FRENKEL

Selenium in Biology and Medicine

Proceedings of the Third International Symposium on Selenium in Biology and Medicine

Held May 27–June 1, 1984
Xiangshan (Fragrance Hills) Hotel
Beijing, People's Republic of China

Sponsored by the
International Selenium Symposium Organizing Committee
in cooperation with the
Chinese Academy of Medical Sciences

International Selenium Symposium Organizing Committee

Dr. S. P. Yang, Coordinator
Texas Tech University
Lubbock, Texas

Dr. J. E. Spallholz, Business Manager
Texas Tech University
Lubbock, Texas

Dr. G. F. Combs, Jr., Program Chairman
Cornell University
Ithaca, New York

Dr. C. G. Hames
Evans County Health Department
Claxton, Georgia

Dr. Jin Daxun
Institute of Health
China National Center for Preventive
  Medicine
Beijing, People's Republic of China

Dr. O. A. Levander
U.S. Department of Agriculture
Human Nutrition Research Center
Beltsville, Maryland

Dr. Niu Shiru
Institute of Health
China National Center for Preventive
  Medicine
Beijing, People's Republic of China

Dr. J. E. Oldfield
Oregon State University
Corvallis, Oregon

Selenium in Biology and Medicine
Part B

Edited by

Gerald F. Combs, Jr.
Department of Poultry and
Avian Sciences, and
Division of Nutritional Sciences
Cornell University
Ithaca, New York

Orville A. Levander
Human Nutrition Research Center
U.S. Department of Agriculture
Agricultural Research Center
Beltsville, Maryland

Julian E. Spallholz
Department of Food and Nutrition
Texas Tech University
Lubbock, Texas

James E. Oldfield
Department of Animal Science
Oregon State University
Corvallis, Oregon

An AVI Book Published by Van Nostrand Reinhold Company
New York

An AVI Book
(AVI is an imprint of Van Nostrand Reinhold Company Inc.)
Copyright © 1987 by Van Nostrand Reinhold Company Inc.

Library of Congress Catalog Card Number

ISBN 0-442-22108-8

Van Nostrand Reinhold Company Inc.
115 Fifth Avenue
New York, New York 10003

Van Nostrand Reinhold Company Limited
Molly Millars Lane
Wokingham, Berkshire RG11 2PY, England

Van Nostrand Reinhold
480 La Trobe Street
Melbourne, Victoria 3000, Australia

Macmillan of Canada
Division of Canada Publishing Corporation
164 Commander Boulevard
Agincourt, Ontario M1S 3C7, Canada

16 15 14 13 12 11 10 9 8 7 6 5 4 3 2 1

Library of Congress Cataloging-in-Publication Data

Selenium in biology and medicine.

Proceedings of the 3rd International Symposium on Selenium in Biology and Medicine, held in Beijing, May 27–June 1, 1984.
Includes bibliographies and index.
1. Selenium—Physiological effect—Congresses.
2. Selenium—Metabolism—Congresses. 3. Selenium in human nutrition—Congresses. 4. Selenium in animal nutrition—Congresses. 5. Selenium—Therapeutic use— Congresses. I. Combs, Gerald F. II. International Symposium on Selenium in Biology and Medicine (3rd : 1984 : Peking, China) [DNLM: 1. Selenium—congresses. QV 138.S5 S464 1984]
QP535.S5S443   1987   599'.019'214   87-2140
ISBN 0-442-22108-8 (set)

Contents

Contributors to Part B

Erling Aadland
Jan Aaseth
B. Ahlrot-Westerlund
B. Åkesson
Nancy W. Alcock
An Ru-Guo
Eiji Araki
Bai Jin
Bai Qian-Fu
Bai Shi-Cheng
D. Behne
B. G. Bennett
P. Brätter
John Bray
Fred Buddingh
Janet S. Butel
Judy A. Butler
Cai Yun
J.-W. Chen
Chen Quan-Guang
W.-W. Chen
Chen Wen
Chen Xiao-Shu
Chen Xue-Cun
Chen Zeng-Cheng
Cheng Dai-Zong
Cheng Yao-Hua
Cheng Yun-Yu
Larry C. Clark
Den Yin-Jie
Deng Jia-Qi
Deng Xue-Jun
Duan You-Jin

W. Elger
Daniel S. Feldman
Maxine E. Fico
Fu Ping
Y.-G. Fu
Fu Zhao-Lin
Gao Fen-Min
Gao Fu-Zheng
Gao Tai
Branko Gavrilović
Ge Ke-You
G. Gissel-Nielsen
Gloria F. Graham
Gu Lu-Zhen
Guan Jin-Yang
B.-G. Guo
Markku Halme
Curtis G. Hames
Han Cong
Han Yue-Ai
T. Höfer-Bosse
Toshiyuki Hosokawa
Hou Chong
Hou Shao-Fan
Jeng M. Hsu
Hu Guo-Gang
Huang Chang-Zhi
F. Huang
Huang Jia-Hong
Huang Jing-Rong
Barbara S. Hulka
M. Ihnat
Clement Ip

Keizo Ito
Ji Lian-Fang
Jin Qun
U. Johansson
Han K. Kang
K. Kasperek
A. Kauppila
Ke Yang
S. Kennedy
P. Koivistoinen
Masaru Kondo
Vesna Kornet
H. Korpela
Klaus Kuehn
Helen W. Lane
Walter M. Lewko
Li Cai-Xia
Li Chong-Zheng
Li De-Hua
Li Fang-Sheng
Li Guang-Sheng
Li Guang-Yuan
Li Ji-Yun
Li Jian-Ye
Li Li
Li Ri-Bang
S.-G. Li
Li Shen-Si
Li Wen-Xian
Liang Shu-Tang
Z.-H. Lin
Liu Bian-Sheng
Liu Jun
Liu Ren-Wei
Liu Sheng-Jie
Liu Su-Mei
Liu Xiu
Liu Yang-Gang
Ingrid Lombeck
Lu Min-De
Luo Xian-Mao
U. M. Mäkilä
R. Masironi

Dubravka Matešić
Kenneth P. McConnell
C. H. McMurray
Sherri Mead
Daniel Medina
Marianne Melcher
Meng Guang-Shan
H. Menzel
John A. Milner
Mo Dong-Xu
Mu Si-Zhang
S. Negretti
Niu Ying-Dou
Pan Bim
Pan Zhong-Ming
R. Parr
Barbara C. Pence
M. Perry
P. J. Peterson
Mary Frances Picciano
L. O. Plantin
Qi Zhi-Ming
Qian Peng-Chu
Lesley S. Reisbord
Ren Ji-Zho
Ren Shang-Xue
D. A. Rice
M. F. Robinson
U. Rösick
Kazuo Saito
Takeshi Saito
Seppo Sarna
I. Savic
Mark D. Schluchter
Gerhard N. Schrauzer
Tammy Schrauzer
Shang Xuan
B.-S. Shi
Carl M. Shy
Å. Siden
Anne M. Smith
Diane K. Smith
H. B. Stockhausen

Su Cheng-Qin
S. Sun
J. Svensson
Tan Jian-An
Tan Wu-Hong
Yngvar Thomassen
C. D. Thomson
Matti Tolonen
Bruce W. Turnbull
Jane L. Valentine
John F. Van Vleet
A. van Leeuwenhoekhuis
P. Varo
C. Veillon
L. N. Vernie
L. Viinikka
Wan Hen-Jun
Wang Fan
Wang Kang-Ning
Wang Shu-Qin
Wang Wu-Yi
Wang Yu-Wen
J. H. Watkinson
We Ting-Gao
Wei Feng-Qun
Bernhard Welz
Wen Zhi-Mei
Philip D. Whanger
W.-H. Wo
M. S. Wolynetz
Wu Ke-Ming
Wu Shi-Quan
Wu Ting-Guo
Xiang Shou-Xian
Xiao Wen-Da
Xie Yu-Hong
Q.-R. Xing

Xu Guang-Lu
Xu Hui-Bi
Xue An-Na
Xue Wen-Lan
Hiroshi Yamamoto
Yan Xiao-Fen
F.-Y. Yang
Yang Feng
Yang Guang-Qi
Yang Jian-Guo
Yang Tong-Shu
E. Yrjänheikki
Yu Shu-Yu
Aniece A. Yunice
Zhai Xu-Jiu
Zhang Fu-Jin
Zhang Fu-Jing
Zhang Fu-Zheng
Zhang Giang-Zhu
Zhang Jiong
Zhang Ju-Chang
Zhang K.
L.-P. Zhang
R.-Q. Zhang
Zhao Yu-Hua
Zheng Da-Xian
Zhou Lan-Hua
J. Zhou
Zhou Zhi-Yu
Zhu Lian-Zhen
Zhu Ping
Zhu Shi-Ying
Zhu Wen-Yu
Zhu Ya-Jun
Zhu Zhen-Yuan
Zou Kang-Nan
Zou Li-Ming

Affiliations of Senior Contributors

Jan Aaseth   Aker Hospital, Oslo, Norway

B. Ahlrot-Westerlund   Department of Neurochemistry and Neurotoxicology, University of Stockholm, Stockholm, Sweden

B. Åkesson   Department of Clinical Chemistry, University Hospital, Lund, Sweden

Nancy W. Alcock   Memorial Sloan-Kettering Cancer Center, New York, NY 10021

D. Behne   Hahn-Meitner-Institut, Berlin, Federal Republic of Germany

B. G. Bennett   Monitoring and Assessment Research Center, London, England

P. Brätter   Hahn-Meitner-Institut für Kernforschung, Berlin, Federal Republic of Germany

Judy A. Butler   Department of Agricultural Chemistry, Oregon State University, Corvallis, OR 97331

Chen Quan-Guang   Beijing Institute for Cancer Research, Chinese Academy of Medical Sciences, Beijing, People's Republic of China

Chen Xiao-Shu   Institute of Health, China National Center for Preventive Medicine, Beijing, People's Republic of China

Cheng Yun-Yu   Sichuan Provincial Sanitary and Antiepidemic Station, Chengdu, Sichuan, People's Republic of China

Larry C. Clark   Department of Preventive Medicine and Division of Nutritional Sciences, Cornell University, Ithaca, NY 14853

Daniel S. Feldman   Multiple Sclerosis Clinic, Departments of Neurology and Section of Nutrition, Department of Medicine, Medical College of Georgia, Augusta, GA 30901

Branko Gavrilović   Agriculture and Food Processing Plant, Kutjevo, Yugoslavia

Ge Ke-You   Institute of Health, Chinese Academy of Medical Sciences, Beijing, People's Republic of China

G. Gissel-Nielsen   Risø Research Establishment, Roskilde, Denmark

Curtis G. Hames   Evans County Medical Department, Claxton, GA 30417

Hu Guo-Gang   Cancer Institute, Chinese Academy of Medical Sciences, Beijing, People's Republic of China

Clement Ip   Department of Breast Surgery and Cancer, Roswell Park Memorial Institute, Buffalo, NY 14263

A. Kauppila   Department of Obstetrics and Gynecology, University of Oulu, Oulu, Finland

S. Kennedy   Veterinary Research Laboratories, Stormont, Belfast, Northern Ireland

P. Koivistoinen   Department of Food Chemistry and Technology, University of Helsinki, Helsinki, Finland

Masaru Kondo   Research Institute of Biological Aging, Tokyo, Japan

Helen W. Lane   Nutrition and Dietetics Program, University of Texas Health Science Center, Houston, TX 77025

Walter M. Lewko   Tumor Cell Biology, Biotherapeutics, Inc., Franklin, TN 37064

Li Chong-Zheng   Endemic Diseases Research Institute of Gansu Province, Lanzhou, People's Republic of China

Li Fang-Sheng   Liaoning Academy of Medicine, Shenyang, People's Republic of China

Li Guang-Yuan   Department of Keshan Disease, Xian Medical College, Xian, People's Republic of China

Li Ji-Yun   Northwestern Institute of Soil and Water Conservation, Shaanxi Province, People's Republic of China

Liang Shu-Tang   Research Institute of Endemic Diseases of Shaanxi Province, Xian, People's Republic of China

Liu Bian-Sheng   Liyau Hospital, Hubei Province, People's Republic of China

Liu Yang-Gang   Sichuan Agricultural College, Chengdu, People's Republic of China

Ingrid Lombeck   University Children's Hospital, Düsseldorf, Federal Republic of Germany

R. Masironi   World Health Organization, Geneva, Switzerland

Dubravka Matešić   Veterinary Faculty of Zagreb, Center for Poultry Science, Zagreb, Yugoslavia

John A. Milner   Department of Food Science and Division of Nutritional Sciences, University of Illinois, Urbana, IL 61801

Mo Dong-Xu   Kaschin–Beck Disease Research Laboratory, Xian Medical College, Xian, People's Republic of China

Barbara C. Pence   Department of Pathology, Texas Tech University Health Sciences Center, Lubbock, TX 79409

Ren Ji-Zho Gansu Grassland Ecological Research Institute, Gansu, People's Republic of China

M. F. Robinson Department of Nutrition, University of Otago, Dunedin, New Zealand

Kazuo Saito Department of Hygiene and Preventive Medicine, Hokkaido University School of Medicine, Sapporo, Japan

Gerhard N. Schrauzer Department of Chemistry, University of California at San Diego, San Diego, CA 92037

Anne M. Smith Department of Foods and Nutrition, Division of Nutritional Sciences, University of Illinois, Urbana, IL 61801

Tan Jian-An Institute of Geography, Academia Sinica, People's Republic of China

Yngvar Thomassen Institute of Occupational Health, Oslo, Norway

Matti Tolonen Department of Public Health, University of Helsinki, Helsinki, Finland

Jane L. Valentine Center for Health Sciences, University of California, Los Angeles, CA 90024

John F. Van Vleet Department of Pathology, School of Veterinary Medicine, Purdue University, West Lafayette, IN 47907

L. N. Vernie Department of Biophysics, The Netherlands Cancer Institute, Amsterdam, The Netherlands

Wang Fan Institute of Endemic Diseases, Bethune Medical College, Changchun, Jihin, People's Republic of China

J. H. Watkinson Ruakura Soil and Plant Research Station, Hamilton, New Zealand

Bernhard Welz Department of Applied Research, Bodenseewerk Perkin-Elmer & Co. GmbH, Überlingen, Federal Republic of Germany

Wen Zhi-Mei Institute of Health, China National Center for Preventive Medicine, Beijing, People's Republic of China

Xiao Wen-Da Northwest Research Institute of Mining and Metallurgy, Lanzhou, People's Republic of China

Xu Guang-Lu Research Laboratory of Keshan Disease, Xi'an Medical College, Xi'an, Shaanxi, People's Republic of China

Xu Hui-Bi Huazhong University of Science and Technology, Huazhong, People's Republic of China

F.-Y. Yang Institute of Biophysics, Academia Sinica, Beijing, People's Republic of China

Yang Guang-Qi Institute of Health, China National Center for Preventive Medicine, Beijing, People's Republic of China

Yu Shu-Yu Department of Biochemistry, Cancer Institute, Chinese Academy of Medical Sciences, Beijing, People's Republic of China

Aniece A. Yunice   Veterans Administration Medical Center, Department of Physiology and Biophysics and Pathology, University of Oklahoma College of Medicine, Oklahoma City, OK 73104

Zhai Xu-Jiu   Qingshui Animal Husbandry and Veterinary Station, Nanjing Agricultural College and Lanzhou Veterinary Institute, People's Republic of China

Zhang Fu-Zheng   Shanghai Research Institute of Environmental Protection, Shanghai, People's Republic of China

Zhang Jiong   Qingshui Animal Husbandry and Veterinary Station, Nanjing Agricultural College and Lanzhou Veterinary Institute, People's Republic of China

Zhu Shi-Ying   Research Laboratory of Keshan Disease, Xian Medical College, Shaanxi, People's Republic of China

Preface

This is a volume of ideas, the ideas of many scientists from several countries working in various areas of biomedical research concerned with selenium. Their ideas are linked with each other and with those of the past by the communication which is the life blood of scientific inquiry. The 225 scientists who came from 17 nations and gathered in Fragrance Hills outside of Beijing in May of 1984 presented and discussed the thoughts and concepts contained in these proceedings; yet, those individuals represented a much larger body of their colleagues, both contemporaries and predecessors, with whom they were joined by these ideas.

The world is getting smaller, and therefore our newly discovered common problems find solutions when we continually seek ways to promote an exchange of our ideas, which generates mutual understanding and respect. The Third International Symposium on Selenium in Biology and Medicine has been such a way. Its beginning was at the Second International Symposium on Selenium in Biology and Medicine held on the campus of Texas Tech University in May, 1980. As that meeting concluded, its organizers (Dr. Leon Hopkins, Dr. J. E. Spallholz, and Dr. S. P. Yang) recognized the need for a third international meeting to continue the exchange. In view of the exciting discoveries of hitherto unrecognized roles of selenium in human health then recently publicized by Chinese scientists, the Texas Tech group realized the tremendous contributions to our collective understanding of the biomedical functions of this trace element that would be made if it were possible to convene the third symposium in the People's Republic of China. They also recognized that such a venue would afford the maximum opportunity for scientific interchange among scientists from China and other countries with long-standing interests in selenium biology. Thus, the incentive for the third symposium was born of the second one; its actualization came about by the efforts of many individuals and groups.

The chapters in these proceedings are organized in nine sections according to subject area and corresponding to the session topics of the symposium. Each section contains the contributions presented as oral reports at that session as well as related presentations given as posters at the symposium. The assignment of presentations was made to provide uniformity with respect to the coverage of the section topic; however, some arbitrary assignments were required, as the reader will observe.

In going through these proceedings, the reader is asked to search, to evaluate, and, where possible, to use the concepts it contains. Only when these ideas are employed by minds that will appropriately challenge and apply them to answer questions and to generate new hypotheses will these proceedings be of continuing value. To the extent that this happens, the work and contributions of the persons and organizations cited above will be rewarded, and the Third International Symposium on Selenium in Biology and Medicine will be not just a meeting of scientists but rather a continuing source of stimulation to further progress in this exciting field of study.

Acknowledgments

The efforts of Dr. S. P. Yang should be recognized, for without his work over a period of almost 3 years, the arrangements for the symposium would not have materialized. Sharing much of his responsibility in organizing the meeting was Dr. Julian E. Spallholz, who ultimately assumed the role of manager of all financial responsibilities, a task frought with difficulties of every kind. These two were assisted by Dr. Curtis Hames, whose sincerely humanitarian nature prompted him to work as a major facilitator of communications between the organizing committee and officials in the host country. Ms. Bette Jones' tireless efforts, which resulted in the logistic arrangements for the meeting, were a major factor enabling the symposium to be held. Dr. Niu Shiru and Dr. Jin Daxun worked as important members of the organizing committee in relations with the Chinese Minstry of Health and the State Science and Technology Council of China, both of which were instrumental in arranging the symposium. The many colleagues of Dr. Niu and Dr. Jin, too numerous to name here, were indispensable to the success of the symposium by virtue of their arrangement of every detail of the meeting, from the construction of poster display boards to the provision of projection equipment. Dr. Orville A. Levander and Dr. James E. Oldfield carried out important functions in the several meetings held in 1983 to develop program plans and financial strategies. Ultimately, it was Dr. Spallholz, Dr. Levander, and Dr. Oldfield who assisted the program chairman in the monumental task of editing the manuscripts submitted for these proceedings. A great debt of gratitude is owed to these individuals and to the many secretaries, assistants, and graduate students who aided them in their respective contributions to the success of the symposium. The editors and their assistants donated their time and talents to the preparation of these proceedings; all royalties will be used as working capital for future International Symposia on Selenium in Biology and Medicine.

Financial support for this symposium was provided by the following groups:

BASF Wyandotte Corporation, Wyandotte, Michigan
The Coca-Cola Company, Atlanta, Georgia
General Foods Corporation, White Plains, New York
Grow Company, Inc., Hackensack, New Jersey
Hames Foundation, Claxton, Georgia
Johnson & Johnson Company, New Brunswick, New Jersey
Ms. Bette Jones, International Center for Travel, Atlanta,
    Georgia
Lederle Laboratories, American Cyanamid Company, Pearl River,
    New York
LyphoMed, Inc., Chicago, Illinois
Mead Johnson & Company, Nutrition Division, Evansville, Indiana
New York State College of Agriculture and Life Sciences, Cornell
    University, Ithaca, New York
Nutrition 21, San Diego, California
The Procter & Gamble Company, Cincinnati, Ohio
Quaker Oats Company, Barrington, Illinois
Avery and Anna Rockefeller, Greenwich, Connecticut
Ross Laboratories Division, Abbott Laboratories, Columbus, Ohio
Schering Corporation, Kenilworth, New Jersey
Shaklee Corporation, San Francisco, California
Texas Tech University, College of Home Economics, Lubbock, Texas
Charles F. Warnell and family, Pembroke, Georgia

Contents of Part A

Section IV
BIOAVAILABILITY OF SELENIUM

Section V

Aspects of Selenium Analysis in Biological Materials

A Simple Rapid Digestion Method for Tissue Analysis of Selenium and Other Trace Metals

Nancy W. Alcock

INTRODUCTION

Methods of tissue preparation for mineral and trace metal analysis which are destructive involve either wet or dry ashing. Wet ashing usually is carried out in strong acid solution at elevated temperatures. Dry ashing requires temperatures in excess of 400°C. Both techniques require the use of glassware or metal vessels which have a potential for contamination. In addition, the environment in which the ashing is carried out is subject to contributing contamination. Recently Clegg *et al.* (1) compared various acid digestions and also compared their wet ashing method of choice with dry ashing (2) prior to measurement of iron, zinc, manganese, and copper. The method most commonly used for tissue digestion prior to the determination of selenium utilizes nitric-perchloric acid as the oxidizing agent (3). A special hood with washing facilities required to eliminate the potential hazards from perchloric acid digestion is a distinct disadvantage of this procedure.

A method described by Chang and Bloom (4) for tissue digestion using hydrogen peroxide as the oxidizing agent, polyethylene vials, and a temperature of 75°C prior to measurement of calcium and magnesium seemed ideal for sample preparation for trace metal analysis. Application, with slight modification to digestion of liver, kidney, gut, and brain prior to the measurement of zinc by flame or flameless atomic absorption, has been reported previously (5). The technique has been further modified and its application to the measurement of selenium and zinc is described here.

MATERIALS AND METHODS

Reagents

Deionized distilled water (DDW) (Millipore Corp., MilliQ system)
Hydrogen peroxide, 30% (Fisher Scientific, catalog no. H-325)
Analytical reagent
Nitric acid—Ultrapure (G. Frederick Smith Chemicals,
    catalog no. 621)
Selenium reference standard, 1 mg/ml (Fisher Scientific)
Zinc reference standard, 1 mg/ml (Fisher Scientific)
Nickel nitrate hexahydrate, analytical reagent (Fisher Scientific)
A solution of 42.3 mg/ml in DDW; 7 µl of this solution contained
    60 µg Ni

Materials

Polyethylene vials, 20-ml capacity with cap (Fisher Scientific, cata-
    log no. 3-337-12); a small hole was made in the vial cap by piercing
    with a 25-gauge needle. A piece of parafilm was firmly pressed
    over the outside of the cap so that the hole was covered.
Automatic pipets, Eppendorf or Finpipette, capacity as required
Parafilm (American Can Co.)
Liver powder (NBS reference material no. 1577a)
All plastic syringe, 10-ml capacity (Aldrich Chemical Co., catalog
    no. Z11,687-4), with a "selectapette" pipet tip (Clay-Adams, cata-
    log no. 4696) attached via 0.5-in. length of Tygon tubing, ⅛-in. i.d.

Instrumentation

Atomic absorption spectrophotometer, Perkin–Elmer, model 5000
    with Zeeman 5000 Furnace and Autosampler As-40, printer
    PR100 and dual pen recorder
Pyrolytically coated graphite tubes and pyrolytic platform
Electrodeless discharge selenium lamp
Atomic absorption spectrophotometer, Perkin–Elmer, model 372
    with single-slit burner; absorption from a hollow cathode zinc
    lamp was measured using an air–acetylene flame

PROCEDURE

1. Digestion

Dried liver powder (50 mg) was weighed into a polyethylene vial. The vial was placed on a glass plate in the oven, which was maintained at a temperature of 75 ± 2°C. After temperature equilibration was reached 1 ml of hydrogen peroxide was added using the plastic syringe with pipet tip attached. When not in use the hydrogen peroxide and syringe were kept in a closed plastic bag to protect them from the possibility of environmental contamination.

After the addition of the hydrogen peroxide the cap was placed on the vial, and it was maintained at 75 ± 2°C for 1 hr. The contents were then examined to ascertain whether oxidation of organic matter was complete, which was apparent from the absence of particles. Incubation was continued until a clear solution remained. The cap of the vial was then removed and the vial continued to be heated to dryness. If there was evidence of undigested matter, an additional 0.5 ml of hydrogen peroxide was added, the vial again capped, and the digestion continued as previously.

When digestion was complete and the solution evaporated to dryness, 2 ml of approximately 0.3 N HNO$_3$ was added, and the vial capped and shaken vigorously. The contents were allowed to stand at room temperature for at least 1 hr with occasional mixing prior to analysis for selenium and zinc by flameless and flame atomic absorption spectrophotometry, respectively.

For recovery studies, separate samples of liver powder, to which selenium and zinc standard solutions in the amounts indicated were added, were treated exactly as described above.

2. Flameless Atomic Absorption

Selenium was measured in the solutions obtained from the digestion of liver using pyrolytically coated graphite tubes with a solid pyrolytic platform. The source was an electrodeless discharge selenium lamp. Peak area absorbance at 196.0 nm was measured over a 5-sec interval at 2500°C. Ashing temperature was 1200°C. The program is shown in Table 1.

An aqueous solution of known selenium content was run daily to check instrument reproducibility. Slit width was 2.0 nm. The internal gas was argon. The sample volume was 15 μl with 7 μl alternate volume of the nickel nitrate solution. The absolute amount of nickel in

TABLE 1. Program for Selenium Determination

Step	1	2	3	4	5
Temperature (°C)	230	1100	2500	2600	20
Ramp (sec)	15	5	0	1	20
Hold (sec)	60	45	5	6	20
Inert gas	300	300	0	300	300
Recorder			−2	0	
Read			−1		
Baseline				15	

the 7-μl volume of nickel nitrate was 60 μg. Triplicate readings of all samples were taken. The method of additions was used to prepare a calibration curve.

3. Flame Atomic Absorption

Zinc was measured by flame atomic absorption spectrophotometry using standards prepared in 0.3 N NOH$_3$. If required, dilution of specimens was performed using 0.3 N NOH$_3$. Integrated absorbance was measured over a 3-sec interval with an aspiration rate of 6 ml/min.

RESULTS

Maintenance of the temperature at 75 ± 2°C produced a satisfactory digestion with minimum frothing during the first step. Higher temperatures caused increased frothing with some deposition of sediment near the top of the vial. Occasionally evidence of frothing into the cap was evident—these samples were considered to be unsatisfactory and were discarded.

Digestion for 5–6 hr was usually adequate, but was continued for 8 hr to assure complete destruction of organic matter. Although the process has been carried out with a cap without the pressure release hole, there has been evidence of expansion of the parafilm, and hence this practice seemed desirable. The presence of particulate matter in the initial reconstituted digests suggested incomplete oxidation. Additional heating with hydrogen peroxide resulted in a clear solution.

Blank readings for both selenium and zinc from reagents subjected to the same procedure as the samples were insignificant. The coefficient of variation of a solution of 37 μg/liter selenium analyzed daily for 10 days was 3%. Peak area absorption by selenium was less in the dissolved digest than in standards prepared in 0.3 N HNO$_3$ when nick-

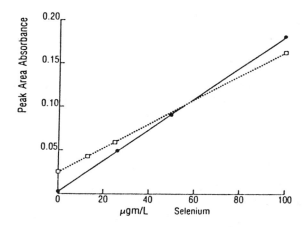

FIG. 1. Graphite furnace analysis for selenium. (●), Aqueous standards; (□), liver digest.

el nitrate was used as matrix modifier, as shown in Fig. 1. Background absorption was not significant, but was less at an ashing temperature of 1200°C than at 1050°C. Since the absorption was not diminished at the higher temperature, analyses were performed at 1200°C. Triplicate readings for selenium in the standards and samples were repeated if they differed by more than 5%.

Values obtained for replicate analyses of NBS liver sample number 1577a for selenium and zinc were in good agreement with those certified (Table 2). The recoveries of 25 ng Se and 5 μg Zn added to the liver samples prior to digestion or to the dissolved digest are shown in Table 3. These are not significantly different from 100%. Agreement between the recovery of additions before and after digestion indicates that losses did not occur during the digestion.

DISCUSSION

The use of polyethylene vials eliminates the necessity for the tedious and extensive acid washing procedures commonly described by investigators performing trace metal analyses (1,2). While most plas-

TABLE 2. Concentration of Se and Zn in NBS Liver (μg/g)

	Se	Zn
Expected concentration	0.71 ± 0.07	123 ± 8
Obtained concentration ($n=10$)	0.69 ± 0.08	120 ± 7

TABLE 3. Recovery of Se and Zn Added to Liver before and after Digestion ($n=10$)

	Se	Zn
Concentration ($\mu g/g$)	0.69 ± 0.08	120 ± 7
Addition before digestion	25 ng	6 μg
Recovery (%)	24.4 ± 0.5 (98%)[a]	6.1 ± 0.1 (102%)
Addition after digestion	25 ng	6 μg
Recovery (%)	24.8 ± 0.4 (99%)	6.1 ± 0.1 (102%)

[a] Mean ± SD.

ticware used by the author has been found to be free of all trace metals examined, a recommended precaution of rinsing the vials with DDW prior to use may be time-saving. The minimum 12-hr predigestion periods used with acid digestions described by Clegg *et al.* (1) followed by a further 4-hr digestion at 100°C are lengthy compared to the time required with hydrogen peroxide described here. In addition, environmental contamination is minimized by carrying out the 75°C digestion in a closed oven, contrasted to heating flasks covered with a glass plate on a hot plate, as used with nitric acid digestion. The system described here offers further improvement over the use of a jet of air, a potential contaminant, used by Chang and Bloom (4). A further advantage of the method is the performance of the entire digestion and dissolution process in the one container.

The low digestion temperature, together with the fact that the procedure is carried out in a closed vessel, minimizes the risk of loss of the volatile selenium, as suggested by Koh *et al.* (6). Recovery of added selenium indicated that losses did not occur. Dry ashing procedures indicated that there was loss of zinc (2), and lower levels of selenium were reported in liver samples following dry ashing than in procedures where wet ashing was employed (6).

Low recoveries for zinc and iron were reported when wet acid digestion was carried out in Vycor or porcelain crucibles. It was suggested that increased binding of the metals to the materials occurred with crucible aging (2). Thorough examination of the polyethylene vials at the completion of analyses failed to reveal any reaction with the surface. One-time use of the vials, the absence of contamination, and hence avoidance of the tedious and hazardous acid washing procedures required for other vessels are distinct advantages of the procedure.

Although the vials are tolerant to temperatures up to 85°C (unpublished observations), the higher temperature encourages increased frothing. Preliminary observations indicate that the frothing may be minimized or eliminated by the presence of a trace of 1-octanol (caprylic alcohol). Hence, higher temperature accompanied by shorter digestion time may be possible.

The procedure described lends itself to trace metal analyses in general and is currently being extended to other elements.

ACKNOWLEDGMENT

The work was supported in part by NIH Grant 5PO1CA29502.

REFERENCES

1. Clegg, M. S., Keen, C. L., Lönnerdal, B., and Hurley, L. S. 1981. Influence of ashing techniques on the analysis of trace elements in animal tissue. I. Wet ashing. Biol. Trace Elem. Res. 3, 107–115.
2. Clegg, M. S., Keen, C. L., Lönnerdal, B., and Hurley, L. S. 1981. Influence of ashing techniques on the analysis of trace elements in biological samples. II. Dry ashing. Biol. Trace Elem. Res. 3, 237–244.
3. Watkinson, J. H. 1966. Fluorimetric determination of selenium in biological materials with 2,3-diaminonaphthalene. Anal. Chem. 38, 92–97.
4. Chang, C., and Bloom, S. 1983. A simple ashing method for the determination of Mg and Ca in laboratory animal feed and tissues. J. Am. Coll. Nutr. 2, 149–155.
5. Alcock, N. W. 1986. Flame and flameless atomic absorption spectrophotometry: Application to the measurement of tissue zinc. In Neurobiology of Zinc. Alan R. Liss, Inc., New York (in press).
6. Koh, T. S., Benson, T. H., and Judson, G. J. 1980. Trace element analysis of bovine liver: Interlaboratory survey in Australia and New Zealand. J. Assoc. Off. Anal. Chem. 63, 809–813.

59

Catalytic Polarographic Determination of Trace Selenium in Biological Materials

Wen-Da Xiao

Methods for estimating trace amounts of Se have been discussed in the literature (1). Neutron activation analysis and spark mass spectroscopy have the highest sensitivity, but they are expensive, time-consuming, and not readily available for use. In contrast, the fluorescence analysis not only has high sensitivity, selectivity, and reproducibility, but also can be done with less expensive equipment.

The author proposed an $SeSO_3^{2-}-IO_4^-(IO_3^-)$ catalytic polarography (2) for trace Se analysis which has a relatively high sensitivity and selectivity in the measurements using ordinary instruments and reagents. It could be applied to the analysis of Se in ore, grain, steel, and iron (2–5). The sensitivity of the method reaches 0.00005 μg/ml and a single determination can be finished within 2 hr. A small sampling was enough for samples with Se concentration at 0.0X–0.X ppm. The relative standard deviation is less than 10%.

EXPERIMENTAL AND RESULTS

Instruments and Reagents

A JP-1A oscillopolarograph with three electrodes and a differential mode was used for current potential curve measurements.

Perchloric acid, nitric acid, hydrochloric acid, sodium chloride, and all other chemicals were of analytical reagent grade.

Na_2SO_3 solution: 20% water solution (treated with Zn powder)

Mixture solution A: 40% (v/v) NH_4OH; 20% (m/v) NH_4Cl; 5% (m/v) EDTA; 0.002% (m/v) arabic gel

Mixture solution B: 1% (m/v) KIO_4 (or KIO_3); 1% (v/v) NH_4OH

Digestion mixture $HClO_4$: HNO_3 = 3:2 (v/v)

Se standard solution: 0.02 μg Se (IV) and 1% HNO_3 after dissolving Se powder with nitric acid

Component Functions of Mixture Solutions and Effects on Catalytic Current

Catalytic Wave Formation and Wave Height Measurement. The catalytic wave is produced through the reaction between $SeSO_3^{2-}$ and IO_4^- (IO_3^-) at pH 9.5–12.0. However, no peak will be observed when Se concentration is lower than 0.0001 μg/ml. The catalytic wave formation and wave height measurement are shown in Fig. 1.

The catalytic wave is divided into a and b sections and a + b represents wave height. However, when Se concentration is too low, b will approximate zero, and only section a is taken as the wave height in this case.

Through digestion all valent Se except Se (VI) is converted to Se (IV), after which the catalytic wave is produced. NH_4OH–NH_4Cl buffer solution is used to control the system at about pH 10. EDTA is used to eliminate the interferences of Cd, Zn, and In, and arabic gel to improve the linearity between Se concentration and catalytic current. Stability of the catalytic wave thus produced is good.

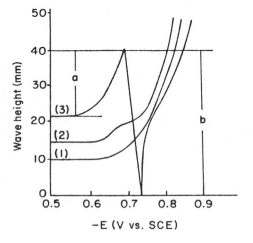

FIG. 1.   Catalytic wave and wave height measurement. (1) Blank solution; (2) Se 0.00005 μg/ml; (3) Se 0.001 μg/ml. Sensitivity: 1.0.

With increasing NH_4Cl concentration, the catalytic current apparently increases, but decreases rapidly while the arabic gel is added. The optimum composition of the mixture solution is 2–5% (v/v) $HClO_4$, 1.0–1.6% (m/v) NaCl, 2% (m/v) Na_2SO_3, 6% (m/v) NH_4Cl, 12% (v/v) NH_4OH, 1.5% (m/v) EDTA, 0.006% (m/v) arabic gel, and 0.5% (m/v) KIO_4 (or KIO_3). The effects of components on catalytic current are given in Fig. 2 (1,2).

Se (VI) Treatment

A pretreatment with NaCl–HCl was necessary before adding Na_2SO_3 in order that Se (VI) could be quantitatively converted to Se (VI) by the formation of $SeSO_3^{2-}$ with Na_2SO_3. A complete recovery of Se (VI) was obtained by the treatment with 0.1–1.3 ml of concentrated HCl and 0.5 ml of 20% NaCl after evaporation with $HClO_4$–HNO_3.

Selection of Digestion Reagent

Various methods had been proposed for the digestion of biological samples (1). $HClO_4$–HNO_3 was found to be the best. The effects of different proportions of $HClO_4$–HNO_3 on the sample digestion are given in Table 1. The results suggested that the proportion of $HClO_4$ to HNO_3 should be higher than 1.

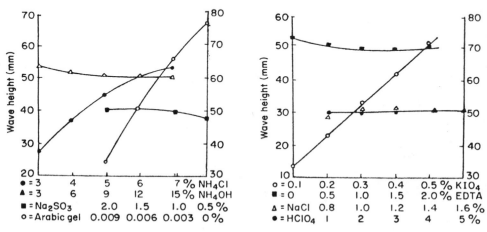

FIG. 2. Effects of mixture components on catalytic current. Sensitivity: 1.0; temperature: 15 ± 0.1°C.

TABLE 1. Effect of Different Proportions of $HClO_4$ to HNO_3 on Sample Digestion

Mixed acids	1:3	1:2	1:1	1.5:1	2:1
Wave height[a]	11.5	31.5	36.0	39.0	39.0

[a] The same sample and the sample volume of mixed acids.

Analytical Procedure and Data Processing

A 10- to 100-mg sample was weighed in a 50-ml beaker, and 5 ml of mixed acids were added. The sample was evaporated to near 1 ml of $HClO_4$ on the heater with medium temperature, then cooled down before adding 0.5 ml of 20% NaCl. Then the sample was evaporated to near dry, and 0.5 ml of concentrated hydrochloric acid was added. Again the solution was heated to near dry, and 2 ml of 20% Na_2SO_3 were added. It was then left for 15–30 min before adding 3 ml of mixture solution A and 5 ml of mixture solution B. The solution was shaken after each addition. It was then put in the water bath at a selected temperature between 15° and 20°C, with a variation within ± 1°C. The cathodized differential polarographic graph was made by measuring the current and potential, starting from −0.5 V at a proper sensitivity position.

A 0.002–0.04 μg Se standard was put into a 50-ml beaker, and 0.5 ml of $HClO_4$ and 0.5 ml of 20% NaCl was added. The solution was then evaporated to near dry. The rest of the procedure was carried out as mentioned above. The working curve is shown on Fig. 3. The results

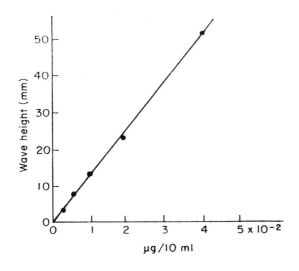

FIG. 3. Calibration curve of Se. Sensitivity: 40; temperature: 15°C.

TABLE 2. Precision of the Method

Sample name	Average value ($\bar{X}$, ppm)	Standard variation ($\times 10^{-3}$ ppm)	RSD (%)
Apple	0.034	2.27	6.68
Eggplant	0.056	3.74	6.68
Chinese cabbage	0.161	9.44	5.86
Grass	0.176	5.04	2.87
Leaves	0.581	2.02	3.48
Dangshen (herb)	0.031	2.82	9.10
Huangqi (herb)	0.035	1.49	4.26

were calculated by interpolation or comparison with standard. Precision of the method is shown in Table 2. The average recovery of added standard in 3 sets of experiments was 93%.

DISCUSSION

This method was applied to measure the trace Se in hair, animal muscle, soil, and wastewater, with satisfactory results. More than 0.1 ppm of selenium content was found in the hair of healthy people, while in Keshan disease and Kaschin–Beck disease patients only 0.05 ppm of Se was found.

With increasing temperature the catalytic current decreases at a range of 4°–40°C. The temperature coefficient is usually larger than −5%/°C.

It is important to complete the digestion process, and care has to be taken to avoid any organic contamination.

No sample preseparation was needed unless the mixture solution was modified. No interferences from inorganic ions were found, but this will happen if any undestroyed organic materials are present.

REFERENCES

1. Hou, S.-F., and Wang, W.-Y. 1980. Fenxi Huaxue 8 (2), 183.
2. Xiao, W.-D., and Wang, B.-Z. 1975. Acta Lanzhou Univ., Nat. Sci. Ed. 1, 9.
3. Xiao, W.-D., Fang, H.-Y., and Wang, S. 1976. Fenxu Huaxue 4 (6), 419.
4. Xu, M.-S., Xiao, W.-D. 1981. Gansu Metall. 1, 59.
5. Liu, C.-Y., and Xial, W.-D. 1982. Metall. Anal. 3, 25.

IUPAC Interlaboratory Trial for the Determination of Selenium in Clinical Materials

Y. Thomassen
M. Ihnat
C. Veillon
M. S. Wolynetz

The element Se has been found to play important roles in human and animal health and disease. Prolific research efforts have been and are at present being devoted to the investigation of many facets of clinical, biochemical, and environmental manifestations of Se. Almost invariably, analytical chemistry and measurement of Se concentrations form an essential background to these research studies. In order to ensure reliability and among-laboratory comparability of analytical results, two prerequisites are appropriate analytical methodologies and reference materials. Although many publications have appeared dealing with research and application studies of analytical methods for selenium, little has been done with respect to the development of chemical reference materials for selenium. Pertinent products currently available are kale (1), orchard leaves, bovine liver, wheat flour and rice flour (2), fish flesh, copepod, human hair, animal muscle and lyophilized animal blood (3), and human urine (4). An earlier interlaboratory trial under the auspices of the IUPAC Subcommittee on Selenium has led to recommended values for Se in a human blood serum proposed reference material (5).

In an attempt to make available additional clinical reference materials, a second IUPAC trial is under way. The objectives of this undertaking are as previously, viz. (a) to establish total Se concentrations in clinical materials to be available as reference materials, and (b) to further assess the performance of various analytical methodologies used to analyze sera and urine with normal levels of Se. It was also decided to extend these considerations to a range of other elements of clinical and environmental significance. This preliminary report will summarize progress to date on the Se aspect of the undertaking.

EXPERIMENTAL

Two clinical materials, a lyophilized human serum designated Seronorm for Trace Elements and lyophilized human urine were prepared in quantity by Nyegaard and Co., AS, Oslo, Norway and kindly made available for this study. Se and nutrient elements were present as endogenous elements at normal concentrations, whereas several other elements of clinical and environmental interest in urine were spiked to higher levels. Twelve vials of each of the two materials were submitted to a population of laboratories similar to that in the previous trial (5) comprising those interested in Se and trace metal determinations in clinical fluids. Analysts were requested to reconstitute the serum and urine with 3 and 10 ml of pure water, respectively, and to analyze the samples by methods of their choice, for primarily Se, but also a range of elements and constituents from the list:

Serum: Na, K, Mg, Ca, Fe, Cu, Zn, Al, total protein

Urine: Na, K, Mg, Ca, Fe, Cu, Zn, Al, As, Cr, Mn, Co, Ni, Cd, Sb, Hg, Pb, F, Tl, Bi, V, Be, Sn, creatinine

One analysis of each of six vials of serum and urine was requested, with analytical results to be submitted on report forms provided.

RESULTS AND DISCUSSION

Of the more than 20 laboratories who agreed to participate in this second IUPAC trial, 14 and 12 reports of analysis for Se in serum and urine, respectively, were received in time for preliminary assessment. Four different analytical methods were employed as summarized in Table 1. Results received are summarized in Table 2 as ranges of Se concentrations, μg/liter, reported in the reconstituted materials. Neither individual results nor means from the laboratories nor overall means are included in this preliminary report in order not to prejudice

TABLE 1. Analytical Methods Used for Determining Se in Human Blood Serum and Urine Materials—Preliminary Data Set

Method	Code	Number of laboratories for	
		Serum	Urine
Acid decomposition–fluorometry	ADF	5	5
Electrothermal atomization, atomic absorption spectrometry	EAAS	3	1
Acid decomposition–hydride generation atomic absorption spectrometry	ADHAAS	5	5
Acid decomposition–isotope dilution mass spectrometry	ADIDMS	1	1

additional information expected in the near future from other cooperators.

Preliminary statistical analyses were carried out to calculate mean concentrations and within-laboratory variance components. In general, statistically significant differences were observed among mean concentrations reported by the various laboratories using different methods. However, for serum data good agreement was evident among means from EAAS, ADHAAS, and ADIDMS, with an indication of somewhat higher values from ADF. One important distinction between the method comparisons here and in the first interlaboratory

TABLE 2. Concentration Ranges (μg/liter) of Se Reported by Cooperating Laboratories for Seronorm Protein/Trace Elements and Urine

Method	Range, all data	Range, after exclusion of outliers (number of labs excluded)[a]
	Seronorm trace element 105	
ADF	79.3–122	79.3–104 (1)
EAAS	80–94.3	80–94.3 (0)
ADHAAS	73.5–117	73.5–102.6 (0)[b]
ADIDMS	86.4–98.7	86.4–98.7 (0)
Overall	73.5–117	73.5–102.6 (1)
	Urine trace element 108	
ADF	33.1–72	49.0–54 (2)
EAAS	11–15	— (1)
ADHAAS	38.3–81.3	38.3–54.0 (1)
ADIDMS	46.4–49.6	46.4–49.6 (0)
Overall	33.1–81.3	38.3–54.0 (4)

[a] Generally exclusion of all data from laboratories deemed to report outlying data, although in one instance only one aberrant datum was excluded from the set.
[b] One datum excluded.

trial (5) is that the technique of ADHAAS demonstrates much better performance. For the case of Se in urine, very good agreement was apparent among ADF, ADHAAS, and ADIDMS; EAAS data from the one and only laboratory analyzing urine by this technique were very low and judged out of line with respect to the other 11 sets of data.

The preliminary mean concentration of Se in this lot of Seronorm protein is in the vicinity of the values reported previously for Seronorm 102 and 103 and within the concentrations reported in the literature for normal adult serum (6–9). The preliminary mean concentration of Se in the urine material is in the vicinity of the value reported for the recently issued National Bureau of Standards standard reference material toxic metals in freeze-dried urine (4).

Performance of the analytical methods with regard to within-laboratory standard deviation is presented in Table 3. Within-laboratory standard deviations varied considerably over laboratories, with heterogeneity of variances being the rule within methods; in fact, only ADF results for serum from four laboratories exhibited homogeneity of variance. The standard deviations were nevertheless encouragingly favorable for these clinical matrices, with Se levels in the 40–100 μg/liter range.

Of the 12 laboratories that analyzed both serum and urine samples, 9 had a serum/urine mean concentration ratio in the narrow range 1.75–1.93. The other three laboratories showed ratios of 1.23 (ADHAAS), 2.42 (F), and 6.52 (EAAS). It was felt that low values for urine were responsible for elevated ratios and that the different methods had different sensitivities to sample matrix. That urine analysis for Se by EAAS may not be as easy as serum analysis may be suggested by

TABLE 3. Performance of Analytical Methods with Regard to Within-Laboratory Variation

	Within-laboratory standard deviation (μg/liter)	
Method	Individual laboratory values	Averaged over all laboratories in method
	Seronorm trace element 105	
ADF	1.51, 1.63, 1.75, 2.66, 9.80	1.94[a]
EAAS	0.84, 1.07, 2.86	1.88
ADHAAS	0.96, 1.30, 4.95, 5.67, 7.26	4.79
ADIDMS	4.40	4.40
	Urine trace element 108	
ADF	0.57, 1.51, 1.54, 3.71, 5.77	2.16[a]
EAAS	1.79	1.79
ADHAAS	1.26, 1.46, 1.99, 2.97, 4.84	2.86
ADIDMS	1.19	1.19

[a] Excluding one laboratory.

the very low results for urine from one laboratory and the fact that no analytical data for urine were reported by two other laboratories who reported for serum.

On account of the several months' cooling period required prior to counting in instrumental neutron activation analysis, INAA, results by this technique were unavailable in the short time allotted. Analytical results by INAA and other techniques together with additional data by methods already discussed are forthcoming. Detailed treatment of the complete data set will be made to arrive at a detailed assessment of method performances with serum and urine materials and to establish recommended concentrations of Se in these proposed clinical reference materials. It is anticipated that analytical information on levels of other constituents and trace elements will also be available.

ACKNOWLEDGMENTS

The authors appreciate the participation and interest in this study of a number of analysts who conducted analyses and submitted results on very short notice.

REFERENCES

1. Bowen, H. J. M. 1965. Proc. Int. Conf. Mod. Methods Act. Anal. pp. 58–60.
2. National Bureau of Standards, Certificate of Analysis: Orchard Leaves SRM1571, Bovine Liver SRM1577a, Wheat Flour SRM1567, Rice Flour SRM1568. NBS, Washington, DC.
3. International Atomic Energy Agency. 1984–1985. Analytical Control Services Programme, Intercomparison Runs, Certified Reference Materials, Reference Materials. IAEA, Vienna.
4. National Bureau of Standards 1984. Certificate of Analysis, Toxic Metals in Freeze-Dried Urine, SRM2670. NBS, Washington, DC.
5. Ihnat, M., Wolynetz, M. S., Thomassen, Y., and Verlinden, M. 1983. IUPAC Interlaboratory Trial on the Determination of Selenium in Lyophilized Human Blood Serum Reference Material, Programme, Int. Conf. Clin. Chem. Chem. Toxicol. Metals, 2nd, COMTOX '83 (also in preparation for publication).
6. Westermarck, T., Raunu, P., Kirjarinta, M., and Lappalainen, L. 1977. Selenium content of whole blood and serum in adults and children of different ages from different parts of Finland. Acta Pharmacol. Toxicol. *40*, 465–475.
7. Saeed, K., Thomassen, Y., and Langmyhr, F. J. 1979. Direct electrothermal atomic absorption spectrometric determination of selenium in serum. Anal. Chim. Acta *110*, 285–289.
8. Kumpulainen, J., and Koivistoinen, P. 1981. Interlaboratory comparison of selenium levels in human serum. Kem—Kemi *8*, 372–373.
9. Oster, O., and Prellwitz, W. 1982. A methodological comparison of hydride and carbon furnace atomic absorption spectroscopy for the determination of selenium in serum. Clin. Chim. Acta *124*, 277–291.

61

Accuracy of Selenium Determination in Human Body Fluids Using Hydride-Generation Atomic Absorption Spectrometry

Bernhard Welz
Marianne Melcher

INTRODUCTION

Selenium has gained substantial interest as an essential trace element for some animals as well as for humans. Its accurate determination, however, is still a major challenge for the analyst.

Hydride-generation atomic absorption spectrometry (HG AAS) offers the advantages of excellent sensitivity and relatively simple instrumentation, and would therefore be well suited for routine purposes. There is some doubt, however, about the accuracy of the results obtained with this technique. In 1983 an IUPAC interlaboratory trial was organized on the determination of selenium in lyophilized human blood serum reference materials (Ihnat *et al.*, 1984). A mean value of 91 ± 7 µg/liter was finally accepted as the "true" selenium concentration in both serum samples. Of the 27 participating laboratories, 4 applied HG AAS for the determination of selenium after acid decomposition of the serum, and their results are shown in Fig. 1. With the exception of one laboratory, the values obtained with this technique exhibit large negative deviations relative to the accepted mean as well as substantial within-laboratory variances. These results indicated the existence of systematic errors.

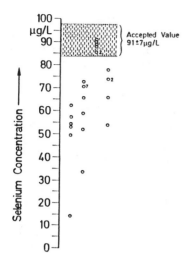

FIG. 1.   Cooperative results obtained for selenium in lyophilized serum, Seronorm (Batch 103), using hydride-generation AAS. Four laboratories, 6 or 12 independent determinations each. IUPAC interlaboratory trial, 1983.

There are several reports in the literature that the sample decomposition technique applied is of great importance for the accuracy of the selenium values obtained in biological materials (Nève et al. 1982; Verlinden 1982; Kotz et al. 1972; Lloyd et al. 1982; Brown et al. 1982: Clinton 1977; Robbins et al. 1979; Oster and Prellwitz 1982). In the hydride-generation technique, by the addition of sodium tetrahydroborate solution to the acid sample digest, gaseous hydrogen selenide (H_2Se) is formed, stripped from the solution, and atomized in a heated quartz cell. For this reaction, it is essential that selenium is present in its tetravalent, ionic form, the selenite. Organic selenium compounds must therefore be decomposed completely for the successful application of HG AAS.

Many selenium compounds are volatile and can be lost during a poorly controlled decomposition procedure. On the other hand, acid-resistant organoselenium compounds, such as selenomethionine, selenocysteine, and the trimethylselenonium ion are often not fully decomposed, hence causing low results (Nève et al. 1982; Verlinden 1982).

Kotz et al. (1972) investigated a decomposition procedure for biological materials, including blood, with nitric acid under pressure in a polytetrafluoroethylene (PTFE) bomb and found quantitative recoveries for selenium. This could not be confirmed by Verlinden (1982), however, who reports only 60% recovery for selenium in blood after a pressure decomposition with nitric acid. Lloyd et al. (1982) applied a digestion with nitric and sulfuric acid at 155°C for the determination of selenium in whole blood, plasma, and erythrocytes. Brown *et al.* (1982) analyzed whole blood, plasma, and urine for selenium after a digestion with nitric and perchloric acids at temperatures up to 200°C. A mixture of nitric and perchloric acids at 210°C was also applied by Clinton (1977) to decompose blood samples prior to the determination of selenium with the hydride technique. A graded destruction with nitric and perchloric acids to a final temperature of 210°C was also recommended by Verlinden (1982). This author found it necessary, however, to use digestion times of 15–30 hr for complete decomposition of human blood and plasma. Robbins *et al.* (1979) applied a mixture of nitric, sulfuric, and perchloric acids (4 + 4 + 1) for digestion of blood samples. The same acids but a more elaborate temperature program were proposed by Oster and Prellwitz (1982). Serum was heated to 180°C with a mixture of nitric and perchloric acids in an automatic digester. After the addition of sulfuric and nitric acids, the temperature was raised to 300°C and kept at this level until the solution was clear. Whenever perchloric acid is applied, selenium is partly oxidized to the hexavalent state. A reduction step is therefore included after the digestion, and most authors simply heat the solution with 5–6 M hydrochloric acid for 15–30 min for this purpose.

In the present work we have investigated the influence of the sample decomposition technique on the accuracy of the selenium values obtained by hydride-generation AAS. A lyophilized human blood serum reference material was investigated as well as whole blood, plasma, and erythrocyte samples. Nitric acid digestion under normal and under elevated pressure, respectively, was compared with a nitric, sulfuric, and perchloric acid decomposition which has already been applied successfully for the decomposition of marine biological tissue samples (Welz and Melcher 1984). The recovery of selenomethionine was tested from aqueous solutions, separately, and added to whole blood.

EXPERIMENTAL

Apparatus

A Perkin–Elmer model 4000 atomic absorption spectrophotometer, equipped with an electrodeless discharge lamp for selenium, operated from an external power supply at 6 W, was used throughout this work. A spectral bandpass of 2.0 nm was used to isolate the 196.0 nm resonance line. The signals were recorded with a Perkin–Elmer model 56 strip chart recorder. A Perkin–Elmer model MHS-20 mercury/hydride system was operated with a quartz cell temperature of 900°C. The time settings at the controller were 40 sec PURGE I, 8 sec REACT (during this time, sodium tetrahydroborate solution is injected into the sample solution continuously), and 30 sec PURGE II.

A Perkin–Elmer Autoclave-3, heated on a hot plate with temperature control, was used for pressure decompositions. Nitric, sulfuric, and perchloric acid digestions were carried out in graduated quartz flasks with long necks and a nominal volume of 40 ml (see Fig. 2). The flasks were heated in a self-made aluminum heating block holding six digestion flasks.

Reagents

Sodium tetrahydroborate (III) solution, 3% (m/v) was prepared by dissolving sodium tetrahydroborate powder (Riedel-de Haen) in de-

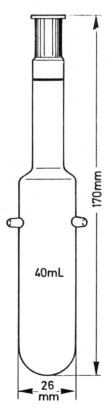

FIG. 2.   Quartz digestion flask for nitric, sulfuric, and perchloric acid decompositions of biological materials (according to Kaiser et al. 1978).

ionized water and stabilizing with 1% (m/v) sodium hydroxide. The
solution was filtered before use and could be stored for only a few days.

Hydrochloric acid (32% m/v), perchloric acid (70% m/v), and sulfuric
acid (96% m/v) were of "suprapur" quality (Merck). Nitric acid (65%
m/v) was further purified by subboiling distillation.

Selenium (IV) stock standard solution (1000 mg/liter) was prepared
from Titrisol concentrates (Merck) by diluting to volume with deionized water. Working reference solutions were obtained by further
dilution with 0.5 M hydrochloric acid.

Selenomethionine, $H_2N\text{-}CH(COOH)\text{-}CH_2\text{-}CH_2\text{-}Se\text{-}CH_3$: 16.4 mg are
dissolved in 50 ml deionized water/methanol (1 + 1). This solution
contains 132 mg/liter Se and is diluted 1:100 with deionized water
before use. Selenomethionine was obtained as described by Nève et al.
(1982).

Preparation of Samples

Lyophilized human blood serum reference material, Seronorm,
Batch 103 (Nyegaard and Co., Oslo, Norway) was dissolved in deionized water according to the manufacturer's recommendations.

Heparinized whole blood samples were obtained from healthy adult
individuals by collecting the blood in heparinized PS tubes, 10 ml
(Greiner no. 160 151). To obtain blood plasma samples, the heparinized
whole blood is centrifuged carefully for 10 min and the supernatant
plasma removed with a Pasteur pipette. To obtain pure erythrocytes,
the residue is washed by mixing with about 5 ml of isotonic saline;
after centrifugation the upper phase is discarded and the washing
procedure repeated. After the last separation, the erythrocytes are
diluted to the original 10 ml with isotonic saline and mixed well before
analysis. The results for selenium in erythrocytes are expressed in
micrograms selenium per kilograms hemoglobin (Hb). Hemoglobin
concentrations were determined by the hemoglobin cyanide method
(Merck no. 9405).

Nitric Acid Digestion

Serum or whole blood (0.5 ml) and 0.5 or 1.0 ml nitric acid are placed
in 10 ml volumetric flasks and heated in a boiling water bath for 6 hr.
After cooling, the solution is diluted to volume with deionized water.
For the determination of selenium, 50 ml of 0.5 M hydrochloric acid
are placed into the reaction flask of the hydride system and 0.5 ml of
the diluted digestion solution (corresponding to 25 µl serum or whole

blood) is added. It is not possible to use larger volumes of the digestion solution because of excessive foam formation during the reaction.

For some experiments a nitric acid digestion was carried out under pressure. The same quantities of sample and acid as described are placed into the PTFE beaker of an Autoclave-3 and heated to 160°C for 1 hr. After cooling, the solution is transferred to a 10-ml volumetric flask, diluted to volume with deionized water, and further analyzed as described.

Nitric, Sulfuric, and Perchloric Acid Digestions

Serum, whole blood, plasma, or erythrocytes (0.5 ml) and 1.0 ml nitric acid are placed in quartz digestion flasks with long necks (Fig. 2), heated slowly to 140°C in an aluminum heating block, and kept at this temperature for 30 min. After cooling to room temperature, 0.5 ml sulfuric and 0.2 ml perchloric acid are added. The temperature is slowly raised to 140°C, 200°C, 250°C, and 310°C and held at each of these temperatures for 15 min before the next increase. The final temperature of 310°C is held for at least 20 min so that most of the perchloric acid is removed and the volume of the digestion solution reduced to about 0.5 ml. After cooling to room temperature, 20 ml of 5 M hydrochloric acid are added to the decomposition solution, heated to 95°C, and held at this temperature for 20 min to reduce all hexavalent to tetravalent selenium. The solution is allowed to cool to room temperature and then diluted to volume (40 ml) with deionized water. For the determination of selenium, 5 ml of the decomposition solution and 15 ml deionized water are placed into the reaction flask of the hydride system. This corresponds to 62.5 μl serum, whole blood, plasma, or erythrocytes. Calibration standards were prepared to contain approximately the same concentration of hydrochloric acid as the samples (0.5 M HCl in the solution for measurement). Foam formation is not a problem when this nitric, sulfuric, and perchloric acid decomposition is used. Larger aliquots or even the total volume of the digestion solution can therefore be transferred to the reaction flask when higher sensitivity is required.

RESULTS AND DISCUSSION

In a first set of experiments a nitric acid digestion in a boiling water bath was compared with a nitric, sulfuric, and perchloric acid decomposition at a maximum of 310°C. The results obtained for a lyophilized

TABLE 1. Determination of Selenium in Lyophilized Human Blood Serum Reference
Material Seronorm, Batch 103[a]

Decomposition number	HNO₃ digestion (6 hr water bath)	HNO₃, H₂SO₄, HClO₄ (310°C)
1	64[b]	91
2	64	89
3	60	95
4	64	88
5	68	86
6	64	86
$\bar{x}$	64	89
s	2	3
RSD %	3.7	3.9

[a] Accepted value: 91 ± 7 μg/liter selenium (Iherat et al. 1984).
[b] Se, μg/liter, found.

blood serum (Seronorm, Batch 103) are shown in Table 1. In six independent digestions with nitric acid, an average selenium concentration of 64 ± 3 μg/liter was found which is very close to the average of 67 ± 7 μg/liter obtained by four laboratories for the same material in the IUPAC interlaboratory trial using AD HG AAS (Ihnat et al. 1984).

Using this nitric, sulfuric, and perchloric acid digestions, however, an average selenium concentration of 89 ± 3 μg/liter was found, which compares favorably with the accepted value of 91 ± 7 μg/liter. This suggests that the low mean obtained with AD HG AAS in the interlaboratory comparison for this reference material is not due to systematic errors in the final determination of selenium with this technique, but to improper acid decomposition procedures.

Selenomethionine is among the most acid-resistant compounds found in human body fluids (Nève et al. 1982; Verlinden 1982). To further elucidate the influence of the decomposition procedure used prior to the determination of selenium with HG AAS, the recovery of this organoselenium compound from aqueous solutions as well as from whole blood was investigated. Figure 3 shows the signals obtained from aqueous solutions of selenomethionine after different acid decompositions in comparison to a reference solution prepared from tetravalent ionic selenium (selenite). Without a decomposition, no signal is found for selenomethionine. Using a nitric acid decomposition, 6 hr on a boiling water bath, or 1 hr at 160°C under pressure, the recovery is 10% and 50%, respectively. Only after a nitric, sulfuric, and perchloric acid decomposition with a maximum temperature of 310°C is full recovery obtained for selenium from selenomethionine.

Essentially, the same behavior was found for selenomethionine add-

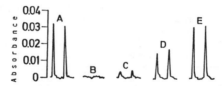

FIG. 3.   Recovery of selenium from selenomethionine in relation to the acid decomposition procedure. (A) Reference solution, 8.0 ng Se (IV). (B–E) Selenomethionine, 7.9 ng Se. (B) No decomposition; (C) nitric acid in boiling water bath; (D) nitric acid under pressure; (E) nitric, sulfuric, and perchloric acid, 310°C.

ed to whole blood (Table 2). The values found for selenium in a whole blood sample after a nitric acid digestion in a boiling water bath or under pressure were 10–15% lower than those obtained after a nitric, sulfuric, and perchloric acid digestion. The recovery of selenomethionine added to this blood sample, however, was only 35% and 77%, respectively, with the two nitric acid digestion procedures, whereas it was always around 100% after a nitric, sulfuric, and perchloric acid digestion.

Finally, samples of heparinized whole blood, blood plasma, and erythrocytes from six healthy adult individuals were analyzed using the recommended procedure. The erythrocyte samples were independently analyzed as well using a solvent extraction-graphite furnace AAS procedure (Welz et al. 1984). All acid decompositions and determinations were done at least in duplicate, and the results (Table 3) were typically in agreement within ±3%. The average values for selenium are 88 μg/liter in whole blood, 75 μg/liter in blood plasma, and 307 μg/kg Hb in erythrocytes. The average value for selenium in erythrocytes found with the independent procedure was 300 μg/kg Hb, which is in good agreement with AD HG AAS.

TABLE 2. Recovery of Selenomethionine (Se-Me) Added to Heparinized Whole Blood after Different Acid Decomposition Procedures[a]

Decomposition procedure	Added Se-Me (μg/liter)	Se found (μg/liter)	Average recovery of Se (μg/liter)	(%)
HNO_3 (6 hr/95°C)	0	90 ± 3	—	—
	79	118 ± 4	28	35
HNO_3 (pressure/160°C)	0	91 ± 5	—	—
	79	152 ± 7	61	77
$HNO_3/H_2SO_4/HClO_4$ (310°C)	0	103 ± 6	—	—
	79	182 ± 6	79	100

[a] Each value is the average of six independent determinations.

TABLE 3. Selenium in Whole Blood, Plasma, and Erythrocytes of 6 Adult Individuals after Nitric, Sulfuric, and Perchloric Acid Decomposition[a]

Subject	Whole blood (μg/liter)	Plasma (μg/liter)	Erythrocytes AD HG AAS	Hb (μg/kg) AD SE GFAAS
BW	94	75	353	345
GS	82	70	279	275
MM	110	91	382	392
SA	78	71	277	267
SG	82	69	287	297
WE	82	71	265	225

[a] Erythrocyte samples were also analyzed using acid decomposition–solvent extraction–graphite furnace AAS (according to Welz et al. 1984).

These experiments show that accurate values for selenium in human body fluids can be obtained with hydride-generation atomic absorption spectrometry when a suitable sample decomposition procedure is applied. The recommended procedure includes a treatment with nitric, sulfuric, and perchloric acid up to a maximum temperature of 310°C.

REFERENCES

Brown, A. A., Ottaway, J. M., and Fell, G. S. 1982. Determination of selenium in biological material: Comparison of three atomic spectrometric methods. Anal. Proc. 19, 321–324.

Clinton, O. E. 1977. Determination of selenium in blood and plant material by hydride-generation and atomic spectroscopy. Analyst 102, 187–192.

Ihnat, M., Wolynetz, M., Thomassen, Y., and Verlinden, M. 1984. IUPAC interlaboratory trial on the determination of selenium in lyophilized human blood serum reference materials. Pure Appl. Chem. (in preparation).

Kaiser, G., Götz, D., Tölg, G., Knapp, G., Maichin, B., and Spitzy, H. 1978. Study of systematic errors in the determination of total Hg levels in the range $<10^{-5}\%$ in inorganic and organic matrices with two reliable spectrometrical determination procedures. Fresenius' Z. Anal. Chem. 291, 278–291.

Kotz, L., Kaiser, G., Tschöpel, P., and Tölg, G. 1972. Decomposition of biological materials for the determination of extremely low contents of trace elements in limited amounts of nitric acid under pressure in a teflon tube. Z. Anal. Chem. 260, 207–209.

Lloyd, B., Holt, P., and Delves, H. T. 1982. Determination of selenium in biological samples by hydride-generation and atomic absorption spectroscopy. Analyst 107, 927–933.

Nève, J., Hanocq, M., Molle, L., and Lefebvre, G. 1982. Study of some systematic errors during the determination of the total selenium and some of its ionic species in biological materials. Analyst 107, 934–941.

Oster, O., and Prellwitz, W. 1982. A methodological comparison of hydride and carbon furnace atomic absorption spectroscopy for the determination of selenium in serum. Clin. Chim. Acta *124*, 277–291.

Robbins, W. B., Caruso, J. A., and Fricke, F. L.   1979.   Determination of germanium, arsenic, selenium, tin, and antimony in complex samples by hydride-generation–microwave-induced plasma atomic emission spectrometry. Analyst *104*, 35–40.

Verlinden, M.   1982.   On the acid decomposition of human blood and plasma for the determination of selenium. Talanta *29*, 875–882.

Welz, B., and Melcher, M.   1984.   Decomposition of marine biological tissues for determination of arsenic, selenium and mercury using hydride-generation and cold vapor atomic absorption spectrometry. Anal. Chem. (submitted for publication).

Welz, B., Melcher, M., and Nève, J.   1984.   Determination of selenium in human body fluids with hydride-generation atomic absorption spectrometry—Optimization of sample decomposition. Anal. Chim. Acta (submitted for publication).

Section VI

Selenium Status of Humans and Human Food

62

Human Selenium Requirements in China

Guang-Qi Yang   *Peng-Chu Qian*
Lian-Zhen Zhu   *Jia-Hong Huang*
Sheng-Jie Liu   *Min-De Lu*
Lu-Zhen Gu

Work has already been done to establish the fact that Keshan disease (KD), an endemic cardiomyopathy widely distributed in China, is closely related to the deficiency of selenium (*1–3*). In order to develop further the role of selenium in KD and to study practical methods of selenium supplementation or fortification to be used in the endemic area for human intervention purposes (*4*) and to acquire more definitive human nutrition standards for the establishment of the recommended dietary allowance (RDA) for selenium, the determination of the selenium requirement of humans has been carried out.

DIETARY STUDIES ON MINIMUM SELENIUM REQUIREMENT

It has been shown that the selenium status of people in KD-affected areas was very poor. The selenium levels in blood and hair and GSHPx activities of blood were extremely low as compared with reports from other low selenium areas of the world (*5,6*). For example, blood selenium levels in KD and KD–Kaschin–Beck disease areas averaged 18 ng/g, of which 22% were less than 10 ng/g and averaged only 7 ng/g (*7*), while it was 60 ng/ml in New Zealand (*5*). Although both selenium levels and enzyme activities in nonaffected areas are consistently

higher than that in adjacent affected areas, the former were still at
lower levels in comparison with data from other parts of China (Fig.
1). It is obvious that selenium intake in a nonaffected area but adjacent to an affected area would be approximate to the minimum Se
requirement for humans.

Sites and Populations Selected

A dietary survey was carried out in two sites in Sichuan Province,
one in the Shiba brigade, Huanglian Commune, and the other in several production teams in Huangshui Commune. The former is the so-
called safety island, a nonaffected site surrounded by the latter affected area. The distance between the two sites is less than 20 km.

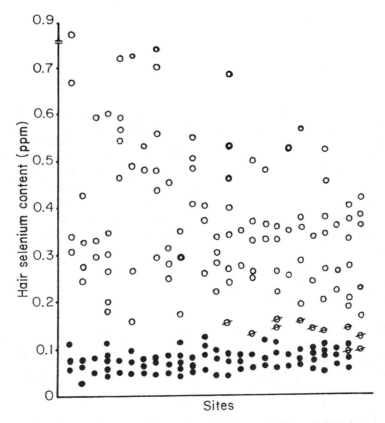

FIG. 1. Distribution of hair selenium in areas of different selenium
status. (●), Affected area; (○) nonaffected; (⌀) nonaffected, close
to affected area.

TABLE 1. Incidence of KD in Huangshui Commune
(Affected) and Huanglian Commune (Not Affected)[a]

	A	NA
Population	294	363
Cases	9	0

[a] In the years from 1974 to 1980.

Prevalence of KD of both sites is shown in Table 1. In the affected
site 9 KD patients were picked out among 294 subjects examined,
while in the safety island none was found among 363 subjects (Table
1). No pronounced nutrition deficiency diseases were found in the subjects examined, and severe cases of protein-calorie deficiency never
appeared among children of susceptible age. The selenium and protein
statuses of male adults from families surveyed are shown in Table 2.
Blood GSHPx activity was determined by the method of Paglia and
Valentine (8). The selenium status of the affected site was within the
range of deficiency previously reported (9), and that of the safety
island was just beyond this range (Table 2). In order to compare the
selenium status with previous data, hair selenium concentrations of
school children were also estimated. They were 0.065 ppm ± SE .009
for the former and 0.160 ppm ± SE .029 for the later. Plasma vitamin
E was estimated before starting the survey and was found at the
marginal level of deficiency, which were 3.8 and 5.0 μg/ml for affected
and nonaffected areas, respectively.

Methods

Dietary surveys were made in May and November, 1982. Eighteen
representative families were selected in each site for 3 consecutive
days of the survey. Food items consumed were collected and the se-

TABLE 2. Se Levels in Hair, Blood, and Urine, and GSHPx Activities in Blood and Other
Nutritive States of Male Adults in KD Affected (A) and Nonaffected (NA) Areas

	A	NA	P
Hair Se levels (ppm)	0.098 (18)[a]	0.186 (18)	<.001
Whole blood Se levels (ppm)	0.017 (16)	0.033 (14)	<.001
Whole blood GSHPx activities (nmoles			
NADPH ox/min/mg Hb)	6.62 (16)	14.36 (14)	<.001
Hemoglobin	15.0 (18)	15.7 (18)	>.05
Serum protein (%)	6.7 (18)	6.6 (18)	>.05
Amount of Se in 12-hr urine (μg)	1.54 (16)	2.77 (21)	<.01

[a] Figures in parentheses indicate number of samples.

TABLE 3. Comparison of the Se Intakes Obtained between Analysis of Food Items and Composite Diet Consumed

Method	Number of subjects	A or NA[a]	Countryside or town	Se intake (μg/day)	Difference[b] (%)
Composite diet	9	A	Countryside	7.4 (5.7– 9.1)[c]	+21.6
Foods	8	A	Countryside	9.0 (5.8–14.1)	
Composite diet	11	NA	Countryside	15.3 (12.1–20.0)	+10.5
Foods	16	NA	Countryside	16.9 (13.6–23.6)	
Composite diet	17	NA	Town	12.4 (5.1–24.0)	−24.2
Foods	18	NA	Town	9.4 (4.4–14.4)	

[a] A, Affected area; NA, nonaffected area.
[b] Composite diet basis.
[c] Numbers in parentheses indicate the range of intake.

lenium and protein contents were analyzed. Daily Se intakes were either obtained by direct estimation of the composite diet or by calculation from the food selenium contents according to the quantities recorded. The latter method needs relatively smaller quantities of foods for use. Results obtained from the two methods are shown in Table 3. They indicate that results obtained by food analysis were 10–20% higher than those by actual composite diet analysis in the countryside, but it was 24% lower when the survey was made in town. It seemed that residents in town are usually somewhat generous with their food for the survey. Therefore, selenium intakes were estimated throughout this survey by food analysis based on careful inquiry of amounts consumed.

Results

Selenium Content of Foods. Selenium content of foods from the safety island is significantly higher than that produced in the nearby affected area (Table 4). Most vegetables contained negligible amounts of selenium, but animal foods such as fish, shrimp, kidney, liver, meat, and eggs are good sources of dietary selenium in endemic areas. The bulk of Se intake is from staple cereals.

TABLE 4. Se Content of Food in KD Area (μg/100 g Fresh Food)

Food items	Affected site	Safety island	Market
Rice	0.6	1.5	0.8
Wheat flour	0.3	1.5	1.1
Soybean	1.2	7.2	—
Peanut	0.780	1.72	—
Vegetables			
Leafy	0.066	0.131	—
Fruity	0.020	0.156	—
Tuberous root	0.041	0.191	—
Fresh beans	0.050	0.126	—
Garlic	0.224	—	—
Garlic green	0.105	—	—
Fish	16.8	12.7	12.0
Shrimp	—	17.1	—
Meat			
Pork	—	—	3.00
Beef	—	3.4	—
Chicken	—	—	3.40
Liver, pig	—	—	5.50
Kidney, pig	—	—	61.3
Egg, chick	—	—	6.1
Egg, duck	—	—	18.4
Egg, alkali preserved	—	—	29.0

TABLE 5. Content of Some Se-Antagonistic Elements and Trace Elements Proposed To Be Related to KD in Staple Foods

Elements	Rice (μg/g)		Wheat flour (μg/g)		Soybean (μg/g)	
	A	NA	A	NA	A	NA
Pb	0.213 (3)	0.061 (4)	0.726 (2)	0.088 (1)	0.146 (1)	0.013 (3)
Cd	0.009 (5)	ND (3)[b]	0.001 (2)	ND (1)	0.005 (1)	ND (1)
Hg	0.001 (5)	0.004 (11)	0.004 (2)	0.007 (1)	0.003 (1)	0.004 (6)
As	0.062 (3)	0.104 (3)	0.172 (2)	0.047 (1)	0.194 (1)	0.224 (3)
Zn	10.6 (3)	8.7 (3)	7.0 (2)	11.1 (1)	36.7 (1)	32.2 (3)
Mo	8.97 (4)	0.52 (4)	1.30 (3)	0.12 (1)	37.5 (4)	5.33 (4)
Cu	3.2 (3)	1.5 (3)	2.4 (2)	1.7 (1)	11.3 (3)	13.4 (3)
Mn	7.5 (3)	11.1 (3)	9.2 (2)	15.6 (1)	23.3 (3)	18.5 (3)

[a] A, Affected area; NA, nonaffected area. Numbers in parentheses are number of samples.
[b] ND, No data.

594

TABLE 6. Amounts of Some Antagonistic Elements Excreted in 12-Hr Urine of Residents in KD Affected (A) and Nonaffected (NA) Areas

Element (μg/liter)	A	NA
Pb	16.4 ± 1.9[b]	13.2 ± 1.6
Cd	0.44 ± 0.14	0.59 ± 0.09
As[a]	40	57
Hg	1.94 ± 0.163	1.13 ± 0.245

[a] Pooled sample.
[b] Mean ± SEM.

Other Trace Element Content of Locally Produced Foods. Elements known to be antagonistic to selenium or that have been proposed to be related to KD were also determined. Pb, Cd (*10*), and Hg (*11*) were determined by flameless atomic absorption methods, Zn, Cu, Mn, and Mg by conventional flame atomic absorption methods, and As (*12*) and Mo (*13*) by oscilloscopic polarography. Results are shown in Tables 5–7. Higher content of Pb and Mo was found in grain samples from affected areas, but only Mo concentration of hair and grain from affected areas was significantly higher than that from nonaffected areas.

Other Nutrient Intakes. Other nutrient intakes such as protein, fat, calories, and vitamins A, B_1, B_2, and C were calculated and none were found to be deficient. Notable signs of deficiency were not found. Dietary protein intake of 3- to 6-year-old children in the affected site was 32 g, of which about 70% came from rice and around 20% from animal foods and soybean; the rest was from other plant sources.

Selenium Intake. Daily selenium intake of different age groups is presented in Table 8. Daily selenium intake of male adults was usually higher than that of females, and averages of both sexes were 7.2 and 16.2 μg/day in affected and nonaffected areas, respectively. Intakes of infants within 1 year of age were calculated from selenium

TABLE 7. Concentrations of Zn and Mo in Hair of Residents in KD Affected (A) and Nonaffected (NA) Areas

Element (μg/g)	A	NA
Zn	138.0 ± 3.7[a]	135 ± 4.4
Mo	0.366 ± 0.016	0.123 ± 0.035

[a] Mean ± SEM.

TABLE 8. Daily Dietary Se Intake of Residents in KD Affected (A) and Nonaffected (NA) Areas

Age (yr)	A			NA	
	Number of subjects	Intakes of Se[b] (μg/day)		Number of subjects	Intakes of Se[b] (μg/day)
1[a]	5	2.0 (1.0– 2.8)		5	3.0 (2.0– 3.4)
3– 6	4	3.1 (2.9– 3.3)		8	5.3 (3.2– 7.6)
7–12	10	4.6 (2.6– 6.3)		8	9.6 (7.6–15.6)
13–17	7	6.1 (3.7– 7.9)		6	14.2 (9.1–19.8)
18–55	26 male	7.7 (4.3–11.3)		22 male	19.1 (9.4–30.3)
	13 female	6.6 (4.2–10.6)		19 female	13.3 (5.4–26.1)
56	10	5.5 (3.6– 8.5)		11	13.8 (8.8–19.4)

[a] Calculated from the Se concentrations of mothers' milk.
[b] Average values from two surveys in May and November, 1982. Numbers in parentheses indicate the range of intake.

TABLE 9. Se Content of Human Milk from Areas of Different Se Levels and Calculated Daily Se Intake of Infants within 1 Year of Age

Sampling site	Se concentration of human milk (ppm)		Se intake calculated (μg/day)
KD-affected area	0.0026	(5)[a]	2.0
Safety island	0.0038	(5)	2.9
Beijing	0.020	(15)	15.7
Enshi	0.283	(3)	220.0

[a] Numbers in parentheses indicate number of subjects.

TABLE 10. Percentage Distribution of Se Intake from Different Dietary Sources in KD Affected (A) and Nonaffected (NA) Areas

Age	A			NA		
	Cereal	Other plant food	Animal food	Cereal	Other plant food	Animal food
3– 6	69.2	22.2	8.6	62.2	18.4	19.4
7–12	71.6	21.1	7.2	70.2	13.4	16.4
13–17	69.4	19.2	11.3	66.0	19.9	14.2
18–55	69.9	21.8	8.4	70.6	16.0	13.5
56	75.6	17.4	6.9	70.0	14.6	15.4

concentration of mother's milk (Table 9) by assuming a daily consumption of 785 ml of milk, which was previously estimated for infants. Milk selenium concentration seemed to be a good indicator of the mother's selenium status.

Dietary Source of Selenium.   Percentage distribution of Se intake from various dietary sources is shown in Table 10. Daily selenium intakes from cereal in both affected and nonaffected areas were about 70%. Less than 10% and about 15–20% of selenium intake came from animal foods in affected and nonaffected areas, respectively.

BIOCHEMICAL STUDY OF THE SELENIUM REQUIREMENT

Site and Population

This study was carried out at the Giaowo farm in a low selenium area of Sichuan Province. Healthy male adults, 18–42 years of age, were selected. Before starting the observations, hair selenium concentrations were estimated and all participants were below 0.2 ppm.

Method

Graded levels, 0, 10, 30, 60, and 90 μg selenium from DL-selenomethionine, were orally supplemented every day and sent to every subject in 5 groups of 8 or 9 subjects each. The experimental diet consisted of foods almost exclusively produced locally. All subjects dined at the public canteen. Daily foods consumed in the canteen were recorded, and individual intakes were calculated during the whole experimental period from April to November in 1983. The average selenium intake for each person was 10.9 ± SD 0.6 μg/day.

Selenium concentrations in erythrocytes, plasma, and 12-hr urine (night), and GSHPx activities in erythrocytes and plasma were determined at 10-day intervals during the first 6 months of observation and once a month afterwards. The observation was continued for 8 months. Blood samples were kept in a low-temperature refrigerator below −20°C before they were transferred to the laboratory for the enzyme assay.

Results

Change of Selenium Concentrations and GSHPx Activities in Erythrocytes.   Selenium concentrations of each group reached a plateau at different levels after the sixth month (Fig. 2). Plateauing of GSHPx activities seemed not to be reached during the whole period (Fig. 3). Only enzyme activity curves of groups 30, 60, and 90 μg declined after the sixth month. Before the sixth month the increase of enzyme activity significantly correlated with selenium concentrations, with the correlation coefficient 0.70. No significant correlation was found after this month.

Change of Selenium Concentration and GSHPx Activities in Plasma. Both selenium concentrations (Fig. 4) and GSHPx activities (Fig. 5) in plasma increased rapidly in the first month, in contrast with what appeared in the erythrocyte, and they tended to reach a plateau after the fourth month. GSHPx activity rose to a maximum in the third month and thereafter decreased gradually, finally reaching a plateau in the sixth month. Groups 30, 60, and 90 μg coincided with each other after the seventh month.

Changes of ratios of plasma GSHPx activity to selenium concentration are shown in Fig. 6. This pattern is somewhat different from the single GSHPx activity change.

Change of Amounts of Selenium Excretion in 12-Hr Urine. Changes of absolute amounts of selenium excreted in 12-hr urines in the first,

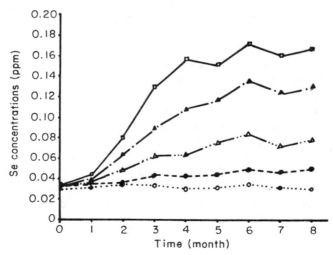

FIG. 2. Change of erythrocyte selenium concentrations of groups supplemented with graded amounts of DL-selenomethionine during the experimental period. (○), 0 μg; (●), 10 μg; (△), 30 μg; (▲), 60 μg; (□), 90 μg.

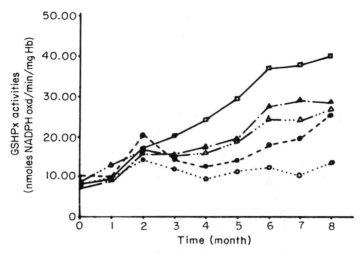

FIG. 3. Change of erythrocyte GSHPx activities of groups supplemented with graded amounts of DL-selenomethionine during the experimental period. See Fig. 2 legend for explanation of symbols.

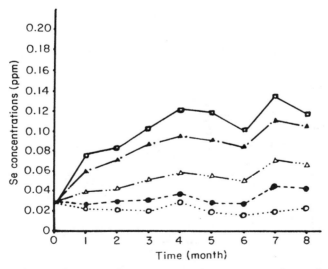

FIG. 4. Change of plasma selenium concentrations of groups supplemented with graded amounts of DL-selenomethionine during the experimental period. See Fig. 2 legend for explanation of symbols.

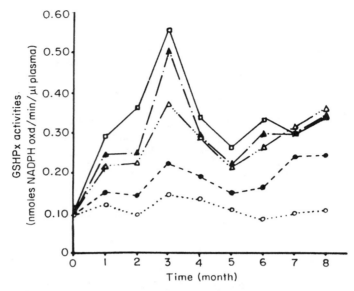

FIG. 5. Change of plasma GSHPx activity groups supplemented with graded amounts of DL-selenomethionine during the experimental period. See Fig. 2 legend for explanation of symbols.

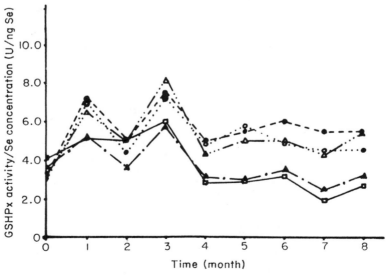

FIG. 6. Change of ratios of plasma GSHPx activity to selenium concentrations of groups supplemented with graded amounts of DL-selenomethionine during the experimental period. See Fig. 2 legend for explanation of symbols.

fourth, and seventh months of each group are shown in Fig. 7. In the first month, almost a linear relationship was obtained between the amount of selenium intake and the selenium excreted. A rise in the 90-µg group in the fourth month may indicate a higher degree of tissue saturation reached earlier in this group. A large increase in the amount of selenium excreted in the same group in the seventh month would in fact indicate that a degree of tissue saturation increased both with time and dose. It appears that within certain limits of selenium intake the longer the supplemented time or larger the dose size, the higher the degree of tissue saturation and hence the larger the amount of selenium appearing in the urine.

As the amount of selenium excreted in the urine was expressed by the percentage of selenium intake (Table 11), it is interesting to find that during the first 3 months, no matter the amount of selenium supplemented, the percentages excreted in all supplemented groups dropped to lower but similar levels, i.e., 12–14% of their intakes. A striking change in the 90-µg group appeared after the fifth month, i.e., the average percentage from the last 3 months rose to a level of 19.7%

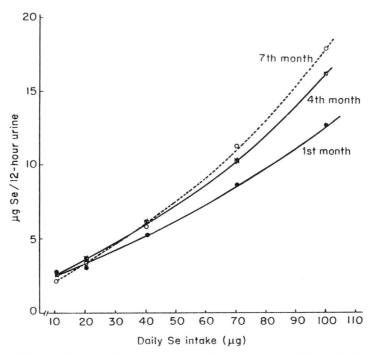

FIG. 7.   Change of amounts of selenium excreted in 12-hr urine in groups supplemented with graded levels of selenium.

TABLE 11. Excretion of Se in 12 Hr Urine of Subjects in Groups Supplemented with Graded Amounts of Se (μg) throughout the Period of Observation

Number of months after supplementation	0		10		30		60		90	
	Amount (μg)	% of intake	Amount (μg)	% of intake	Amount (μg)	% of intake	Amount (μg)	% of intake	Amount (μg)	% of intake
0	2.2 (7)[a]	20.0	2.5 (9)	22.7	2.6 (8)	23.7	2.2 (9)	19.6	2.5 (8)	23.1
1	2.8 (7)	20.6	3.0 (9)	14.0	5.2 (8)	13.3	8.6 (9)	12.1	12.7 (8)	12.6
2	2.0 (7)	20.0	3.4 (9)	15.0	5.8 (8)	14.2	9.1 (8)	13.0	13.0 (8)	13.0
3	2.0 (7)	18.2	3.0 (9)	13.8	5.5 (8)	13.8	8.2 (9)	11.5	13.3 (7)	13.2
4	2.6 (7)	23.4	3.6 (8)	16.9	6.1 (8)	14.7	10.2 (8)	14.3	16.2 (7)	16.0
5	2.8 (7)	27.0	3.8 (9)	17.3	7.8 (7)	18.4	12.0 (9)	16.3	21.7 (7)	21.6
6	1.5 (6)	13.6	1.6 (6)	7.6	5.4 (4)	13.0	10.0 (8)	14.1	20.0 (5)	19.8
7	2.1 (7)	18.0	3.3 (6)	15.0	5.8 (7)	13.9	11.2 (8)	16.9	17.9 (4)	17.6
Average of the first 3 months	2.1	19.4	3.1	14.2	5.5	13.8	8.6	12.2	13.0	12.9
Average of the last 3 months	2.1	19.5	2.9	13.3	6.3	15.1	11.1	15.8	19.8	19.7

[a] Numbers in parentheses indicate number of subjects observed.

compared with the level of the unsupplemented group of 19.5%. Again this indicated a gradual increase in saturation of tissues as the process of supplementation was continued until they were restored to their stable starting level of 23.1%. Saturation levels of other groups were expected to appear if supplementation was continued.

DISCUSSION

Studying the relation between blood selenium concentrations and GSHPx activities in New Zealand indicated (14) that erythrocyte enzyme activity varied with selenium concentrations up to about 0.14 μg/ml, and thereafter the enzyme activity reached a plateau. This plateauing of enzyme activity was interpreted by these authors (14) to mean that the selenium requirement for the enzyme was met and this dietary selenium intake would represent a dietary requirement. In this study we found a plateauing of plasma enzyme activities among groups supplemented with 30, 60, and 90 μg selenium daily, and a coincidence of activity among these three groups was obtained after the seventh month (Fig. 5). Based upon the above suggestion, if plateauing of enzyme activity either in erythrocyte or in plasma is with the same mean, the selenium requirement for adults would be around 40 μg selenium per day.

Plasma selenium levels increased rapidly once selenium supplementation began (Fig. 4), in agreement with observations in Finland (15). Plasma enzyme activities also rose rapidly until they reached a maximum. After a gradual decrease, maximal levels were never restored again even though the enzyme activity reached a plateau. This probably indicates that the increase of plasma enzyme was too fast for the body to deal with the excess enzyme by its homeostatic mechanisms during the preceding stage of selenium supplementation, thus a peak appears in the curve (Fig. 5). This peak activity could not be explained by an unusual change of dietary selenium intake, since this high level of enzyme activity in plasma is an average of four determinations within 40 days in the third and fourth months after supplementation.

Plateauing of erythrocyte GSHPx activities has not been observed, even though erythrocyte selenium concentrations tended to reach a plateau at a level of 0.17 ppm selenium in the 90-μg group and 0.13 ppm in the 60-μg group. One possibility is that the observed time was still not long enough for the tissues to reach a state of equilibrium.

The daily selenium intake of 16.2 μg estimated in the safety island should approach the minimum requirement. This is the level of marginal intake below which KD may occur. Calculating the selenium content of staple cereals in the nonaffected area indicates that the daily minimum selenium requirement should not exceed 20 μg per day, since samples containing more than 0.02 ppm selenium were exclusively from nonaffected areas (7). Peasants in the endemic area usually had 70% of their selenium intakes from staple cereals, in good agreement with what was obtained in New Zealand by balance studies (16).

The result obtained from the biochemical study in this trial is more than twice the requirement obtained from dietary intake. Large differences in the size of body selenium pools of residents in different areas should not be neglected. A selenium depletion/repletion trial in young North American men (17) indicated an initial urinary excretion of 54 ± SD 11 μg/day during the first 3-day depletion period. In connection with the selenium status of the subjects at the terminal stage of this observation, Table 12 indicates that selenium excretion of the unsupplemented groups was only 4.8 ± SE 1.7 μg/day. This is less than one-tenth the daily urinary excretion of young healthy American men.

In spite of the low level, the human body seems able to handle this level of selenium intake well. This is evidenced by the data in Table 11. The results indicated that among the percentages excreted, only the 90-μg group could be restored to starting levels compared with the unsupplemented group at the final stage of supplementation. Probably the body could keep its urinary selenium as a constant percentage of its intake, even though the intake was extremely low. When the selenium status is poor, the body will retain more selenium to raise its tissue selenium concentration to a new level at the initial stage of supplementation. Thus, a decrease of selenium excretion was expected in the supplemented groups. Once the body reaches a new stable level, it will assign the same percentage of selenium to be excreted by way of

TABLE 12. Se Status of Subjects at the Terminal Stage of the Experimental Period in Groups Receiving Graded Levels of Se Supplementations

Amount of Se supplemented (μg)	Whole blood (ppm)	Hair (ppm)	Nail (ppm)	Urine (μg/24 hr)
0	0.029 ± 0.001[a]	0.164 ± 0.019	0.169 ± 0.019	4.8 ± 1.7
10	0.050 ± 0.002	0.252 ± 0.014	0.241 ± 0.030	8.7 ± 1.8
30	0.089 ± 0.013	0.373 ± 0.040	0.396 ± 0.030	15.9 ± 3.8
60	0.118 ± 0.039	0.503 ± 0.077	0.618 ± 0.102	25.7 ± 4.1
90	0.178 ± 0.006	0.591 ± 0.022	0.796 ± 0.163	39.3 ± 9.4

[a] Mean ± SEM.

the kidney. The controversy in the literature (18,19) over the effect of the dose on the urinary excretion of selenium in experimental animals perhaps could be attributed to the differences in the selenium status of animals, size of dose, or duration of selenium supplementation. Time is particularly important to permit the body to reach the corresponding degree of saturation. It seems that the homeostatic mechanisms of the body could control the excretion of ingested selenium to a constant percentage through the kidney when the selenium intake was at nutritional levels. In this respect, Burk et al. (19) had observed a decreased percentage of selenium excretion in urine in the experimental animals when the intake of selenium was in excess.

Levander et al. (17) suggested that young North American men need a dietary selenium intake of 70 μg/day to replace losses and maintain body store when absorption of selenium in the diet is taken into account. Their result is in fair agreement with what we obtained in this study. In our 90-μg group (daily selenium intake = 100), the amount of selenium excreted in the urine was 39.3 ± SE 9.4 μg (Table 12), compared with 54 ± SD 11 μg obtained in North America. The reason for the little discrepancies in intakes and excretions could be attributed to the difference in original selenium status of populations and in bioavailabilities of the selenium supplement in their traditional diets.

If an adaptation mechanism does exist in humans, it may operate even during infancy. An estimation of the selenium content in mother's milk indicated that there was a great disparity between concentrations from high and low selenium areas (Table 8). Infants within 1 year of age get only 2–3 μg selenium/day in the endemic area, and this intake is far less than the 8 μg/day for infants receiving cow's milk formula in West Germany (20). In Beijing, the intake from mother's milk was 15.7 μg/day, but it increased to 220 μg/day in seleniferous areas without any signs of intoxication. This is 100-fold as high as that from a low selenium area.

Mertz (21) emphasized the influence of environmental pollution on the dietary selenium requirement. Since KD is unlikely to be an uncomplicated selenium deficiency disease and the interactions between selenium and some metals such as Cd, Hg, and As are well known, these metals were studied in this experiment. No significant contamination of the above metals was found in endemic areas. Only environmental contamination of Pb was observed in grain samples taken from affected areas. However, evidence for an antagonism between Pb and nutritional levels of selenium is inconclusive (22). Mo was found to be higher in samples of grain and hair from affected areas, and this may represent an oxidant stress on these residents. Mertz (23) supposed the second factor in the etiology of KD may consist of dietary sulfur-containing amino acids. It has been repeatedly demonstrated that methionine supplementation will increase the availability of dietary selenium in experimental animals (7,24). However, the real meaning of this result needs to be examined in practical diets of humans in the endemic area.

ACKNOWLEDGMENTS

The authors express their gratitude to Dr. Orville A. Levander, Vitamin and Mineral Nutrition Laboratory, Beltsville Human Nutrition Research Center, Beltsville, Maryland, for his enthusiastic assistance during the planning of this study.

REFERENCES

1. Keshan Disease Research Group of the Chinese Academy of Medical Sciences 1979. Epidemiological studies on the etiologic relationship of selenium and KD. Chin. Med. J. *92*, 477.
2. Chen, X.-S., Yang, G.-Q., Chen, T.-S., Chen, X.-C., Wen, Z.-M., and Ge, K.-Y. 1980. Studies on the relations of selenium and Keshan disease. Biol. Trace Elem. Res. *2*, 91.
3. Xian Medical College 1979. Observations on the efficacy of sodium selenite for the prevention of Keshan disease. Chin. Med. J. *59*, 457 (in Chinese).
4. Yang, G.-Q. *et al.* 1982. Approaches to the supplementation of selenium in the prevention of Keshan disease. Acta Nutr. Sin. *4*, 1.
5. Watkinson, J. H. 1981. Changes of blood selenium in New Zealand adults with time and importation of Australian wheat. Am. J. Clin. Nutr. *34*, 936.
6. Lombeck, I., Kasperck, K., Feinendegen, L. E., and Bremer, H. J. 1981. Low selenium state in children. *In* Selenium in Biology and Medicine. J. E. Spallholz, J. L. Martin, and H. E. Ganther (Editors), p. 269. AVI Publishing Co., Westport, CT.
7. Yang, G.-Q. Research on selenium-related problems in human health in China. This volume, Chapter 2.
8. Paglia, D. E., and Valentine, W. N. 1979. Studies on the quantitative and qualitative characterization of erythrocyte glutathionine peroxidase. J. Lab. Clin. Med. *70*, 158.
9. Yang, G.-Q. *et al.* 1982. Relationship between the distribution of Keshan disease and selenium status. Acta Nutr. Sin. *4*, 191 (in Chinese).
10. Legotte, P. A. *et al.* 1980. Determination of Cd and Pb in urine and other biological samples by graphite furnace atomic absorption spectrometry. Talanta *27*, 39.
11. Campe, A. *et al.* 1978. Determination of inorganic mercury, phenylmercury and total mercury in urine. At. Absorp. Newsl. *17*, 100.
12. Zhou, L.-J. *et al.* 1977. Determination of trace amounts of As in hair by oscilloscopic polarography. Proc. Clin. Polaro. Anal. Chem. Conf. (in Chinese).
13. Gua, X.-S. 1974. Application of catalytic wave polarography in water analysis. Bull. Chem. Chin. Acad. Sci. *1*, 15 (in Chinese).
14. Rea, H. M., Thomson, C. D., Campbell, D. R., and Robinson, M. F. 1979. Relation between erythrocyte selenium concentrations and glutathionine peroxidase activities of New Zealand residents and visitors to New Zealand. Br. J. Nutr. *42*, 201.

15. Levander, O. A., Huttunen, J. K., Kataja, M., Koivistoinen, P., and Pikkarainen, J. 1983. Bioavailability of selenium to Finnish men as assessed by platelet glutathionine peroxidase activity and other blood parameters. Am. J. Clin. Nutr. *37*, 887.

16. Stewart, R. D. H., Griffiths, N. M., Thomson, C. D., and Robinson, M. F. 1978. Quantitative selenium metabolism in normal New Zealand women. Br. J. Nutr. *40*, 45.

17. Levander, O. A., Sutherland, B., Morris, V. C., and King, J. C. 1981. Selenium balance in young men during selenium depletion and repletion. Am. J. Clin. Nutr. *34*, 2662.

18. Hopkins, L. L., Jr., Pope, A. L., and Baumann, C. A. 1966. Distribution of microgram quantities of selenium in the tissues of the rat, and effects of previous selenium intake. J. Nutr. *88*, 61.

19. Burk, R. F., Seely, R. J., and Kiker, K. W. 1973. Selenium: Dietary threshold for urinary excretion in the rat. Proc. Soc. Exp. Biol. Med. *142*, 214.

20. Lombeck, I., Kasperek, K., and Menzel, H. Selenium status in pediatric patients. This volume, Chapter 64.

21. Mertz, W. 1976. Present status and future development of trace element analysis in nutrition. In Ultratrace Metal Analysis in Biological Science and Environment. H. R. Terence (Editor), No. 172. Adv. Chem. Ser., Washington, D.C.

22. Levander, O. A. 1979. Lead toxicity and nutritional deficiencies. Environ. Health Perspect. *29*, 115.

23. Mertz, W. 1982. Personal communication.

24. Gu, L.-Z. *et al.* Influence of dietary constituents on the bioavailiability of selenium. This symposium, Chapter 49.

Assessment of Human Exposure to Environmental Selenium

B. G. Bennett
P. J. Peterson

INTRODUCTION

Selenium is an element that occurs naturally in the environment and is released from industrial sources, particularly from fossil fuel combustion. Harmful effects in animals and man may result from both deficient or excessive amounts of selenium intake.

Intake of selenium by man occurs primarily from ingestion of selenium-containing foods. Inhalation of selenium in air is a secondary intake pathway. With understanding of the environmental and metabolic behavior of selenium, the pathways of selenium transfer to man can be evaluated. Representative behavior relationships have been formulated into an exposure commitment assessment of environmental selenium (Bennett 1983). Following this procedure, exposures to selenium and the levels in the body can be anticipated for various conditions of occurrence of environmental selenium. This will be demonstrated in the discussion to follow.

SOURCES AND ENVIRONMENTAL OCCURRENCE

Selenium is released into the environment from several natural sources, including volcanoes, wind-blown dusts, and volatile releases from plants, animals, soils, and from microorganisms in soils and sediments. Emissions from anthropogenic sources arise from fossil fuel combustion, estimated to be around 60% of the total man-made release [National Academy of Sciences (NAS) 1976], and secondarily from metal smelting and refining and glass and ceramics manufacturing.

Concentrations of selenium in air are of the order of a few nanograms per cubic meter, the levels depending particularly on coal-burning power plants in the vicinity. A good correlation can be expected between the concentrations of selenium and sulfur in air, reflecting the occurrence ratio in fossil fuels whenever fuel combustion is a dominant source. Higher concentrations of selenium in air are reported in industrialized and urban areas and lower concentrations at more remote background sites. Natural emissions of alkylselenides from aquatic environments with high biological activity may reach levels encountered in air near anthropogenic sources (Jiang et al., 1983).

Selenium occurs naturally in rocks and soils. The average concentration in the earth's crust is estimated to be 0.05 $\mu g\ g^{-1}$ (Taylor 1964); however, there are wide geographic variations. Higher concentrations of selenium are found in some sedimentary rocks, leading to higher concentrations in soil. There are notable seleniferous areas, such as in the United States, Australia, and Ireland, and important selenium-deficient areas in China, New Zealand, and elsewhere.

Concentrations of selenium in soil are extremely variable, ranging from 0.1 $\mu g\ g^{-1}$ in selenium-deficient areas to over 1000 $\mu g\ g^{-1}$ in seleniferous areas. Based on several reviews, Berrow and Burridge (1980) suggest a normal range of selenium in cultivated surface soils of 0.1–2 $\mu g\ g^{-1}$, with a typical concentration of 0.4 $\mu g\ g^{-1}$.

The uptake of selenium by plants varies widely. Some plants grow only in soils of high selenium content and accumulate concentrations up to a few thousand $\mu g\ g^{-1}$. These are generally nonfood plants but, occasionally, very unusually high concentrations of selenium are reported in some legumes, nuts, and mushrooms. Secondary selenium-absorbing plants contain 50–500 $\mu g\ g^{-1}$. Most crop plants, grains, and grasses rarely contain more than 30 $\mu g\ g^{-1}$, the general range being 0.05–1 $\mu g\ g^{-1}$.

Relatively low levels of selenium are found in most surface and ground waters, generally less than 10 $\mu g\ liter^{-1}$. Irrigation drainage from seleniferous soils may result in water concentrations up to 400 $\mu g\ liter^{-1}$ (NAS 1976). Selenium may be found in wastewater streams of ore refineries, coal-fired power plants, and industrial plants in which selenium is utilized. Precipitation of selenium with metal hydroxides usually greatly reduces the levels in larger rivers and lakes.

INTAKE AND METABOLISM

Most dietary intake of selenium occurs via plant foods, the main exception being fish and certain seafoods. There is considerable variability in levels in plants due to soil conditions, whereas most consistent levels are contained in animal tissues to meet the nutritional needs of the animal.

Some volatile selenium compounds escape from foods during cooking, but reductions in total selenium content for most foods are not significant (Ganapathy et al., 1978). Dry heating of cereals can result in a more significant reduction in selenium content (Higgs et al. 1972).

The chemical forms of selenium in foods have not been well established; however, approximately 40% of the selenium in wheat has been shown to be associated with methionine (Olson et al. 1970). In animal tissues, selenium is present as selenocysteine, selenomethionine, selenotrisulfide, and selenopersulfides, as well as other forms, such as metal selenides (Burk 1976).

Availability of selenium in foods following ingestion is higher for those of plant origin than for those of animal origin (Cantor et al. 1975). Selenium in fish is found in stable complexes with metals and is much less available.

Estimates of average daily intake of selenium in diet range from 6 to 220 μg day^{-1} (Bennett 1983). In the lower range are areas of China, New Zealand, and Italy, in midrange are Finland and the United Kingdom, and in the upper range are the United States, Canada, and Japan. In some large countries, such as China and the United States, certain regions of both high and low dietary intake can be found.

The minimum dietary selenium requirement to avoid deficiency effects has been tentatively estimated to be 20–120 μg day^{-1} (NAS 1976; Stewart et al. 1978). Toxic effects of selenium may occur at intakes in excess of 500 μg day^{-1} (Sakurai and Tsuchiya 1975).

Ingested selenium is readily absorbed, though with some dependence on chemical form. Organic selenium compounds, such as selenomethionine, may be incorporated directly into tissue and accumulate to high levels (Scott 1973). Absorption of inorganic selenite may be limited by available selenium-binding sites in tissues, although biotransformation occurs to organic forms which are then incorporated directly into tissue proteins (Glover et al. (1979). In general, absorption of selenites, selenates, and organic selenium compounds is efficient. Metal selenides and elemental selenium are poorly absorbed. It is estimated that, on average, 80% of dietary selenium is absorbed (Stewart et al. 1978).

Absorbed selenium is widely distributed by blood to organs and tissues, the highest concentrations occurring in liver and kidneys. In a study of selenium in human tissues from Japan, somewhat higher concentrations were found in lungs, with about equal concentrations in kidneys, spleen, pancreas, muscle, and brain (Yukawa et al. 1980). Regional differences in tissue concentrations are expected, reflecting variable dietary selenium intake. Whole-body selenium content has been determined to be approximately 3–6 mg in the low selenium region of New Zealand (Stewart et al. 1978) and is estimated to be about 13 mg elsewhere [International Commission on Radiological Protection (ICRP) 1975].

There are several components of selenium retention in the body, a rapid first-phase elimination, an intermediate phase, and a longer retention phase representing slower whole-body turnover of selenium. Animal studies indicate that particularly the longer-term phase retention may increase where dietary intake of selenium is very low (Griffiths et al. 1976). For general retention estimation, suggested retention half-times for man are 3, 30, and 150 days, with component fractions of 0.1, 0.4 and 0.5, respectively (ICRP 1980). The two longer-term components correspond to an effective half-time of 100 days and a mean retention time of 140 days.

PATHWAYS ANALYSIS—AVERAGE INTAKE

Pathways analysis of selenium transfers to man, utilizing the exposure commitment method (Bennett 1981), illustrates the contribution to body concentrations from the various sources. General transfer occurs via the inhalation and ingestion pathways. Compartmental arrangements representing the transfer pathways are formulated and transfer factors describing intercompartmental transfers are evaluated. The pathways analysis for selenium, based on representative values of environmental concentrations and intake rates, is presented here.

For the inhalation pathway, the main assumptions are the air-breathing rate of 22 m^3 day^{-1} (8000 m^3 $year^{-1}$), retention in the lungs, and absorption to blood of 20% of the intake amount, distribution of 90% of the absorbed selenium to body tissues, with a residence time in the body of 140 days.

Retention of inhaled ambient airborne particles in the lungs is of the order of 35%. Absorption of selenium compounds from the retained particles is uncertain. Tentatively, the combined retention–absorption transfer factor is taken to be 0.2. The mean retention time of selenium in the body (140 days) corresponds to the longer-term components of the retention function (ICRP 1980).

For the ingestion pathway, the analysis may begin with estimated deposition of airborne selenium onto soil. Alternatively, the estimated transfer may be made from background concentrations in soil. More certainty can be had by beginning directly with estimates of dietary intake.

The transfer factor relating soil concentrations of selenium to dietary intake can be inferred from representative values—0.4 μg g^{-1} in agricultural soil and 70 μg day^{-1} intake. The values will differ for various environmental conditions, but the ratio may be more constant and thus generally applied.

Dietary selenium is transferred to the gastrointestinal tract, from where fractional absorption of selenium to blood has been assumed to be 80%. Retention of selenium in the body is the same as following absorption of inhaled amounts.

The product of the transfer factors along pathways of transfer gives the estimated contribution to the concentrations of selenium in the body. For the inhalation pathway, the result is 0.008 ng g^{-1} in the body for each 1 ng m^{-3} of selenium in air. For the ingestion pathway, the result is 0.3 μg g^{-1} in the body for each 1 μg g^{-1} of selenium in soil. For direct estimates of dietary intake, the estimated relationship is 1.4 ng g^{-1} in the body from each 1 μg day^{-1} of dietary intake.

The contributions to selenium concentrations in the body for representative environmental conditions are listed in Table 1. For urban air concentrations of selenium of the order of 3 ng m^{-3}, the contribution to the body concentration is 0.02 ng g^{-1}. From soil concentrations of 0.4 μg g^{-1} or from dietary intake of 70 μg day^{-1}, the estimated contribution to the body concentration is 0.1 μg g^{-1}. The inhalation pathway is therefore negligible compared with the ingestion pathway. An additional contribution from drinking water intake may be included with total dietary intake, should this be of any significance.

The estimated mean concentration of selenium in the body of 0.1 μg g^{-1} under these representative conditions corresponds to a whole-body

TABLE 1. Representative Levels of Selenium in the Environment and Man

Pathway	Air	Soil	Diet	Body
Inhalation	3 ng m^{-3}			0.00002 μg g^{-1}
Ingestion		0.4 μg g^{-1}	70 μg day^{-1}	0.1 μg g^{-1}
			Total	0.1 μg g^{-1}

content of 7 mg. Concentrations up to an order to magnitude greater could apply to the main organs of accumulation of selenium, according to reported measurements, and concentrations somewhat lower than the mean would be expected in other tissues.

PATHWAYS ANALYSIS—EXTREME LEVELS OF DIETARY INTAKE

The transfer of environmental selenium to man, which is dominated by the ingestion pathway, has been evaluated above for "representative" conditions. The estimation of transfer in more specific circumstances is of interest, particularly at the extremes of dietary intake from selenium-deficient areas and from areas with nearer to toxic levels. There are various factors which can influence selenium transfer, including availability and retention of dietary forms, physiological adaptations in metabolism, genetic variations (ethnic differences), and other mineral or vitamin deficiencies.

Estimates of dietary intake for various regions of the world are given in Table 2. Selenium intake ranges from less than 10 μg day^{-1} to several thousand μg day^{-1}. The selenium content of cereals may vary up to 1000-fold, for example, from wheat grown on the low selenium soils in China and New Zealand (<0.01 μg g^{-1}) to Australian wheat (0.15 μg g^{-1}) and some United States wheat (25 μg g^{-1}) grown

TABLE 2. Estimated Dietary Intakes in Various Countries

Country	Dietary Se intake (μg day^{-1})	Reference
China: Deficiency area	10	Levander (1982)
Italy: Deficiency area	20	Rossi et al. (1976)
New Zealand	30	Thompson and Robinson (1980)
	60	Watkinson (1974)
Egypt	30	Maxia et al. (1972)
Finland	30	Varo and Koivistoinen (1980)
United Kingdom	60	Thorn et al. (1978)
Japan	100	Sakurai and Tsuchiya (1975)
Netherlands	110	Cresta (1976)
Italy: Countrywide	140	Cresta (1976)
Canada	170	Thompson et al. (1975)
France	170	Cresta (1976)
United States: Countrywide	170	Levander (1982)
South Dakota	220	Olson et al. (1978)
Venezuela	220	Mondragon and Jaffé (1976)
China: Selenosis area	4990	Levander (1982)

on the seleniferous soils of South Dakota (Robinson 1982). The influence of cereals on selenium nutrition in a general population has been demonstrated in New Zealand. Occasionally, Australian wheat has to be imported to supplement supplies in the North Island and is reflected in higher blood selenium levels (Watkinson 1981).

Animal studies have shown a relatively poor availability of selenium from soybean meal (Gabrielsen and Opstvedt 1980), a major dietary component in low selenium areas of China. Soybean proteins are low in sulfur amino acids, and this may be a partial explanation of reduced selenium availability from this food source (Young et al. 1979). The dietary form of selenium may also affect tissue retention (Griffiths et al. 1976; Robinson et al. 1978). The vitamin B_6 status of animals has been shown to affect the bioavailability of different selenium compounds (Yasumoto et al. 1979), while copper deficiency affects tissue retention of selenium and reduces glutathione peroxidase activity (Jenkinson et al. 1982). It is unlikely that vitamin E intake is implicated in Keshan disease as plasma α-tocopherol was adequate and not different between affected and nonaffected groups (Robinson 1982).

Experiments with rats have demonstrated an adaptation to chronic selenium intake (Jaffé and Mondragon 1975) and that urinary selenium excretion increases proportionally with selenium intake when threshold dietary levels are exceeded (Glover et al. 1979). The exhalation of volatile selenium compounds through the lungs becomes a significant route of excretion in subacute selenium poisoning (Glover et al. 1979).

The transfer of selenium to man for representative exposure conditions has been given above as 14.10^{-4} μg g^{-1} average concentration in the body per μg day^{-1} dietary intake. Representative levels of selenium in tissues and blood have been estimated as 0.1 μg g^{-1} and 0.2 μg ml^{-1}, respectively (Bennett 1983). This would indicate a relationship of 28 μg ml^{-1} in blood per μg day^{-1} intake. Based on a survey of selenium in diet and blood of several individuals, Schrauzer and White (1978) have derived an empirical relationship of the form 9.10^{-4} μg ml^{-1} selenium in blood per μg day^{-1} dietary intake, plus a constant value of 0.06 μg ml^{-1}. This suggests a somewhat lower transfer of ingested selenium to blood, but with a minimum maintained level. For representative exposure conditions (70 μg day^{-1} dietary intake), a blood selenium concentration of 0.12 μg ml^{-1} would be expected, giving a ratio of blood to dietary selenium of 1.7 μg ml^{-1} per μg day^{-1}. Comparisons of calculated and observed blood selenium concentrations are given in Table 3. Deviations in the above average value of the selenium blood-to-diet ratio are indicative of the altered

TABLE 3. Relationships between Selenium Levels in Man

Location	Dietary selenium (µg day^{-1})	Blood selenium (ng ml^{-1})		Blood Se[b]/ diet Se	Inference
		Actual	Calculated[a]		
Representative exposure	70	—	120	1.7	Low availability
China (low intake)	11	8	70	0.7	"Normal" absorption
New Zealand (low intake)	28	60	85	2.1	Slightly increased excretion
South Dakota, United States (elevated intake)	216	265	250	1.2	Low availability and increased excretion
China (endemic selenosis)	4990	3180	4550	0.6	—

[a] Empirical relationship: 9×10^{-4} µg ml^{-1} per µg day^{-1} intake + 0.06 µg ml^{-1} (Schrauzer and White 1978).
[b] Blood selenium: actual value, except for representative exposure.

transfers occurring in the areas of unusually high or low selenium exposure.

The degree of selenium transfer from areas of unusual occurrence depends, therefore, on local conditions. Stewart *et al.* (1978) have indicated that a dietary intake of 20 μg day^{-1} in the New Zealand population is adequate. In comparison, a dietary intake of 11 μg day^{-1} in China produces pathological changes. This indicates that, in addition to total dietary intake of selenium, the availability of dietary forms and the metabolic characteristics of the populations have to be taken into account. As these particular influences are not readily known, it is of paramount importance to define and measure the parameters which influence the metabolism of selenium, especially in populations which are susceptible to deficiencies or are exposed to toxic levels.

REFERENCES

Bennett, B. G. 1981. Exposure commitment concepts and application. *In* Exposure Commitment Assessments of Environmental Pollutants I. 1. MARC Rep. No. 23. Monitoring and Assessment Research Centre, Chelsea College, Univ. of London.

Bennett, B. G. 1983. Exposure of man to environmental selenium—An exposure commitment assessment. Sci. Total Environ. *31*, 117–127.

Berrow, M. L., and Burridge, J. C. 1980. Trace element levels in soils: Effects of sewage sludge. *In* Inorganic Pollution and Agriculture, MAFF Ref. Book 326, pp. 159–183. Her Majesty's Stationery Office, London.

Burk, R. F. 1976. Selenium in man. *In* Trace Elements in Human Health and Disease. A. S. Prasad (Editor), Vol. 2, pp. 105–133. Academic Press, New York.

Cantor, A. H., Scott, M. L., and Noguchi, T. 1975. Biological availability of selenium in foodstuffs and selenium compounds for prevention of exudative diathesis in chicks. J. Nutr. *105*, 96–105.

Cresta, M. 1976. Food Nutr. *2*, 8.

Gabrielsen, B. O., and Opstvedt, J. 1980. Availability of selenium in fish meal in comparison with soybean meal, corn gluten meal, and selenomethionine relative to selenium in sodium selenite for restoring glutathione peroxidase activity in selenium-depleted chicks. J. Nutr. *110*, 1096–1100.

Ganapathy, S. N., Joyner, B. T., Sawyer, D. R., and Häfner. K. M. 1978. Selenium content of selected foods. *In* Trace Element Metabolism in Man and Animals—3. M. Kirchgessner (Editor), p. 322. Technische University, Munich.

Glover, J., Levander, O., Parizek, J.. and Vouk, V. 1979. Selenium. *In* Handbook on the Toxicology of Metals. L. Friberg, G. F. Nordberg, and V. B. Vouk (Editors), pp. 555–577. Elsevier/North-Holland Biomedical Press, Amsterdam.

Griffiths, N. M., Stewart, R. D. H., and Robinson, M. F. 1976. The metabolism of [^{75}Se]selenomethionine in four women. Br. J. Nutr. *35*, 373–382.

Higgs, D. J., Morris, V. C., and Levander, O. A. 1972. Effect of cooking on selenium content of foods. J. Agric. Food Chem. *20*, 678–680.

International Commission on Radiological Protection (ICRP) 1975. Task Group Report on Reference Man, ICRP Publ. 23. Pergamon Press, Oxford, UK.

International Commission on Radiological Protection (ICRP) 1980. Limits for Intakes of Radionuclides by Workers, ICRP Publ. 30, Part 2. Pergamon Press, Oxford, UK.

Jaffé, W. G., and Mondragon, C. 1975. Effects of ingestion of organic selenium in adapted and nonadapted rats. Br. J. Nutr. *33*, 387.

Jenkinson, S. G., Lawrence, R. A., Burk, R. F., and Williams, D. M. 1982. Effects of copper deficiency on the activity of the selenoenzyme glutathionine peroxidase and on excretion and tissue retention of $^{75}SeO_3^{2-}$. J. Nutr. *112*, 197–204.

Jiang, S., Robberecht, H., and Adams, F. 1983. Identification and determination of alkylselenide compounds in environmental air. Atmos. Environ. *17*, 111–114.

Levander, O. A. 1982. Selenium: Biochemical actions, interactions, and some human health implications. *In* Clinical, Biochemical and Nutritional Aspects of Trace Elements. A. S. Prasad (Editor), pp. 345–368. Alan R. Liss, New York.

Maxia, V., Meloni, S., Rollier, M. A. *et al.* 1972. Selenium and chromium assay in Egyptian foods and in blood of Egyptian children by activation analysis. *In* Nuclear Activation Techniques in the Life Sciences, pp. 527–550. IAEA, Vienna.

Mondragon, M. C., and Jaffé, W. G. 1976. Ingestion of selenium in Caracas, compared with some other cities. Arch. Latinoam. Nutr. *26*, 341.

National Academy of Sciences (NAS) 1976. Medical and Biological Effects of Environmental Pollutants—Selenium. NAS, Washington, DC.

Olson, O. E., Novacek, E. J., Whitehead, E. I., and Palmer, I. S. 1970. Investigations of selenium in wheat. Phytochemistry *9*, 1181–1188.

Olson, O. E., Palmer, I. S., and Howe, M. 1978. Selenium in foods consumed by South Dakotans. Proc. S.D. Acad. Sci. *58*, 113.

Robinson, M. F. 1982. Clinical effects of selenium deficiency and excess. *In* Clinical, Biochemical and Nutritional Aspects of Trace Elements. A. S. Prasad (Editor), pp. 325–343. Alan R. Liss, New York.

Robinson, M. F., Rea, H. M., Friend, G. M., Stewart, R. D. H., Snow, P. C., and Thomson, C. D. 1978. On supplementing the selenium intake of New Zealanders. II. Prolonged metabolic experiments with daily supplements of selenomethionine, selenite, and fish. Br. J. Nutr. *39*, 589–600.

Rossi, L. C., Clemente, G. F., and Santaroni, G. 1976. Mercury and selenium distribution in a defined area and its population. Arch. Environ. Health *31*, 160–165.

Sakurai, H., and Tsuchiya, K. 1975. A tentative recommendation for maximum daily intake of selenium. Environ. Physiol. Biochem. *5*, 107–118.

Schrauzer, G. N., and White, D. A. 1978. Selenium in human nutrition: Dietary intakes and effects of supplementation. Bioinorg. Chem. *8*, 303–318.

Scott, M. L. 1973. Nutritional importance of selenium. *In* Organic Selenium Compounds, Their Chemistry and Biology. D. L. Klayman and W. H. H. Günther (Editors), pp. 629–661. John Wiley, New York.

Stewart, R. D. H., Griffiths, N. M., Thompson, C. D., and Robinson, M. F. 1978. Quantitative selenium metabolism in normal New Zealand women. Br. J. Nutr. *40*, 45–54.

Taylor, S. R. 1964. Abundance of chemical elements in the continental crust: A new table. Geochim. Cosmochim. Acta *28*, 1273–1285.

Thompson, C. D., and Robinson, M. F. 1980. Selenium in human health and disease with emphasis on those aspects peculiar to New Zealand. Am. J. Clin. Nutr. *33*, 303.

Thompson, J. N., Erdody, P., and Smith, D. C. 1975. Selenium content of food consumed by Canadians. J. Nutr. *105*, 274.

Thorn, J., Robertson, J., and Buss, D. H. 1978. Trace nutrients. Selenium in British food. Br. J. Nutr. *39*, 391.

Varo, P., and Koivistoinen, P.   1980.   Mineral element composition of Finnish foods: N, K, Ca, Mg, P, S, Fe, Cu, Mn, Zn, Mo, Co, Ni, Cr, F, Se, Si, Rb, Al, B, Br, Hg, As, Cd, Pb, and ash. XII. General discussion and nutritional evaluation. Acta Agric. Scand., Suppl. *22*, 165.

Watkinson, J. H.   1974.   The selenium status of New Zealanders. N. Z. Med. J. *80*, 202.

Watkinson, J. H.   1981.   Changes of blood selenium in New Zealand adults with time and importation of Australian wheat. Am. J. Clin. Nutr. *34*, 936.

Yasumoto, K., Iwami. K., and Yoshida, M.   1979.   Vitamin B_6 dependence of selenomethionine and selenite utilization for glutathione peroxidase in the rat. J. Nutr. *109*, 760–766.

Young, V. R., Scrimshaw, N. S., Torún, B., and Viteri, F.   1979.   Soybean protein in human nutrition. An overview. J. Am. Oil Chem. Soc. *56*, 110–120.

Yukawa, M., Suzuki-Yasumoto, M., Amano, K., and Terai, M.   1980.   Distribution of trace elements in the human body determined by neutron activation analysis. Arch. Environ. Health *35*, 36–44.

Selenium Status of Pediatric Patients

Ingrid Lombeck
K. Kasperek
H. Menzel

INTRODUCTION

During childhood, the Se content of blood shows a marked age relationship (*18*). Low serum or plasma Se values at birth were found in West Germany (*18*), Finland (*40*), the United States (*29*), Norway (*13*), Italy (*27*), Spain (*9*), and Belgium (*37*). No significant differences were measured between cord blood samples of preterm and full-term newborns (*1,13,40*). The Se content of erythrocytes or whole blood or erythrocyte glutathione peroxidase (GSHPx) activity was also reduced in some investigations of cord blood samples (*11,27,29,40,42*), but comparable to adult values in others (*12,18,35*).

During early infancy the Se content of serum, plasma, or whole blood and erythrocyte GSHPx activity decreases further in term and preterm infants (*12,18,30,35,37*). In the second half year of life and during later childhood the values slowly increase to the adult level.

In Libyan infants the hair Se content was estimated during infancy. It decreased rapidly in the first half year of life from 1071 ± 75 ng/g in newborns to 301 ± 99 ng/g in infants from North Libya, to 557 ± 204 ng/g in infants from South Libya (Fig. 1) (*25*). The high hair Se content at birth probably reflects Se storage during pregnancy, as does the high liver Se content (*4,41*). In newborns Se seems to exhibit another distribution pattern between different tissues than in adults.

The decrease in the Se state during infancy is due to the low Se

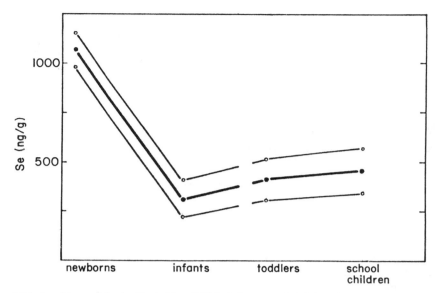

FIG. 1. The content of hair ($\bar{x} \pm$ SD) in infants and children from North Libya.

intake. Human milk and cow's milk are rather low in Se. In West Germany (*20*) mature human milk contains 17–38 ng/ml and cow's milk 19–29 ng/ml. The Se content of colostrum is higher than that of mature human milk. Data of West Germany and the United States (*31,32*) show comparable values. In contrast the Se content of cow's milk infant formulae is much lower. The average Se content of 10 different German formulae (*20*) amounts to less than one-third of that of mature human milk. American casein-based infant formulae and soy-based infant formulae were as low as cow's milk infant formulae (*43*). A 2-month-old formula-fed infant therefore receives usually less than 8 μg Se/day, while a breast-fed infant receives about 22 μg/day.

In the second half year of life the Se intake increases, the values vary between 8 and 80 μg/day, with a median value of 33 μg/day (*22*). The median Se intake, 34 μg/day, is similar in the third half year of life. The amount of different foods added to the child's diet varies considerably. The choice of food has an important influence on daily Se intake. Fresh fruit contains less than 5 ng/g (*8,24*), with the exception of bananas, which exhibit a broad variation, with a median level of 32 ng/g. Vegetables or vegetables plus meat or chicken, typical infant meals, do not contain more Se than cow's milk in West Germany (*8*). Cereals enriched with milk, or biscuits and eggs are the richest source of Se for small children. The addition of infant cereals to a milk formula leads to a severalfold increase of the Se content of the respective meal. Bioavailability studies of infant cereals have not yet been performed. As in adults from some countries (34,36,39), about 43% of the daily Se intake is derived from cereals, mainly from milk paps (Table 1). The Se intake is not equally distributed over the day. About one-quarter is eaten in the morning or at midday, and the other half at supper.

Children at risk with respect to low Se intake are those who do not get cereals rich in Se, either by getting no whole grain products or by being fed with locally grown cereals with low Se content. Besides young formula-fed infants (20,32), patients with Kwashiorkor (2), patients on semisynthetic diets (19), children from low Se areas in Finland with neuronal lipofuscinosis (41), children with poor diets (7), and children from Keshan areas in China (15) account for risk groups. Controversial results about the Se status of patients with malabsorption syndromes like cystic fibrosis (5,16,26) or celiac disease (38) have been published. No correlation of low Se state and sudden infant death syndrome (28) or reduced erythrocyte GSHPx activity and muscular dystrophy (3) could be proved. Despite increased erythrocyte GSHPx activities and superoxide dismutase (10), a reduced Se content of plasma was found by Neve et al. (26) in patients with Down's syndrome.

The reduced Se status in pediatric patients was measured by estimating the Se content of serum, whole blood, erythrocytes, hair, or erythrocyte GSHPx activity. Seldom was plasma GSHPx activity measured, which is generally low (33), or platelet GSHPx activity, which is, in agreement with platelet Se content (14), about three times higher than the erythrocyte values (Fig. 2). Few data about Se content and GSHPx activity (4,41) of other tissues in children exist.

In patients with phenylketonuria or maple syrup urine disease, the Se state is markedly reduced (17). This is due to the very low Se intake,

TABLE 1. Median Se Intake with Different Food Constituents in an 11-Month-Old Infant

Foods	Se intake (μg/day)	Se intake (%)
Cereals (enriched with milk)	13.8	41.2
Cereals (dry)	1.0	2.9
Milk + milk products	5.8	17.2
Vegetable + meat	4.8	13.4
Egg	1.7	5.0
Vegetable	1.3	4.0
Beverages	0.3	0.8
Fruit	5.2	15.5
	33.9	100.0

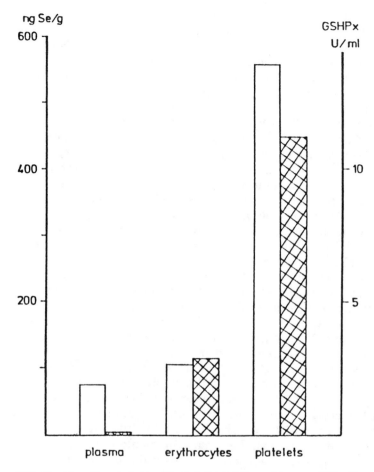

FIG. 2.   The Se content and glutathione peroxidase activity of
plasma, erythrocytes, and platelets in healthy young children.
Open bar, ng Se/g; cross-hatched bar, U/ml.

which amounts to 3–11 μg/day in young dietetically treated patients
and is practically not much higher in older treated patients (22). This
very low Se intake derives from the composition of their food, which
contains vegetables, fruit, and protein-poor cereals. Whole grain products, eggs, meat, and fish are avoided in the diet, and milk is only
allowed in small quantities. From 60 to 80% of their daily protein is
provided by amino acid mixtures or protein hydrolysates, which are
low in Se. This diet is started usually in the newborn period when the
diagnosis of the inborn error of amino acid metabolism is confirmed.

Within a few months the Se content of the serum drops to 15–25% of age-matched reference values (*19*), of erythrocytes to 35%, and of platelets to 47%. The GSHPx activity of plasma erythrocytes and platelets is also markedly reduced (Fig. 3).

In these patients two supplementation studies were performed with yeast rich in Se. In the first trial 5 patients, age 1.3–6.0 years, got 45 μg Se/day for 4 months, and the Se content of serum, whole blood, and erythrocyte GSHPx activity was measured. In a second trial 75–100 μg Se/day were given to 5 patients (age 4–8 years) for 4 to 5 months, and plasma and platelet GSHPx activity and platelet glutathione *S*-

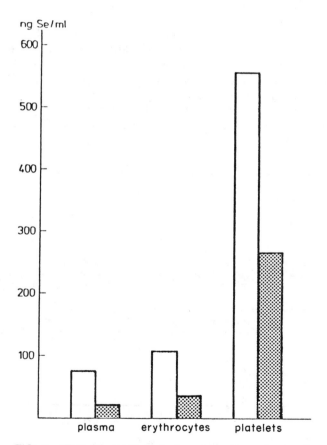

FIG. 3.   The Se content of plasma, erythrocytes, and platelets in healthy children (□) and dietetically treated (▨) patients with phenylketonuria or maple syrup urine disease.

transferase were measured. In addition, platelet GSHPx activity and
the Se content of serum and whole blood were estimated after the end
of the supplementation period.

In patients with a reduced Se state, the Se content of serum or
GSHPx activity of plasma increased to values of healthy children
within less than 4 weeks. The Se content of whole blood needed from 4
to 8 weeks (21). In contrast to plasma GSHPx activity (33), which
increased within 4 weeks to a plateau, erythrocyte GSHPx activity
needed 2 to 3 months, the life span of erythrocytes (Fig. 4). Two adults
with normal Se values did not show any change of plasma GSHPx
activity during a 3-month supplementation period with 200 μg Se/day.
In platelets GSHPx activity was also estimated with 2 to 3 substrates:
hydrogen peroxide, t-butyl hydroperoxide, and cumene hydroperoxide.
In addition, glutathione S-transferase was measured with 1-chlo-
ro-2,4-dinitrobenzene in platelets. Glutathione S-transferase was not
detectable in plasma. Platelet GSHPx activity was reduced in the dietetically treated patients to 30–40% of control values. During Se
supplementation it also increased rapidly and reached a plateau within 2 to 3 weeks. The increase of platelet GSHPx activity was slightly
slower than in plasma, but markedly more rapid than in erythrocytes.

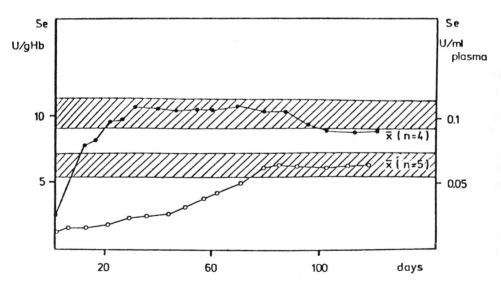

FIG. 4.   The erythrocyte GSHPx activity in 5 dietetically treated patients with
phenylketonuria (PKU) or maple syrup urine disease during 4 months supplementation with 45 μg Se/day (Se-rich yeast) (○), and plasma GSHPx activity in 4
dietetically treated patients with PKU during 4 months of supplementation with 75–
100 μg Se/day (Se-rich yeast) (●).

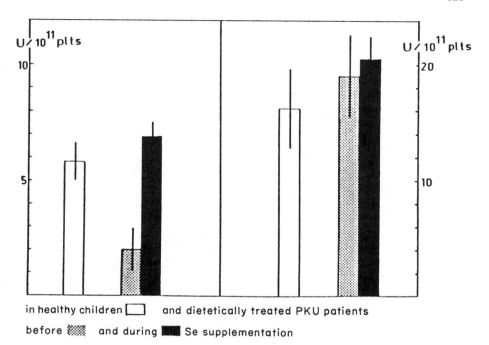

FIG. 5. Platelet GSHPx (left) and glutathione-S-transferase activity (right) in healthy children and dietetically treated patients with phenylketonuria and maple syrup urine disease before and after 2 weeks of supplementation with 75–100 μg Se/day.

The increase of the activity was similar with organic hydroperoxides (accounting for Se- and non-Se-peroxidase) and hydrogen peroxide (accounting for Se-GSHPx) (Fig. 5). There was no change in the ratios of the activities with t-butyl hydroperoxide and hydrogen peroxide and no change of the glutathione-S-transferase activity. This suggests that there was no significant non-Se-GSHPx activity in platelets, although we detected glutathione-S-transferase activity. Platelet glutathione S-transferase therefore does not metabolize organic hydroperoxides either in the Se repleted or in the depleted state (23).

After the end of about 4 months of Se supplementation, platelet GSHPx activity dropped slowly and needed about 24 weeks to reach a low level plateau again. One year later one of the patients received the same Se supplementation for another 34 days. The increase was as rapid as during the first supplementation period, but the platelet GSHPx activities reached a higher plateau.

Not only was the Se content and GSHPx activity of blood constituents reduced, but also the Se content of hair. It amounted to less than

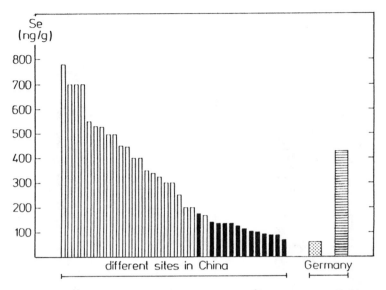

FIG. 6. The Se content of hair in Chinese children and healthy German children, and dietetically treated patients with phenylketonuria or maple syrup urine disease. (□), Nonaffected sites; (■) affected sites in China (from Chen, Ref. 6); (▦), dietetically treated patients; (▤), healthy children.

20% of the values of healthy children of the same age. In healthy preschool or school children from Libya, Iran, or West Germany, the median hair Se content is between 390 and 460 ng/g (*25*). In the dietetically treated patients the median value is 72 ng/g, with a range from 13 to 140 ng/g (*19*). In these dietetically treated patients, the hair Se content is as much reduced as in Chinese children from areas where the endemic cardiomyopathy—Keshan disease—occurs (Fig. 6) (*6*). In the dietetically treated patients with low Se status no clinical or biochemical deficiency syndromes were observed. Plasma proteins, vitamin E content, and liver function tests were normal. Although a semisynthetic diet is used, the intake of proteins, carbohydrates, fats, minerals, vitamins, and most of the trace elements was adequate. Besides the lower Se intake sometimes there were periods of a reduced zinc or rubidium intake. The dietetically treated patients had no clinical signs of myopathy of the heart or skeletal muscles, and electrocardiograms and electromyograms were normal. No increased hemolysis rate or bleeding abnormalities were observed. At the moment we do not find any hint for a long-term effect of low Se status. In the phenylketonuria patients the diet poor in natural proteins and Se lasts for at least 10–12 years, in the maple syrup urine disease patients for their entire life. The clinical course of the patients is monitored in regular intervals from 1 week to 3 months.

It remains to be discovered what other factors in food or the environment, which are not present in patients with inborn errors of amino acid metabolism, cause in addition to the Se deficiency the cardiomyopathy in Chinese children from Keshan areas.

CONCLUSIONS

The Se intake of young children is influenced by regional differences. Even more than in adults, it depends on the amount of cereals ingested and the Se content of cereals, which can be naturally low or are low by processing.

At birth the Se content of plasma is low, whereas it may be low, normal, or high in erythrocytes, hair, and liver. The Se content of serum, whole blood, and the respective GSHPx activities decrease during early infancy, reflecting the low Se intake. During phases of rapid growth they also show a rapid change. Platelets and hair belong to the tissues with the highest Se content.

Groups at risk for a low Se intake are formula-fed infants, children from Se-poor areas, children fed semisynthetic diets, or who are on parenteral nutrition. Besides endemic cardiomyopathy, a reduced Se state was observed in a variety of different clinical entities: chromosomal aberration, malabsorption, and neurologic disorders. Only in pediatric patients with endemic cardiomyopathy did Se supplementation change the clinical course.

In patients with a reduced Se state it is easy to monitor Se supplementation. The Se content and GSHPx activity of plasma increases within days, as does the platelet GSHPx activity. Long-term changes of the Se intake can be seen by measuring GSHPx activity of erythrocytes or in field studies by head hair analysis.

An Se depletion can be easily monitored in infants only during rapid growth. In older children or repleted patients a reduced Se intake needed months to show in GSHPx activities of plasma, erythrocytes, or platelets.

At the moment, it is difficult to define the necessary amount of Se intake which is optimal for health and growth for children under different conditions.

REFERENCES

1. Amin, S., Chen, S. Y., Collipp, P. J., Castro-Magana, M., Maddaiah, V. T., and Klein, S. W. 1980. Selenium in premature infants. Nutr. Metab. *24,* 331–340.
2. Burk, R. F., Pearson, W. N., Wood, R. P., and Viteri, F. 1967. Blood-selenium levels and *in vitro* red blood cell uptake of Se in Kwashiorkor. Am. J. Clin. Nutr. *20,* 723–733.
3. Burri, B. J., Chan, S. G., Berry, A. J., and Yarnell, S. K. 1980. Blood levels of superoxide dismutase and glutathione peroxidase in Duchenne muscular dystrophy. Clin. Chim. Acta *105,* 249–255.
4. Casey, C. E., Guthrie, B. E., Friend, G. M., and Robinson, M. F. 1982. Selenium in human tissues from New Zealand. Arch. Environ. Health *37,* 133–135.
5. Castillo, R., Landon, C., Lewiston, N., Morris, V., and Levander, O. 1981. Selenium levels in patients with cystic fibrosis. *In* Selenium in Biology and Medicine. J. E. Spallholz, J. L. Martin, and H. E. Ganther (Editors), pp. 464–467. AVI Publishing Co., Westport, CT.
6. Chen, X., Guang-Qi, Y., Chen, J., Chen, Y., Wen, Z., and Ge, K. 1980. Studies on the relations of selenium and Keshan disease. Biol. Trace Elem. Res. *2,* 91–107.
7. Collipp, P. J., and Chen, S. Y. 1981. Cardiomyopathy and selenium deficiency in a 2-year-old girl. N. Engl. J. Med. *304,* 1304–1305.
8. Ebert, K. H. *et al.* 1984. Selenium content of infant food. Z. Ernaehrungswiss. (in press).
9. Fraga, J. M., Cocho, J. A., Alvela, M., Alonso, J. R., Pena, J., and Tojo, R. 1983. Selenium state of children. The selenium content of the serum of normal children and children with inborn errors of metabolism. J. Inherited Metab. Dis. *6,* 99–100.
10. Frants, R. R., Eriksson, A. W., Jongbloet, P. H., and Hamers, A. J. 1975. Superoxide dismutase in Down syndrome. Lancet *2,* 42–43.
11. Glader, B. E., and Conrad, M. E. 1972. Decreased glutathione peroxidase in neonatal erythrocytes: Lack of relation to hydrogen peroxide metabolism. Pediatr. Res. *6,* 900–904.
12. Gross, S. 1976. Hemolytic anemia in premature infants: Relationship to vitamin E, selenium, glutathione peroxidase, and erythrocyte lipids. Semin. Hematol. *13,* 187–199.
13. Hågå, P., and Lunde, G. 1978. Selenium and vitamin E in cord blood from preterm and full-term infants. Acta Paediatr. Scand. *67,* 735–739.
14. Kasperek, K., Lombeck, I., Kiem, J., Iyengar, G. V., Wang, Y. X., Feinendegen, L. E., and Bremer, H. J. 1982. Platelet selenium in children with normal and low selenium intake. Biol. Trace Elem. Res. *4,* 29–34.
15. Lancet 1979. Selenium in the heart of China. Lancet *2,* 889–890.
16. Lloyd-Still, J. D., and Ganther, H. E. 1980. Selenium and glutathione peroxidase levels in cystic fibrosis. Pediatrics *65,* 1010–1012.
17. Lombeck, I., Kasperek, K., Feinendegen, L. E., and Bremer, H. J. 1975. Serumselenium concentrations in patients with maple syrup urine disease and phenylketonuria under dietotherapy. Clin. Chim. Acta *64,* 57–61.
18. Lombeck, I., Kasperek, K., Harbisch, H. D., Feinendegen, L. E., and Bremer, H. J. 1977. The selenium state in healthy children. I. Serum selenium concentration at different ages; activity of glutathione peroxidase of erythrocytes at different ages; selenium content of food of infants. Eur. J. Pediatr. *125,* 81–88.
19. Lombeck, I., Kasperek, K., Harbisch, H. D., Becker, K., Schumann, E., Schröter, W., Feinendegen, L. E., and Bremer, H. J. 1978. The selenium state of children. II. Selenium content of serum, whole blood, hair and the activity of erythrocyte glu-

tathione peroxidase in dietetically treated patients with phenylketonuria and maple syrup urine disease. Eur. J. Pediatr. *128*, 213–223.

20. Lombeck, I., Kasperek, K., Bonnermann, B., Feinendegen, L. E., and Bremer, H. J. 1978. Selenium content of human milk, cow's milk, and cow's milk infant formulas. Eur. J. Pediatr. *129*, 139–145.

21. Lombeck, I., Kasperek, K., Bachmann, D., Feinendegen, L. E., and Bremer, H. J. 1980. Selenium requirements in patients with inborn errors of amino acid metabolism and selenium deficiency. Eur. J. Pediatr. *134*, 65–68.

22. Lombeck, I. *et al.*, unpublished data.

23. Menzel, H., Steiner, G., Lombeck, I., and Ohnesorge, F. K. Glutathione peroxidase and glutathione S-transferase activity of platelets. Eur. J. Pediatr. *140*, 244–247.

24. Morris, V. C., and Levander, O. A. 1970. Selenium content of foods. J. Nutr. *100*, 1383–1388.

25. Musa Al-Zubaidy, I., Lombeck, I., Kasperek, K., Feinendegen, L. E., and Bremer, H. J. 1983. Zinc status of Libyan children—A pilot study. Z. Ernaehrungswiss. *22*, 1–5.

26. Neve, J., van Geffel, R., Hanocq, M., and Molle, L. 1983. Plasma and erythrocyte zinc, copper, and selenium in cystic fibrosis. Acta Paediatr. Scand. *72*, 437–440.

27. Perona, G., Guidi, G. C., Piga, A., Cellerino, R., Milani, G., Colautti, P., Moschini, G., and Stievano, B. M. 1979. Neonatal erythrocyte glutathione peroxidase deficiency as a consequence of selenium imbalance during pregnancy. Br. J. Haematol. *42*, 567–574.

28. Rhead, W. J., Schrauzer, G. N., Saltzstein, S. L., Cary, E. E., and Allaway, W. H. 1972. Vitamin E, selenium, and the sudden infant death syndrome. J. Pediatr. *81*, 415–416.

29. Rudolpf, N., and Wong, S. L. 1978. Selenium and glutathione peroxidase activity in maternal and cord plasma and red cells. Pediatr. Res. *12*, 789–792.

30. Rudolph, N., Preis, O., Bitzos, E. J., Reale, M. M., and Wong, S. L. 1981. Hematologic and selenium status of low-birth-weight infants fed formulas with and without iron. J. Pediatr. *99*, 57–62.

31. Shearer, T. R., and Hadjimarkos, D. M. 1975. Geographic distribution of selenium in human milk. Arch. Environ. Health *30*, 230–233.

32. Smith, A. M., Picciano, M. F., and Milner, J. A. 1982. Selenium intakes and status of human milk and formula-fed infants. Am. J. Clin. Nutr. *35*, 521–526.

33. Steiner, G., Menzel, H., Lombeck, I., Ohnesorge, F. K., and Bremer, H. J. 1982. Plasma glutathione peroxidase after selenium supplementation in patients with reduced selenium state. Eur. J. Pediatr. *138*, 138–140.

34. Thompson, J. N., Erdody, P., and Smith, D. C. 1975. Selenium content of food consumed by Canadians. J. Nutr. *105*, 274–277.

35. Thomson, C. D., and Robinson, M. F. 1980. Selenium in human health and disease with emphasis on those aspects peculiar to New Zealand. Am. J. Clin. Nutr. *33*, 303–323.

36. Thorn, J., Robertson, J., Buss, D. H., and Bunton, N. G. 1978. Trace nutrients: Selenium in British food. Br. J. Nutr. *39*, 391–396.

37. Verlinden, M., van Sprundel, M., van der Auwera, J. C., and Eylenbosch, W. J. 1983. The selenium status of Belgian population groups. II. Newborns, children, and the aged. Biol. Trace Elem. Res. *5*, 103–113.

38. Ward, K. P., Arthur, J. R., Russell, G. R., and Aggett, P. J. 1986. Blood selenium content and glutathione peroxidase activity in children with cystic fibrosis, celiac disease, asthma, and epilepsy. In press.

39. Watkinson, J. H. 1974. The selenium status of New Zealanders. N. Z. Med. J. *523*, 202–205.

40. Westermarck, T., Raunu, P., Kirjrinta, M., and Lappalainen, L. 1977. Selenium content of whole blood and serum in adults and children of different ages from different parts of Finland. Acta Pharmacol. Toxicol. *40*, 465–475.

41. Westermarck, T. 1977. Selenium content of tissues in Finnish infants and adults with various diseases, and studies on the effects of selenium supplementation in neuronal ceroid lipofuscinosis patients. Acta Pharmacol. Toxicol. *41*, 121–128.

42. Whaun, J. M., and Oski, F. A. 1970. Relation of red blood cell glutathione peroxidase to neonatal jaundice. J. Pediatr. *76*, 555–560.

43. Zabel, N. L., Harland, J., Gourmican, A. T., and Ganther, H. E. 1978. Selenium content of commercial formula diets. *In* Trace Element Metabolism in Man and Animals—3. M. Kirschgessner (Editor). Technische University, Munich.

Status of the Food Supply and Residents of New Zealand

M. F. Robinson
C. D. Thomson

The low Se environment and the low Se status of New Zealand (NZ) residents is well known, but as yet no obvious signs of deficiency or increased liability to diseases such as cancer and cardiovascular conditions have been detected (*1*). At the last Se symposium in 1980, we heard about Keshan disease and the low Se environment in the Keshan areas of China (*2*). It is intriguing that Keshan disease has not yet been seen in NZ. Our Se status might be not quite as low as that in the Keshan areas, but it could well be that interacting factors are of greater importance. Our search continues toward identifying these interacting factors and for a greater understanding of our low Se status.

This chapter reports on (a) progress regarding the description of the Se status of NZ residents, (b) the dietary Se intake and how it is metabolized by NZ residents, (c) identification and implications of other nutrients in the NZ diet, and finally, (d) progress in identifying any factor which might be protecting NZ residents from becoming marginally deficient in Se.

LOW Se STATUS OF NEW ZEALAND RESIDENTS

Our laboratory is situated in Dunedin on the east coast of the province of Otago on South Island and is surrounded by low soil Se agricultural areas such as in West Otago at Tapanui where, with P. Snow, we have carried out Se dosing trials (*3*). Further north are the

Canterbury plains where NZ cereals low in Se (<0.01 μg Se/g) are mainly grown. South Island supplies most of the cereals for the NZ market, and at times of poor harvest, Australian wheat is imported to supplement NZ wheat, but this happens only for North Island (4). J. Watkinson has shown that such importations of Australian wheat have raised periodically the Se status of North Island residents. At Lubbock (5) we showed that in 1977 Auckland residents had higher blood Se (83 ± 13 ng ml^{-1}) than Dunedin residents (59 ± 12 ng ml^{-1}), with Auckland vegetarians (mainly lacto-ovo-vegetarian) having higher still (101 ± 19 ng ml^{-1}), presumably from consuming imported vegetarian foods with a higher Se content.

Many more techniques are in use for assessing Se status (Table 1), which indicates that nothing is entirely satisfactory. However, with all techniques our local residents always show a low Se status, lower than in most other countries (1).

Our values for Se might even be still lower. We participated with 26 other laboratories in the IUPAC interlaboratory trial for determination of Se in lyophilized human serum. The results were reported at a meeting in Lyngby (1983) of the IUPAC Subcommittee on Selenium and they showed a remarkably close agreement between almost all laboratories. However, we were in the upper part of the range (about 10% above the mean), and this could mean that our values for Se status might even be a little lower than reported (6).

Almost identical blood Se levels to those for our control group of Otago blood donors were obtained recently in a health survey in Milton (7). For other tissues, Morris et al. showed low levels for toenails from Tapanui (8), but hair has not yet been followed because of the periodic use of Se-containing medicated shampoos. The Se content of NZ foods reflects our low Se environment (9), and our daily intakes are usually less than 30 μg Se day^{-1} (10). Blood Se for 10 Dunedin mothers was

TABLE 1. Se Status of Otago, New Zealand Residents

Se in	$\bar{x} \pm$ SD
Blood (ng ml^{-1})	59 ± 12
Plasma (ng ml^{-1})	48 ± 10
Erythrocytes (ng ml^{-1})	74 ± 16
Toenails (μg g^{-1})	0.26 ± 0.09[a]
Intake (μg day^{-1})	$6 - 70$
Urine (μg day^{-1})	$6 - 20$
Milk (ng ml^{-1})	7.6 ± 2
Autopsy tissue, e.g., liver (μg g^{-1} dry matter)	0.72 ± 0.18

[a] Morris et al. (8).

low (46 ± 11 ng ml^{-1}), and their milk was the lowest ever reported for mature human milk (11). A relationship was observed between blood and milk Se concentrations (P <.01). Autopsy tissues gave values similar to those in other low Se areas (12).

Despite the growing disenchantment with the glutathione peroxidase (GSHPx) assay as a measure of Se status because of interference from hemoglobin and the non-Se-dependent peroxidase activity, we continue to use this assay because of the linear relationships for NZ residents between erythrocyte GSHPx and blood Se levels, where whole blood Se is less than 100 ng SE ml^{-1} (6). Following recommendations of Levander (13), we have now established platelet GSHPx assays, and this also shows that NZ residents have a lower platelet activity (14). Enzyme activities are also measured for those other enzymes involved in the prevention of lipid peroxidation and protection from free radicals: glutathione S-transferase (GST; EC 2.5.1.8) (15), catalase (CAT) (16), and superoxide dismutase (SOD; EC 1.15.1.1) (17). α-Tocopherol measured by the high-performance liquid chromatography (HPLC) technique of Bieri et al. (18), was 10 ± 1.2 μg α-tocopherol ml^{-1} plasma for Dunedin residents, well within the accepted range. These measurements give only static indices of Se status, and we are now hoping to introduce functional indices as urged by Solomons and Allen (19).

A. van Rij and his group in the Department of Surgery, Otago Medical School, have established techniques for malondialdehyde (20) and thromboxane (21), and such measurements are used in our collaborative studies of patients and other subjects suspected of having a low Se status.

DIETARY Se INTAKE AND ITS METABOLISM

In the summer of 1982–1983, we participated in Dunedin in a collaborative study with O. A. Levander and C. Veillon, U.S. Department of Agriculture Human Nutrition Center, Beltsville, MD, on the effect of supplementing the daily diet of four NZ women for 8–13 weeks with 200 μg Se provided as United States high Se wheat bread. Wheat was kindly donated by O. E. Olson, Brookings, South Dakota. Such a supplement had been used by Levander and his group in Beltsville (22) and in Finland (22a). The subjects ate their normal diet, excluding the high Se foods fish, liver, and kidney, until later in the bread-supplementation period. At intervals the dietary intake was measured by collection of duplicate diets for 3 consecutive days. Throughout the study, urine and fecal samples were collected daily and then weekly later in the study, while blood samples were collected weekly throughout.

Stable isotopes of Se were used at the beginning of the control and supplementation periods and then 5 weeks later, and their metabolism was followed by the Beltsville group (23).

A preliminary report has also been given of the increase in Se concentrations and GSHPx activities in whole blood, erythrocytes, and plasma and also in platelet GSHPx of all subjects (24). On the other hand, activities of GST, SOD, or CAT in erythrocytes and plasma did not change throughout the study, nor did the levels of plasma vitamin E. This suggested that there had been no compensatory changes in the antioxidant enzymes which might have been expected to fall during supplementation when GSHPx had increased, or alternatively to rise again when bread supplement had stopped and GSHPx activity had decreased again.

Clearly this study has yielded a wealth of information about the NZ situation, but this chapter is mainly about the metabolism of food Se by the subjects. It was a surprise to find that the mean Se intake was 11.4 μg day^{-1} (range 4.5–17.5 μg day^{-1}), which is roughly half the intake of 24 μg/day in our 1973 study of another four women (25) (Table 2). The similarity between the urine and fecal outputs in the two studies suggested a possible error, but reexamination of the diets in the two studies confirmed the low Se intake, which incidentally was similar to that in low Se areas of China (1, 25–27).

In the 1973 study, fish, liver, or kidney were consumed on 13 of the 56 days (23% days), giving a mean intake of 45 μg Se day^{-1} (Table 3). Higher intakes were given for days with fish than with liver. Poultry feeds in NZ are supplemented with Se, and eggs and chicken were eaten on 33% of the days, giving a mean intake of 27 μg Se. For the remaining time (41% days), mean intake was 11.3 μg Se day^{-1}, similar to that for the 1983 study when subjects were requested not to eat fish, liver, or kidney, and eggs and chicken were eaton on 33% of the days. It is not clear why the mean intake for these days was half that for the 1973 study, but two eggs were eaten more frequently than in the 1983 study. This might in part account for the difference, but it seems unlikely that the level of Se supplement in poultry mash had been changed.

Urinary outputs were similar for the two studies, as also were their plasma Se levels which covered similar ranges. Fecal output was only 2 μg Se day^{-1} less on the lower intake of the 1983 study, resulting in the subjects losing apparently about 10 μg Se, whereas the other four women were more or less in balance.

Apparent absorption was also low (11–14% intake) for three of the

TABLE 2. Metabolic Studies of New Zealand Women

Year	Subjects	Length of study (days)	Intake	Feces (µg Se day⁻¹)	Urine (µg Se day⁻¹)	Balance	Plasma (ng Se/ml)	Apparent absorption (%)
1973	4 females	14	24.2[a] (20.1–33.5)	10.8 (8.7–13.4)	13.1 (8.0–18.7)	+0.3 (−2.2 to +4.4)	58 (48–76)	55 (50–60)
1983	4 females	3	11.4 (9.1–13.7)	9.0 (7.8–9.4)	12.7 (10.2–14.9)	−10.3 (−12.5 to −8.8)	61 (52–78)	20 (11–43)

[a] Mean, with range of means for each subject in parentheses.

TABLE 3. Dietary Intakes of New Zealand Women

| Year | Subjects | Length of study (days) | Intake (μg Se day^{-1}) | | | +Fish liver, kidney |
			Mean	−Eggs	+Eggs, chicken	
1973	4 females	14	24.2	11.3[a] (5–17) 41%[b]	27.0 (11–42) 36%	44.7 (18–102) 23%
1983	4 females	3	11.4	9.5 (4–13) 67%[b]	13.3 (9–17.5) 33%	—

[a] Mean, with range in parentheses.
[b] Percentage of days taken.

four subjects. It could be that the endogenous fecal component had become proportionately an even greater contributor to the fecal output. The use of radiotracer ^{75}Se permitted the estimation of endogenous fecal output in the 1973 study by assuming absorbed ^{75}Se is partitioned for excretion between urine and feces in the same proportion as is absorbed Se (28). This gave a mean of 5.6 μg Se day^{-1}, amounting to half the total fecal output. By comparing the two sets of results for plasma and endogenous fecal output, it seems that there might be some kind of relationship between them for the four subjects, even though it is not statistically significant ($y = 0.03 \times +3.64$; $r = .35$). Because the mean plasma Se and ranges were similar in the two studies, endogenous fecal outputs for the 1983 study were derived from the graph for the 1973 study to yield unabsorbed food Se and thereby an estimate of true absorption, 71% (range 61–84%), which is similar to 80% for the 1973 study (Table 4). This obviously needs to be confirmed, and also the extent to which plasma Se influences the endogenous fecal output.

The selenite tracer [74]Se gave a mean apparent percentage of absorption of 64% (range 55–72%) (23) for an intake of 40 μg [74]Se. Apparent

TABLE 4. Estimate of Unabsorbed Food Se and "True" Absorption

| Year | Intake | Total | Fecal output (μg Se day^{-1}) | | True absorption (%) |
			Endogenous	Unabsorbed food Se	
1973	24.2	10.8	5.6	5.2	80
1983	11.4	9.0	5.7[a]	3.3[a]	71

[a] Assuming relationship between plasma Se/endogenous fecal output in 1973 study.

absorption remained relatively constant throughout the study when 57 and 70% of 200 μg doses of [76]Se and [74] Se had been given on day 1 and day 36 of the supplementation period (23).

It was extraordinarily timely to receive the FASEB Abstract (29) from the group at the Cancer Institute, Chinese Academy of Medical Science, Beijing and Endemic Diseases Institute, Werchang, about Se intake and metabolic balance for 3 consecutive days in 10 men consuming self-selected diets in an Se-deficient area of Hebei Province, People's Republic of China (PRC) (Table 5). The intakes are almost identical with those in our 1983 study for control and post-bread periods. The difference was that the Chinese subjects were in positive balance in both studies, whereas NZ subjects were in negative balance. Fecal output was half the NZ value, and the urinary excretion was proportionately less still, only 20–30% NZ value for 1983. This would suggest that plasma Se was also considerably less.

We have been comparing the renal plasma clearances of Se in residents in NZ and in North America. Clearances were calculated using the conventional formula

$$\frac{[Se]_u}{[Se]_p} \times V$$

where $[Se]_u$ and $[Se]_p$ are Se concentrations in urine and plasma, and V the rate of production in ml min^{-1}. It seems that the kidneys of NZ subjects excreted Se more sparingly than those of North American subjects. Perhaps the kidneys of the Chinese subjects are even thriftier still.

The positive balance for the Chinese men and negative balance for NZ women indicated that for both groups the intakes had been changed and the subjects had apparently not yet adapted to the changes from a habitually lower intake for the Chinese men and a greater intake for the NZ women.

The four women in the 1973 study appeared to be in balance with their intake, which was strengthened because the means (±SD) for monthly samples of plasma Se and daily urine Se for the successive 7–10 months were almost identical for each subject, respectively, with those for samples taken during the 14-day experimental period.

Fish, liver, and kidney are not usually major food items in the NZ diet, but excluding them from the diets in 1983 had reduced (possibly halved) their habitual Se intake. A meal of fish provides 50–100 μg Se day^{-1}, which would double the weekly and daily intakes. In the Milton health survey, few men and women reported eating liver or kidney regularly, and almost half ate fish weekly. Blood Se and

TABLE 5. Metabolic Studies of Subjects Resident in People's Republic of China (PRC) and New Zealand (NZ)

	Subjects	Season	Intake	Feces (μg Se day^{-1})	Urine (μg Se day^{-1})	Balance	Apparent absorption (%)
PRC	10 males	Summer	13.3 ± 3.1[a]	4.5 ± 1.8	4.3 ± 1.0	+4.4 ± 4.6	63.1 ± 20.5
		Fall	9.2 ± 1.0[b]	4.3 ± 1.8	2.7 ± 1.0	+2.2 ± 2.4	52.4 ± 21.5
NZ	4 females	Summer	11.4 ± 3.0	9.0 ± 1.2	12.7 ± 2.1	−10.3 ± 1.7	20 ± 16

[a] Luo et al. (29).
[b] Mean ± SD.

638

TABLE 6. Supplement of United States High Se Wheat Bread[a] for 8 to 13 Weeks

Group	Intake	Feces (µg Se day^{-1})	Urine (µg Se day^{-1})	Balance	Apparent absorption (%)	Plasma (ng Se ml^{-1})
Control	11.4	9.0	12.7	−10.3	20	61
Bread						
Week 2	252	83	72	+97	67	103
	(239–271)	(71–98)	(50–86)	(64–147)	(60–73)	(92–118)
Week 8	248	77	112	+59	69	166
	(241–253)	(56–103)	(77–125)	(22–93)	(58–77)	(154–181)
Post-bread						
Week 7	12.5	14	22	−24		107
						(95–121)

[a] Supplement of 221 ± 9 µg Se day^{-1}. Wheat donated by O. E. Olson, Brookings, South Dakota.

GSHPx levels of those who did and those who did not eat fish showed the expected trend, but were not statistically different (7).

Mertz emphasized recently the importance of the "time factor" in trace element nutrition (30). He pointed out that the time factor expressed the buffering effect of homeostatic regulation on short-term changes of dietary intake.

The differences in how the Chinese men and NZ women handle their intakes are intriguing and need to be studied further. It would be extremely fascinating if some collaborative study were to be set up for comparing the metabolism of Se by both groups, and the feasibility of such a study requires further investigation.

Table 6 contains a summary of the metabolic studies of Se during the control, bread, and post-bread periods. Briefly, bread supplementation gave an immediate increase in fecal Se, which seemed to plateau. Fecal Se dropped back down again after Se bread was stopped, but at the end of the study it had not returned to prebread values. Plasma Se rose steeply, as did urinary Se, and then appeared to plateau. When bread supplementation ceased, urine Se dropped dramatically while the plasma Se fell more slowly and had not returned to predosing values by week 7. The significance of our findings is not clearly understood and further work is in progress.

OTHER COMPONENTS IN THE NEW ZEALAND DIET

New Zealand residents consume a Western-type diet comprising foods mainly produced locally. A comprehensive survey of New Zealanders and their diet was carried out in 1977 by J. A. Birkbeck, Department of Nutrition, University of Otago, and the second revised edition is now available (31).

Table 7 compares the nutrient content of the NZ diets with that of United States self-selected diets of Maryland residents (32). Analyses for NZ 1983 diets are not yet completed. Because the energy contents of NZ 1973 and United States diets are so close to 2000 kcal, the nutrients are given as total intakes and not on the basis of nutrient density. The Se contents show the greatest differences, with the NZ diet providing 15–30% Se of the United States diets. Men usually have higher protein intakes than women, and this may account for the greater United States intakes. Protein intakes were adequate as judged by the NZ (33) and United States (34) recommended intakes, respectively. Amino acid composition of the NZ diets was determined by A. Carne and I. Emerson, Department of Biochemistry, University

TABLE 7. Daily Dietary Intakes of New Zealand and United States Residents

Nutrient	NZ 1973 4 females, 14 days	NZ 1983 4 females, 12 days	U.S. 1976[a] 11 male + 11 female, 6 days
Se (μg)	24 ± 19[b]	11 ± 3	81 ± 41
Energy (kcal)	1911 ± 428	—	2027 ± 906
Protein (g)	61 ± 13	61	84 ± 25[c]
Fat (g)	86 ± 30	—	82 ± 22
Carbohydrate (g)	221 ± 43	—	235 ± 63
Methionine (g)	0.91 ± 0.12	0.87 ± 0.11	—
Met (mg g^{-1}N)	94	92	—
Met + Cys (g)	1.6	1.6	—
Cu (mg)	1.8 ± 0.5	1.3 ± 0.5	1.1 ± 0.6
Zn (mg)	7.5 ± 2.0	8.3 ± 1.7	9 ± 5
Mn (mg)	2.4 ± 0.6	—	2.7 ± 1.8
Ca (mg)	511 ± 110	738 ± 235	827 ± 455

[a] Welsh et al. (32).
[b] Mean ± SD.
[c] Values for United States dietary intake are given as mean nutrient level per 2000 kcal.

of Otago, and gave similar methionine intakes, expressed per gram N or per day. There was no difference in the 1983 study between Met intake for the control period and for the entire experiment of four 3-day periods. The combined Met and Cys intake was calculated as 1.7 times the Met intake. The intakes were adequate but not generous, as judged by recommendations of the Food and Agriculture Organization/World Health Organization (35) and The U.S. National Research Council (36). Protein contributed more to the energy intake of United States diets (17%) than to the NZ diets (13%), while fat contributed less (37 and 41%, respectively).

Some other nutrients suggested to be interrelated with Se are also listed in Table 7. The higher Cu content for the 1973 study reflects the inclusion of one meal of liver or kidney by each subject during the 14-day study (37), whereas such foods were excluded in the 1983 study. Such an intake is marginally less than the recommendation for minimum safe intake (MSI) (33), and likewise the Zn intake. Calcium intakes were adequate for all groups. The implications of the intakes of these nutrients and their possible interaction with Se has yet to be assessed. But it must not be forgotten that the data are derived from only 8 women, even though the diets were collected for 14 days in the 1973 study and for 4 periods of 3 days in the 1983 study. When the analyses are completed, the data will be tested for links between Se and other food components.

ARE NEW ZEALAND RESIDENTS MARGINALLY DEFICIENT IN Se?

Our search continued for factors that might be protecting NZ residents from the implications of their low Se status. Mertz pointed out at the trace element conference in Lund (1983) that retrospective diagnosis or normalization of an impaired function seemed the most reliable way to demonstrate a marginal deficiency. Possibly the lower GSHPx activity of most NZ residents which can be raised by Se supplementation would suggest some impairment of function (38). However, such Se regimes have not shown compensatory changes in the other antioxidant enzymes (24). We are now seeking evidence of oxidative damage in subjects with extremely low Se status. Recently, Mertz set out a scheme for four stages of trace element deficiency (30). Our metabolic studies suggest that NZ residents have adapted to their low Se environment by their thriftiness in urinary excretion of Se. They therefore could be considered to be in the "initial depletion" Stage I, but possibly not in Stage II, "compensated metabolic phase," because of the absence of compensatory changes in antioxidative enzyme activities.

A comparison of intake, metabolism, and function of Se in NZ and of the presence of interacting factors with those in Se adequate and other low Se areas where Western and vegetarian-type diets are consumed (such as in the PRC, Scandinavian countries, and the United States) might yield some clues to the extent of involvement of other factors.

ACKNOWLEDGMENT

Work in the authors' laboratory was supported by the Medical Research Council of New Zealand.

REFERENCES

1. Robinson, M. F., and Thomson, C. D.   1983.   The role of selenium in the diet. Nutr. Abstr. Rev. *53*, 1–26.
2. Chen, X., Chen, X., Yang, G., Wen, Z., and Ge, K.   1981.   Relation of the selenium deficiency to the occurrence of Keshan disease. *In* Selenium in Biology and Medicine. J. E. Spallholz, J. L., Martin, and H. E. Ganther (Editors), pp. 171–175. AVI Publishing Co., Westport, CT.
3. Robinson, M. F., Campbell, D. R., Stewart, R. D. H., Rea, H. M., Thomson, C. D., Snow, P. G., and Squires, I. H. W.   1981.   Effect of daily supplements of selenium on patients with muscular complaints in Otago and Canterbury. N.Z. Med. J. *93*, 289–292.

4. Watkinson, J. H. 1981. Changes of blood selenium in New Zealand adults with time and importation of Australian wheat. Am. J. Clin. Nutr. *34*, 936–942.
5. Robinson, M. F., and Thomson, C. D. 1981. Selenium levels in humans vs environmental sources. *In* Selenium in Biology and Medicine. J. E. Spallholz, J. L. Martin, and H. E. Ganther (Editors), pp. 283–302. AVI Publishing Co., Westport, CT.
6. Rea, H. M., Thomson, C. D., Campbell, D. R., and Robinson, M. F. 1979. Relation between erythrocyte selenium concentrations and glutathione peroxidase (EC 1.11.1.9) activities of New Zealand residents and visitors to New Zealand. Br. Nutr. *42*, 201–208.
7. Robinson, M. F., Campbell, D. R., Sutherland, W. H. F., Herbison, G. P., Paulin, J. M., and Simpson, F. O. 1983. Selenium and risk factors for cardiovascular disease in New Zealand. N.Z. Med. J. *96*, 755–757.
8. Morris, J. S., Stampfer, M. J., and Willett, W. 1983. Dietary selenium in humans. Biol. Trace Elem. Res. *5*, 529–537.
9. Thomson, C. D., and Robinson, M. F. 1980. Selenium in human health with emphasis on those aspects peculiar to New Zealand. Am. J. Clin. Nutr. *33*, 303–323.
10. Griffiths, N. M. 1973. Dietary intake and urinary excretion of selenium in some New Zealand women. Proc. Univ. Otago Med. Sch. *51*, 8–9.
11. Williams, M. M. F. 1983. Selenium and glutathione peroxidase in mature human milk. Proc. Univ. Otago Med. Sch. *61*, 20–21.
12. Casey, C. E., Guthrie, B. E., Friend, G. M., and Robinson, M. F. 1982. Selenium in human tissues from New Zealand. Arch. Environ. Health *37*, 133–135.
13. Levander, O. A. 1983. Considerations in the design of selenium bioavailability studies. Fed. Proc., Fed. Am. Soc. Exp. Biol. *42*, 1721–1725.
14. Thomson, C. D., and Duncan, A. 1981. Glutathione peroxidase and selenium status. *In* Trace Element Metabolism in Man and Animals—4. J. Mc.C. Howell, J. M. Gawthorne, and C. L. White (Editors), pp. 22–25. Australian Academy of Sciences, Canberra.
15. Habig, W. H., Pabst, M. J., and Jakoby, W. B. 1974. Glutathione *S*-transferases. The first enzymatic step in mercapturic acid formation. J. Biol. Chem. *249*, 7130–7139.
16. Beers, R. S., and Sizer, I. W. 1952. A spectrophotometric method for measuring the breakdown of hydrogen peroxide by catalase. J. Biol. Chem. *195*, 133–140.
17. McCord, J. M., and Fridovich, I. 1969. Superoxide dismutase. An enzymatic function for erythrocuprein (hemocuprein). J. Biol. Chem. *244*, 6049–6055.
18. Bieri, J. G., Tolliver, T. J., and Catignani, G. L. 1979. Simultaneous determinations of α-tocopherol and retinal in plasma or red cells by high-pressure liquid chromatography. Am. J. Clin. Nutr. *32*, 2143–2149.
19. Solomons, N. W., and Allen, L. H. 1983. The functional assessment of nutritional status: Principles, practice, and potentials. Nutr. Rev. *41*, 33–50.
20. Wade, C. R., Jackson, P. G., van Rij, A. M., and Highton, J. 1984. Measurement of lipid peroxidation in rheumatoid synovial fluid by a new method using ion-pairing reverse-phase HPLC. Proc. Univ. Otago Med. Sch. *62*, 61–62.
21. Loughlin, A. E., Wade, C. R., Kirk, I., and van Rij, A. M. 1983. Platelet thromboxane production in patients with peripheral vascular disease—The effect of a single low dose of aspirin. Proc. Univ. Otago Med. Sch. *61*, 77–79.
22. Levander, O. A., Sutherland, B., Morris, V. C., and King, J. C. 1981. Selenium balance in young men during selenium depletion and repletion. Am. J. Clin. Nutr. *34*, 2662–2669.
22a. Levander, O. A., Alfthan, G., Arvilommi, H., Gref, C. G., Huttunen, J. K., Kataja, M., Koivistoinen, P., and Pikkarainen, J. 1983. Bioavailability of selenium to

Finnish men as assessed by platelet glutathione peroxidase activity and other blood parameters. Am. J. Clin. Nutr. *37*, 887–897.

23. Edmonds, L. J., Veillon, C., Robinson, M. F., Thomson, C. D., Morris, V. C., and Levander, O. A. 1984. Use of stable isotopes to monitor kinetics of selenium (Se) excretion in New Zealand women before and after Se supplementation. Fed. Proc., Fed. Am. Soc. Exp. Biol. *43*, Abstr. 1099.

24. Robinson, M. F., Thomson, C. D., Ong, L. K., and Huemmer, P. 1984. Effects of supplementation with high-selenium wheat bread on glutathione peroxidase and related enzymes in women. Proc. Nutr. Soc. *43*, 17A.

25. Diplock. A. T. 1981. Metabolic and functional defects in selenium deficiency. Philos. Trans. R. Soc. London, Ser. B *294*, 105–117.

26. Chen, X., Yang, G., Chen, J., Chen, X., Wen, Z., and Ge, K. 1980. Studies on the relations of selenium and Keshan disease. Biol. Trace Elem. Res. *2*, 91–107.

27. Levander, O. A. 1982. Selenium: Biochemical actions, interactions, and some human health implications. *In* Clinical, Biochemical and Nutritional Aspects of Trace Elements. A. S. Prasad (Editor), pp. 345–368. Alan R. Liss, New York.

28. Stewart, R. D. H., Griffiths, N. M., Thomson, C. D., and Robinson, M. F. 1978. Quantitative selenium metabolism in normal New Zealand women. Br. J. Nutr. *40*, 45–54.

29. Luo, X. M., Yang, C. L., Wei, H. J., Liu, X., Xing, J., Liu, J., Qiao, C. H., Feng, Y. M., Liu, Y. X., Wu, Q., Guo, J. S., Stoecker, B. J., Spallholz, J. E., and Yang, S. P. 1984. Selenium intake and metabolic balance in 10 men consuming self-selected diets in a selenium-deficient area of Hebei Province, People's Republic of China. Fed. Proc., Fed. Am. Soc. Exp. Biol. *43*, Abstr. 1097.

30. Mertz, W. 1985. Metabolism and metabolic effects of trace elements. *In* Trace Elements in Nutrition of Children. R. K. Chandra (Editor), pp. 107–119. Raven Press, New York.

31. Birkbeck, J. A. 1983. New Zealanders and Their Diet. A report of the National Heart Foundation of New Zealand on the National Diet Survey 1977.

32. Welsh, S. O., Holden, J. M., Wolf, W. R., and Levander, O. A. 1981. Selenium in self-selected diets of Maryland residents. J. Am. Diet. Assoc. *79*, 277–285.

33. Nutrition Advisory Committee, New Zealand 1983. Recommendations for Selected Nutrient Intakes of New Zealanders.

34. National Research Council 1980. Recommended Dietary Allowances, 9th Editon. National Academy of Sciences, Washington, DC.

35. Davidson, S., Passmore, R., Brock, J. F., and Truswell, A. S. 1975. Human Nutrition and Dietetics, 7th Edition, pp. 42–43. Churchill-Livingstone, Edinburgh and London.

36. Anon. 1959. Evaluation of protein nutrition. N.A.S.–N.R.C., Publ. *711*.

37. Guthrie, B. E., and Robinson, M. F. 1977. Daily intakes of manganese, copper, zinc, and cadmium by New Zealand women. Br. J. Nutr. *38*, 55–63.

38. Thomson, C. D., Robinson, M. F., Campbell, D. R., and Rea, H. M. 1982. Effect of prolonged supplementation with daily supplements of selenomethionine and sodium selenite on glutathione peroxidase (EC 1.11.1.9) activities in blood of New Zealand residents. Am. J. Clin. Nutr. *36*, 24–31.

Selenium in Finnish Food

P. Koivistoinen
P. Varo

INTRODUCTION

Finland is one of the world's low Se areas. Its soils are derived mainly from very old magmatic rocks in which Se is scarce. The melt waters of the latest Ice Age thoroughly leached the soils. The current climate, with low temperatures and fairly high humidity, is responsible for the reducing condition of the soil. The pH is generally on the acidic side, most often between 5 and 6 (Kurki 1979).

All these factors mean that Finnish soils have a very low content of available Se (Koljonen 1975; Sippola 1979). Diseases related to Se/vitamin E deficiency used to be common in domestic animals. In 1969, selenium enrichment of commercial animal feeds and mineral concentrates was started. Muscular dystrophy and other Se deficiency disorders are no longer a serious problem in Finland.

A comprehensive study of the mineral and trace element composition of Finnish foods conducted from 1975 to 1979 included close scrutiny of the Se content of foods (mainly locally grown, but also imported). The gross Se intake by the Finnish population was also estimated (Koivistoinen 1980).

SELENIUM IN FOODS OF VEGETABLE ORIGIN

It was found that foods of vegetable origin were all very low in Se. The main staple cereals in Finland, wheat and rye, contained only 10–15 µg of Se/kg dry matter (dm). A comparative study of locally grown and imported grain showed that the Finnish grain samples had the lowest contents of Se (Varo and Koivistoinen 1981). Finnish grain contained about 15 µg Se/kg dm, the general European level was 25–60 µg, and in grain imported from North America the level was 300–700 µg/kg. In years with favorable growing and harvest seasons, Finland is self-supporting in grain production. However, each 10-year period usually includes 4–6 more or less unfavorable years, and abundant grain imports are frequently required.

All other plant products are also low in Se (Table 1). However, very high levels can be found in some wild mushrooms in Finland. Concentrations of up to 35,000 µg/kg dm have been detected in the cep (*Boletus edulis*) (Piepponen *et al.* 1983). This indicates that soils do contain Se, which is available to mushrooms and lichens but not to green plants.

SELENIUM IN FOODS OF ANIMAL ORIGIN

In spite of the Se supplementation of commercial feeds, the Se status of animal products was exceptionally low in the mid 1970s. Both beef and pork contained distinctly less Se in Finland than elsewhere (Table 2). In pig liver and kidney, however, the level was about the same as reported elsewhere. This may reflect the form of Se supplement: It is added as inorganic selenite, which mainly affects the concentrations in these organs.

The Se content of milk was 3 µg/liter, which is also very low. Of the

TABLE 1. Se Content of Some Vegetables and Plants in Finland (µg/kg Dry Matter)

Item	Se
Potato	<10
Carrot	<10
Cabbage	<10
Lettuce	10 (5–10)
Onion	<10
Fruits	<10
Berries	<10
Mushrooms	
Cantharellus cibarius	40 (30–50)
Lactarius trivialis	220 (100–300)
Agaricus bisporus	300
Boletus edulis	9000 (5,000–13,000)
Lichen	200

TABLE 2. Se Content of Beef in Different Countries (µg/kg Dry Matter)

Origin	Se	Reference
Finland	50 (30–100)	Nuurtamo et al. (1980)
Sweden	150 (90–210)	Kolar and Widell (1977)
Norway	230 (150–290)	Hellesnes et al. (1975)
Germany	700 (40–2500)	Hecht (1977)
United States	940	Morris and Levander (1970)
United States (South Dakota)	620	National Academy of Sciences (1976)

agricultural products only eggs were a reasonably rich source. Egg production is almost totally dependent on the use of commercial feeds.

A comparison of the Se concentrations in powdered milk samples of different geographic origin showed a type of variability similar to that in cereals (Varo et al. 1984). The Finnish samples from 1976 were the lowest of all, about 30 µg/kg dm, whereas those from 1982 contained more than twice the amount of Se. The samples from the United States (North Dakota) contained up to 400 µg Se/kg dm.

VARIATION WITH TIME

We have followed the changes in the Se concentrations in some indicator foods for some time. This has revealed that in foods of animal origin the Se levels are generally higher than in the mid-1970s (Table 3). This may be a reflection of intensified use of Se-supplemented feeds. Some of this increase may also be due to the fairly large-scale import of feedstuffs in 1980–1982. However, no sign of a decrease after that period has so far been observed.

Finnish grain is still as low in Se as it was in the mid-1960s (Ok-

TABLE 3. Changes in Se Content of Some Finnish Foods (µg/kg Dry Weight)

Food	1975/1977	Fall, 1981	Spring, 1982	Spring, 1983	Fall, 1983
Wheat bread	10	80	190	50	35
Rye bread	20	25	55	140	25
Beef, top round	55	160	125	180	210
Pork, ham	220	370	330	330	360
Cooked sausage	40	85	75		
Whole milk	30	45	65	65	60
Eggs	415	675	525	610	700
Baltic herring	830		725		

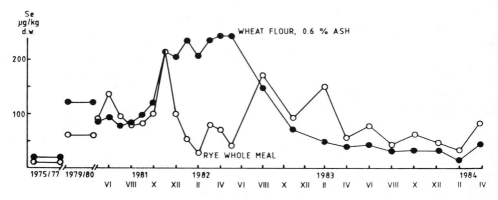

FIG. 1.   Development of Se content in flour.

sanen and Sandholm 1970) or in the mid-1970s (Koivistoinen 1980). The variation in the Se content observed in wheat flour, rye flour, and bread is fully dependent on the geographic origin of the imported grain and on the ratio of domestic and imported grain used in making flour (Fig. 1).

SELENIUM INTAKE AND ITS CONNECTIONS WITH DISEASES

From the above data, we have estimated the average gross Se intake in Finland at different times. The calculations are based on the food consumption patterns indicated by the Finnish food balance sheets (Agricultural Economics Research Institute 1977–1983) and are standardized to an energy intake level of 10 MJ (2400 kcal). The Se intake varies with time (Mutanen and Koivistoinen 1983) and the extent of variation is mainly determined by the volume of the domestic grain harvest (Fig. 2). Only during periods of abundant grain imports from North America does the gross intake reach the lower limit (50 μg/day) of the estimated safe and adequate daily intake defined by the National Academy of Sciences (1980).

The Se status in the Finnish population has been studied with supplementation tests and its possible connection with cardiovascular diseases and cancer with epidemiological studies (Table 4). The supplementation studies indicate that the level of Se is usually suboptimal even in periods of almost maximal dietary intake, as in the Jyväskylä study (Levander *et al.* 1983). Some, but not all, of the epidemiological

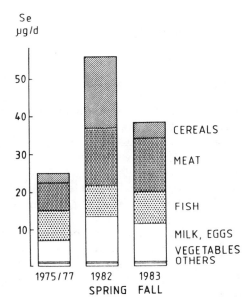

FIG. 2. Daily intake of Se at different times in Finland, estimated according to the food consumption statistics (energy level 10 MJ).

TABLE 4. Se Studies in Human Populations

	Subjects	Indications
Supplementation studies		
Levander et al. (1983)	Healthy, middle-aged men	Platelet GSHPx strongly affected: Se intake suboptimal
Kumpulainen et al. (1983)	Infants, lactating mothers	Se intake suboptimal
Epidemiological studies		
Salonen et al. (1982)	283[a]	Increased risk of cardiovascular deaths at plasma Se levels <45 μg/liter
Miettinen et al. (1983)	33	Plasma Se–cardiovascular deaths: Connection nonsignificant
Virtamo et al. (1984)	1110	Plasma Se–cardiovascular deaths: Connection nonsignificant
Salonen et al. (1984)	128	Increased risk of cancer at plasma Se levels <45 μg/liter

[a] Number of subjects.

studies suggest that very low plasma Se, below 45 μg/liter, may increase the risk of cardiovascular and cancer deaths.

PUBLIC INTERESTS AND OFFICIAL MEASURES

The Se issue has been a focus of public interest for years, and from time to time the discussion has reached nearly hysterical tones. Do-it-yourself supplementation has become popular and the health store pill business has boomed. Paradoxically, some people even consider a good domestic grain harvest almost a national catastrophe.

Against this background, the Finnish authorities decided to increase the Se content of domestic grain nearly 10-fold, to 100 μg/kg, by fertilization. About 10 g of selenate Se/ha is required to reach this level (Yläranta 1982). The main arguments for this action are the attempt to restore consumer reliance in the quality of locally grown food and to calm the exaggerated interest in unsupervised self-medication. The decision is also supported by the accumulating clinical evidence of the great physiological significance of Se.

A strict control system has been planned to monitor the consequences of Se fertilization. The Se content of fertilizers, soils, animal feeds, foods, and human sera will be analyzed regularly and systematically. The Se content in fertilizers can be adjusted or the whole procedure discontinued, if necessary.

REFERENCES

Agricultural Economics Research Institute   1977–1983.   Food Balance Sheet. AERI, Finland, Helsinki.

Hecht, H.   1977.   Der Gehalt des Fleisches an toxischen Elementen. Ber. Landwirtsch. *55*, 796–808.

Hellesnes, I., Underdal, B., Lunde, G., and Haire, G. N.   1975.   Selenium and zinc concentrations in kidney, liver, and muscle of cattle from different parts of Norway. Acta Vet. Scand. *16*, 481–491.

Koivistoinen, P. (Editor)   1980.   Mineral element composition of Finnish foods. Acta Agric. Scand., Supp. *22*, 1–171.

Kolar, K., and Widell, A.   1977.   Untersuchung ueber den Selengehalt in Fleisch, Leber und Nieren von Schwein, Rind and Kalb. Mitt. Geb. Lebensmittelunters. Hyg. *68*, 259–266.

Koljonen, T.   1975.   The behaviour of selenium in Finnish soils. Ann. Agric. Fenn. *14*, 240–247.

Kumpulainen, J., Vuori, E., Kuitunen, P., Mäkinen, S., and Kara, R. 1983. Longitudinal study on the dietary selenium intake of exclusively breast-fed infants and their mothers in Finland. Int. J. Vitam. Nutr. Res. *53*, 420–426.

Kurki, M. 1979. Changes in Fertility of Finnish Soils (in Finnish). Viljavuuspalvelu Oy, Helsinki.

Levander, O. A., Alfthan, G., Arvilommi, H., Gref, C. G., Huttunen, J. K., Kataja, M., Koivistoinen, P., and Pikkarainen, J. 1983. Bioavailability of selenium to Finnish men as assessed by platelet glutathione peroxidase activity and other blood parameters. Am. J. Clin. Nutr. *37*, 887–897.

Miettinen, T., Alfthan, G., Huttunen, J. K., Pikkarainen, J., Naukkarinen, V., Mattila, S., and Kumlin, T. 1983. Serum selenium concentration related to myocardial infarction and fatty acid content of serum lipids. Br. Med. J. *287*, 517–519.

Morris, V. C., and Levander, O. A. 1970. Selenium content of foods. J. Nutr. *100*, 1383–1388.

Mutanen, M., and Koivistoinen, P. 1983. The role of imported grain on the selenium intake of Finnish population in 1941–1981. Int. J. Vitam. Nutr. Res. *53*, 102–108.

National Academy of Sciences 1976. Selenium. Medical and Biologic Effects of Environmental Pollutants. NAS, Washington, DC.

National Academy of Sciences 1980. Recommended Dietary Allowances, 9th Edition. NAS, Washington, DC.

Nuurtamo, M., Varo, P., Saari, E., and Koivistoinen, P. 1980. Mineral element composition of Finnish foods. V. Meat and meat products. Acta Agric. Scand., Suppl. *22*, 57–76.

Oksanen, H. E., and Sandholm, M. 1970. The selenium content of Finnish forage crops. J. Sci. Agric. Soc. Finl. *42*, 250–254.

Piepponen, S., Liukkonen-Lilja, H., and Kuusi, T. 1983. The selenium content of edible mushrooms in Finland. Z. Lebensm.-Unters -Forsch. *177*, 257–260.

Salonen, J. T., Alfthan, G., Huttunen, J. K., Pikkarainen, J., and Puska, P. 1982. Association between cardiovascular death and myocardial infarction and serum selenium in a matched-pair longitudinal study. Lancet *2*, 175–178.

Salonen, J. T., Alfthan, G., Huttunen, J. K., and Puska, P. 1984. Association between serum selenium and the risk of cancer. Am. J. Epidemiol. *120*, 342–349.

Sippola, J. 1979. Selenium content of soils and timothy (*Phleum pratense* L.) in Finland. Ann. Agric. Fenn. *18*, 182–187.

Varo, P., and Koivistoinen, P. 1981. Annual variations in the average selenium intake in Finland: Cereal products and milk as sources of selenium in 1979–80. Int. J. Vitam. Nutr. Res. *51*, 79–84.

Varo, P., Nuurtamo, M., and Koivistoinen, P. 1984. Selenium content of non-fat powdered milk in various countries. J. Dairy Sci. (in press).

Virtamo, J., Valkeila, E., Alfthan, G., Punsar, S., Huttunen, J. K., and Karvonen, M. J. 1985. Serum selenium and the risk of coronary heart disease and stroke in elderly Finnish men. Am. J. Epidemiol. *122*, 276–282.

Yläranta, T. 1982. Uptake by plants of selenium compounds added to soil. 12. Linderstrøm-Lang Conference, IUB Symp. No. 110: Selenium, Glutathione Peroxidase and Vitamin E, Abstract, p. 14.

Development of Selenium Deficiency in the Total Parenteral Nutrition of Infants

P. Brätter
S. Negretti
U. Rösick
H. B. Stockhausen

INTRODUCTION

Nutritional selenium deficiency in man has been shown to be responsible for the development of cardiomyopathy (1,2). It has been shown by various workers (3–5) that during total parenteral nutrition (TPN), besides the trace elements zinc and copper, the serum level of selenium also decreases. A sufficient supply of selenium through nutrient fluids is highly important during the early months of life because after delivery the child has to build up his own antioxidant defense mechanism. Smith *et al.* (6) have shown that feeding practices directly influence selenium status in early infancy. However, comparison between the amounts of trace elements supplied by serum-based infusion solutions and the estimated daily requirements for newborn infants showed that the selenium requirements were not met during TPN (7).

Therefore, in the present work, studies of the variation of the selenium serum level of 38 infants, suffering from various diseases, in the course of long-term TPN have been carried out in order to establish whether selenium supplementation might be indicated. For comparison the trace element content in serum of healthy breast-fed ($n =$

TABLE 1. Determination of Fe, Rb, Se, and Zn in NBS Bovine Liver by INAA, Comparison of Results

| Element | Element concentration (μg/g) | |
	Found ($N = 10$)	Certified
Fe	267.5 ± 8.7[a]	268 ± 8
Rb	18.3 ± 0.35	18.3 ± 1
Se	1.10 ± 0.008	1.1 ± 0.1
Zn	129.6 ± 1.8	130 ± 13

[a] Mean ± SD.

35) and formula-fed ($n = 26$) babies aged from 12 to 14 weeks was determined.

MATERIALS AND METHODS

Assuming that the trace element concentration of blood reflects the nutritional intake, blood constituents might be a suitable monitor for the selenium status of newborns and infants. In order to keep the blood and serum samples free from contamination all materials involved in sampling and storing were tested and if necessary precleaned in the analytical laboratory. Only small sample volumes of about 100–200 μl serum were available for the trace element investigation. Instrumental neutron activation analysis with high-resolution γ-spectrometry has been applied becuase it provides a high sensitive determination of trace elements even in small serum samples. The samples were sealed

TABLE 2. Content of Selenium in Serum of Infants in the Course of Total Parenteral Nutrition[a]

| Duration of TPN (weeks) | No. | Selenium content | | |
		(μg/g dry)	(ng/g wet)	Range
Start	38	0.62 ± 0.16[b]	32.3 ± 11.5	42–15.6
1	13	0.50 ± 0.23	27.0 ± 10.0	44–15.3
2	18	0.39 ± 0.09	21.8 ± 4.1	33–14.0
3	15	0.34 ± 0.09	20.0 ± 6.4	35–10.9
4	12	0.26 ± 0.07	15.7 ± 4.6	24– 9.3
5	4	0.24 ± 0.06	12.2 ± 3.4	16– 7.8
6	5	0.21 ± 0.01	10.7 ± 1.7	13– 8.7
7–12	7	0.22 ± 0.04	12.2 ± 2.9	15– 8.7

[a] Age group: 1–30 days; two-thirds were newborns.
[b] Mean ± SD.

in high-purity quartz ampoules and irradiated for 10 days at a neutron flux density of about 7×10^{13} **n** cm^{-2}sec^{-1} (Berlin Research Reactor BER II). The accuracy of the selenium determination was checked by means of the standard reference material bovine liver (National Bureau of Standards, Washington). The results are presented in Table 1.

RESULTS AND DISCUSSION

The results of the determination of the selenium level in sick infants during intravenous feeding are summarized in Table 2. After two

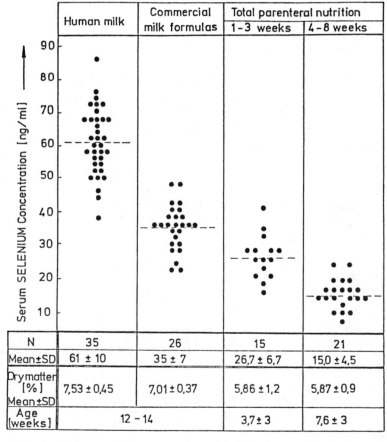

	Human milk	Commercial milk formulas	Total parenteral nutrition 1-3 weeks	Total parenteral nutrition 4-8 weeks
N	35	26	15	21
Mean±SD	61 ± 10	35 ± 7	26,7 ± 6,7	15,0 ± 4,5
Dry matter [%] Mean±SD	7,53 ± 0,45	7,01 ± 0,37	5,86 ± 1,2	5,87 ± 0,9
Age [weeks]	12 - 14		3,7 ± 3	7,6 ± 3

FIG. 1.   Serum selenium concentration of infants. Comparison of breast-feeding, formula-feeding, and total parenteral nutrition.

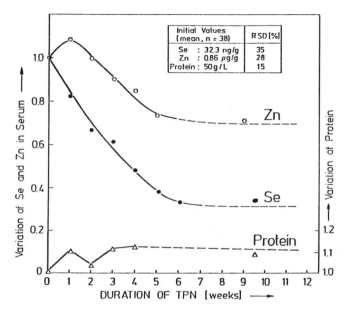

FIG. 2.   Variation of Se, Zn, and protein in serum of infants (mostly newborns) during total parenteral nutrition.

weeks the selenium levels become significantly lower as compared to the values of healthy infants (Fig. 1). After 6 weeks of TPN the serum selenium content of the infants reached levels which have been reported in connection with the development of deficiency syndromes. In the course of TPN the decrease of selenium values were found to be independent of the serum protein content, which was nearly constant after the third week of TPN (Fig. 2). The trace element zinc, which was determined simultaneously, shows after the second week of TPN a dependence similar to selenium, although at the beginning of TPN selenium in the serum drops faster than zinc, which may indicate smaller body stores for selenium and/or faster release from the stores. After 6 weeks of TPN, the variation of both elements in serum tends toward a lower limit of about 9 ppb Se and 0.5 ppm Zn, respectively. It must be noted that in the course of TPN no variation of the selenium-containing enzyme glutathione peroxidase was found within the experimental error [1.54 ± 0.47 U/g hemoglobin ($n = 23$) at start of TPN and 1.49 ± 0.51 ($n = 11$) after 6 weeks of TPN]. This is to be expected, as the lifetime of erythrocytes is much longer than the observation period.

The results of this study suggest the need of selenium supplementation in the parenteral nutrition of newborns and infants. This is further supported by recently published case reports (*8–10*) describing selenium deficiency and cardiomyopathy in patients receiving long-term TPN without selenium supplementation. However, it must be emphasized that intravenous selenium administration is still a problem. The small margin between beneficial and toxic action of this element makes the choice of the suitable chemical form and dose difficult, particularly in the neonatal period.

REFERENCES

1. Keshan Disease Research Group of the Chinese Academy of Medical Sciences 1979. Epidemiological studies on the etiologic relationship of selenium and Keshan disease. Chin. Med. J. *92*, 477–482.
2. Salonen, J. T., Alfthan, G., Pikkarainen, J., Huttunen, J. K., and Puska, P. 1982. Association between cardiovascular death and myocardial infarction and serum selenium in a matched-pair longitudinal study. Lancet *2*, 175–179.
3. Hankins, D. A., Riella, M. C., Scribner, B. H., and Babb, A. L. 1976. Whole blood trace element concentration during total parenteral nutrition. Surgery (St. Louis) *79*, 674–677.
4. Armin, S., Chen, S. Y., Collipp, P. J., Castro-Magna, M., Maddaiah, V. T., and Klein, S. W. 1980. Nutr. Metab. *24*, 331–340.
5. Lombeck, I., Kasparek, K., and Harbisch, H. D. 1978. The selenium state of children. II. Selenium content of serum, whole blood, hair, and the activity of erythrocyte glutathione peroxidase in dietetically treated patients with phenylketonuria and maple syrup urine disease. Eur. J. Pediatr. *128*, 213–223.
6. Smith, A. M., Picciano, M. F., and Milner, J. A. 1982. Selenium intakes and status of human milk and formula-fed infants. Am. J. Clin. Nutr. *35*, 521–526.
7. Brätter, P., Gardiner, P. E., Negretti, V. E., Schulze, G., and von Stockhausen, H. B. 1983. The determination and speciation of essential trace elements in intravenous infusion solutions—Importance in total parenteral nutrition. *In* Trace Element Analytical Chemistry in Medicine and Biology. P. Brätter and P. Schramel (Editors), Vol. 2, pp. 45–59. Walter de Gruyter, Berlin and New York.
8. Fleming, C. R., Lie, J. T., McCall, J. T., O'Brien, J. F., Baillie, E. E., and Thistle, J. L. 1982. Selenium deficiency and fatal cardiomyopathy in a patient on home parenteral nutrition. Gastroenterology *83*, 689–693.
9. Collipp, P. J., and Chen, S. Y. 1981. Cardiomyopathy and selenium deficiency in a 2-year-old girl. N. Engl. J. Med. *304*, 1304–1305.
10. Johnson, R. A., Baker, S. S., Fallon, J. T., Maynard, E. P., Ruskin, J. N., Wen, Z.-M., Ge, K.-Y., and Cohen, H. J. 1981. An occidental case of cardiomyopathy and selenium deficiency. N. Engl. J. Med. *304*, 1210–1212.

68

Role of Selenium in Aging:
An Overview

Aniece A. Yunice
Jeng M. Hsu

Although the process of aging has been described in many different ways, it is generally agreed that it is a time-dependent process in which there is a reduced capacity for self-maintenance and the ability to repair body cells (Alfin-Slater 1979), particularly DNA (Ames 1983). Among the many factors that modulate aging, nutrition is perhaps one of the most important (Moment 1982). Micronutrients, and among them are the trace elements, interact at the structural and functional level of the gene by influencing translational events and/or by modulating posttranslational processes.

Several essential trace metals have been implicated in the aging process. The reader can find excellent reviews on copper, zinc, chromium, selenium, and others (Yunice *et al.* 1976; Hsu 1979; Burch *et al.* 1979; Smith and Hsu 1982; Yunice and Hsu 1984). Perhaps the best candidate among these trace elements that may play a role in the aging processes is selenium in view of its influence on reducing the toxicity of other metals (Hill 1975; Perry and Erlanger 1977; Whanger 1981), in addition to the fact that it is one of several antioxidative defense mechanisms available in the body against free radicals and lipid peroxidation. Other mechanisms of equal importance are such antioxidants as superoxide dismutase, vitamin E, vitamin C, and the sulfur-containing amino acids. Liebovitz and Siegel (1980) have recently reviewed the important role of these antioxidants in the aging process.

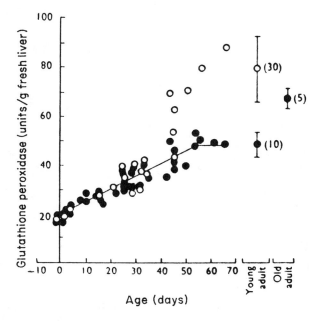

FIG. 1. Activities of hepatic GSHPx in relation to age in male (●) and female (○) rats. The ages of the fetuses are referred to as −1 and −2 days old; young adults and old adults were rats 4 months old and more than 18 months old, respectively. *Source: Pinto and Bartley (1969), reproduced with permission.*

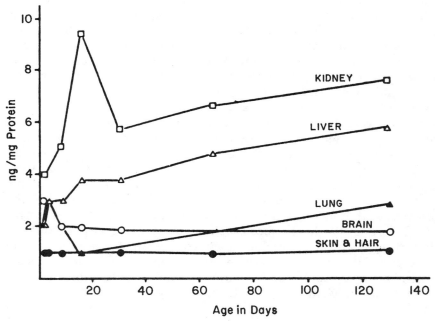

FIG. 2. Selenium levels in different rat organs. *Source: Burch et al. (1979), reproduced with permission.*

DIRECT EVIDENCE: CORRELATION OF SELENIUM CONCENTRATION WITH DIFFERENT AGE GROUPS

Correlation of hepatic glutathione peroxidase (GSHPx) activity with age in the male and female rat is shown in Fig. 1. The activity of this enzyme increased steadily after birth until about 55 days, at which time the activity was similar to young adult males. At 18 months of age, it was 40% higher than in young adults. This increase may represent an autoprotection against free radicals (Pinto and Bartley 1969).

Selenium levels in different rat organs are presented in Fig. 2. In the rat, selenium content increased in the kidney, liver, and lung with

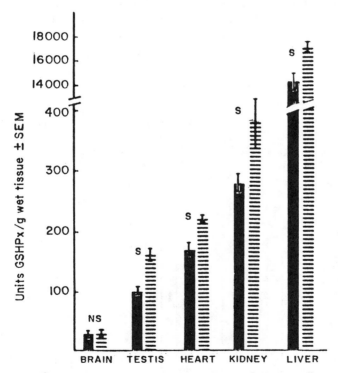

FIG. 3. GSHPx enzyme activity in tissues from weanling (stippled) and 19-month-old (barred) rats. Bar represents the mean of 3 rats ± SEM. Significance between means by the Student's t test ($P<.05$) is indicated with the letter S. NS is not significant.

Source: Csallany et al. (1981), reproduced with permission.

age. Skin, hair, and brain selenium content did not show any significant change up to 128 days of age (Burch *et al.* 1979).

Figure 3 shows GSHPx activity in tissues of weanling rats vs 19-month-old rats. GSHPx activity was significantly higher in heart, kidney, liver, and testes of old rats as compared to weanling rats. No significant changes were observed in the brain tissue (Csallany *et al.* 1981).

Figure 4 shows that elderly people of age range 60–90 had significantly lower blood selenium than the age group comprised of age 16–60 (Robinson and Thomson 1981). Whether the difference is due to reduced intake by the elderly is a question to be answered by data reported by Lane *et al.* (1983) (see later).

Figure 5 depicts changes in serum concentration of selenium in children and adults up to 20 years of age. Serum selenium content at birth amounted to half as much as adults (Lombeck *et al.* 1981).

Table 1 demonstrates the difference in whole blood selenium concentration between children and adults of three different geographic

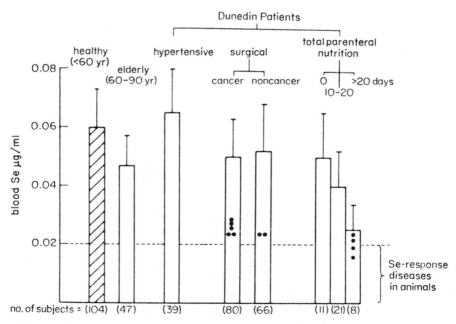

FIG. 4.   Se concentrations in whole blood. Results are shown for healthy Dunedin subjects, elderly subjects, hypertensive patients, surgical patients with and without cancer, and patients on TPN.
Source: Robinson and Thomson (1981), reproduced with permission.

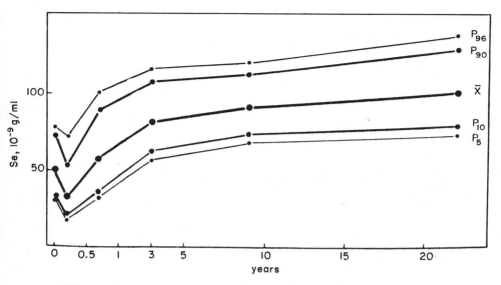

FIG. 5.   Serum selenium concentration in healthy children and adults.
Source: Lombeck et al. (1981), reproduced with permission.

areas in New Zealand. In all areas, the difference is highly significant
(Thomson and Robinson 1980).

Selenium content of whole blood and serum in Finnish children of
various ages is demonstrated in Table II. Results are given in μg/ml.
Again there is a gradual increase from newborn to 12 years of age
(Westermarck *et al.* 1977).

TABLE 1.   Se Concentrations in Whole Blood of New Zealand Residents in Auckland,
Dunedin, and Tapanui

Subjects	Age (yr)	N	Blood Se (μg Se/ml)
Auckland			
Children	7 ± 3[a]	13	0.060 ± 0.012
Adults	36 ± 4	122	0.083 ± 0.012
Dunedin			
Children	9 ± 3	18	0.059 ± 0.011
Adults	33 ± 12	59	0.062 ± 0.012
Tapanui			
Children	11 ± 2	50	0.048 ± 0.010
Adults	35 ± 13	49	0.060 ± 0.012

[a] Values are means ± SD.
Source: Thomson and Robinson (1980), reproduced with permission.

TABLE 2. Selenium Content (Mean ± SD) of Whole Blood and Serum in Finnish Children of Different Ages[a]

District	Age	Whole blood[b]	Serum[b]
Helsinki	1–7 days		
	Deceased prematures	40 ± 12 (13)	ND[c]
	Deceased full-term infants	36 ± 7 (6)	ND
	Living full-term infants	49 ± 12 (3)	ND
	1–6 years	59 ± 11 (14)	50 ± 17 (9)
	7–12 years	60 ± 14 (13)	51 ± 14 (8)
Maarianhamina	4–15 years	ND	73 ± 20 (9)

[a] Results are expressed in ng/ml. The number of subjects is given in parentheses.
[b] Values reported from other countries: Plasma Se, 38.2 ± 11.9 ng/ml in 10 normal French infants, by neutron activation technique; serum Se, 51.2 ng/ml (range: 18.5–112 ng/ml) in 9 healthy infants of the age group from 1 month to 1 year, 81 ng/ml (range: 41–141 ng/ml) in 10 healthy toddlers (1–5 years), and 97 ng/ml (range: 40–125 ng/ml) in 7 healthy children (5–12 years) from Germany, by neutron activation analyses.
[c] ND, Not determined.
Source: Westermarck et al. (1977), reproduced with permission.

Figure 6 depicts changes in lipid peroxidation with age. Lipid peroxidation was measured by the thiobarbituric acid method in the hepatic mitochondria and microsomes of chicks. It is very high in newly hatched selenium and vitamin E-deficient chicks, but decreases rapidly at 6 and 9 days of age (Combs et al. 1975).

Figure 7 demonstrates the inverse relationship between lipid peroxidation (lipofuscin pigment) and GSHPx activity in the tissues of 1- and 19-month-old rats. GSHPx was increased and lipofuscin was decreased in the kidney, heart, and testes of the 19-month-old rats. In the liver, GSHPx was elevated, but the lipofuscin was unchanged in the old rats (Csallany et al. 1981).

In humans, selenium supplementation in patients with neuronal ceroid lipofuscinosis in whom GSHPx activity was depressed was corrected, suggesting a defective uptake of selenium in these patients (Westermarck and Sandholm 1977).

To answer the question as to whether the decrease of selenium observed in the elderly is due to reduced intake or some other causes, Lane et al. correlated selenium levels in two groups of subjects: institutionalized and free-living. Figure 8 shows a positive correlation between dietary selenium and caloric carbohydrate and protein intake in the free-living elderly. Similar results were obtained in the institutionalized subjects (Lane et al. 1983).

In Fig. 9 a positive correlation is demonstrated between erythrocyte

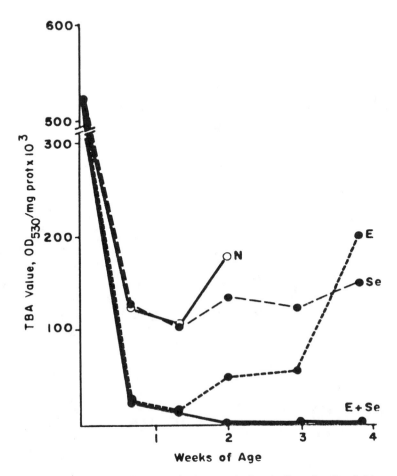

FIG. 6.   Effects of dietary selenium and vitamin E on *in vitro* lipid peroxidation in chick liver mitochondria. N, none.
Source: Combs et al. (1975), reproduced with permission.

selenium and dietary selenium in the free-living elderly [r = .38 (Lane *et al*. 1983)].

In Fig. 10 no correlation is shown between plasma selenium concentrations and dietary selenium in the free-living elderly subjects [r = .13 (Lane *et al*. 1983)].

Table III illustrates the correlation between dietary selenium and blood selenium levels in free-living and institutionalized subjects. A positive correlation was observed between dietary selenium, on one

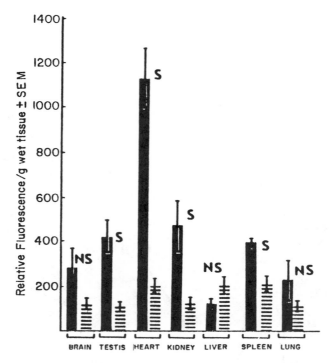

FIG. 7. Organic-solvent-soluble lipofuscin pigment concentrations in tissues from weanling (stippled) and 19-month-old (barred) rats. Bar represents the mean of 3 rats ± SEM. Significance between means by the Student's *t* test (*P*<.05) is indicated with the letter S; NS is not significant.

Source: Csallany et al. (1981), reproduced with permission.

hand, and calorie, carbohydrate, and protein intake for both groups, on the other (Lane *et al.* 1983). The difference between males and females is shown in Table IV. Selenium and GSHPx of erythrocytes were significantly higher in the males than females. The authors concluded from this study that the elderly seem to have adequate selenium intake as judged by their selenium status when compared to levels found in populations at risk for selenium deficiency (Lane *et al.* 1983).

Figure 11 shows positive correlation between GSHPx and selenium content of erythrocytes [*r* = .81 (Lombeck *et al.* 1981)].

Figure 12 demonstrates the wide geographic variability of blood selenium of healthy adults in five countries of the world. This most likely reflects changes in soil selenium (Robinson and Thomson 1981).

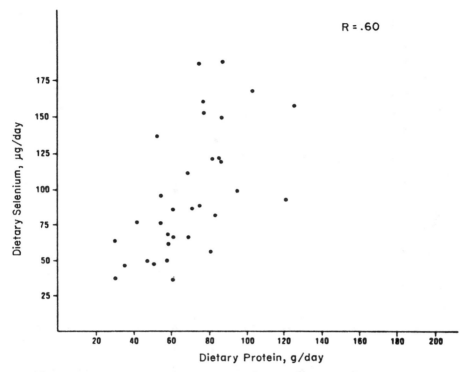

FIG. 8.   Correlation between dietary selenium levels and dietary protein in the free-living elderly subjects.
Source: Lane et al. (1983), reproduced with permission.

INDIRECT EVIDENCE: THE FREE-RADICAL HYPOTHESIS AND THE ROLE OF ANTIOXIDANTS

It is difficult to assign a role for selenium in aging without invoking the antioxidant theory of Tappel (1965, 1980) as delineated by Harman (1965; Harman and Eddy 1979) and others (Yunice *et al.* 1976; Fridovich 1978; Burch *et al.* 1979). The theory stipulates that an increase of free-radical production with aging leads to a decline in cellular integrity due to free-radical chain reactions which cause cross-linking of the membrane.

Free radicals are produced during a variety of cellular activities such as autoxidation, irradiation, mitochondrial respiration, or during environmental stress. Polyunsaturated fatty acids which are located on the cellular membrane are attacked by free radicals formed from

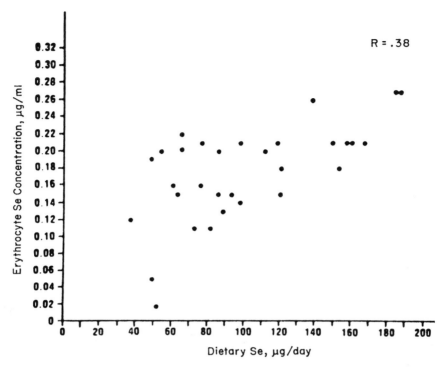

FIG. 9.   Correlation between erythrocyte selenium levels and dietary selenium
in the free-living elderly subjects.
Source: Lane et al. (1983), reproduced with permission.

molecular oxygen. This yields lipid free radicals that add molecular
oxygen to produce peroxy radicals (Tappel 1975; Liebovitz and Siegel
1980). Oxygen radicals as endogenous initiators of degenerative diseases such as cancer and cardiovascular disorders can enhance the
aging processes by virtue of their destructive ability of the cellular
membrane (Sun and Sun 1982).

Although there is no direct evidence that selenium can prevent or
slow down the aging processes in humans or animals, indirect evidence
links selenium and/or its enzyme GSHPx to some manifestations of
aging, such as the fluorescent pigment, lipofuscin, or the products of
lipid peroxidation. Indirect evidence can be derived also from studies
on selenium status and/or GSHPx in different age groups as well as
nutritional status of the elderly and their intake of selenium compared
to other age group populations.

Aging is known to be associated with increased accumulation of lipofuscin pigments in several tissues of the aged (Porta and Hartroft 1969; Reichel 1968). The product of these pigments, which are derived from nucleic acid or protein breakdown, is the substance malonaldehyde, which many investigators have used as an index of the extent of lipid peroxidation and hence cellular membrane damage (Chio and Tappel 1969; Tappel *et al.* 1973; Chen 1973; Tappel 1975). Selenium as a component of GSHPx presumably protects against lipid peroxide-induced cellular damage by destroying the peroxides (Lucy 1972; Diplock and Lucy 1973; Diplock 1981). The selenium-dependent enzyme utilizes reduced glutathione to catalyze the reduction of lipohydroperoxides to alcohol derivatives (Little and O'Brien 1968; Christophersen 1969). This mechanism of action, however, is not recognized by other investigators. For example, McCay *et al.* (1976) postulate that the enzyme acts by preventing the initial free-radical attack on lipid membranes by the rapid removal of hydrogen peroxide.

The demonstration of a protective effect of selenium against perox-

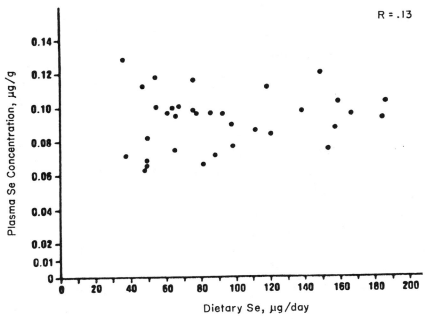

FIG. 10.   Correlation between plasma selenium levels and dietary selenium in the free-living elderly subjects.
Source: Lane et al. (1983), reproduced with permission.

TABLE 3. Correlations between Dietary Selenium and Blood Selenium Measures and GSHPx Activity

Measures	Correlation	
	r	P
Free-living[a]		
Diet		
Calories (kcal/day)/Se (μg/day)	.46	< .05
Protein (g/day)/Se (μg/day)	.60	< .01
Carbohydrate (g/day)/Se (μg/day)	−.19	> .05
Fat (g/day)/Se (μg/day)	.43	< .01
Plasma		
Dietary Se (μg/day)/plasma Se (μg/g)	.13	< .05
Erythrocyte		
Dietary Se (μg/day)/erythrocyte Se (μg/ml)	.38	> .05
Dietary Se (μg/day)/erythrocyte GSHPx (units)[b]	−.15	< .05
Institutionalized[c]		
Diet		
Calories (kcal/day)/Se (μg/day)	.44	< .05
Protein (g/day)/Se (μg/day)	.36	< .05
Carbohydrate (g/day)/Se (μg/day)	.51	< .005
Fat (g/day)/Se (μg/day)	.42	< .005

[a] Thirty-six subjects.
[b] Units expressed as μmole NADPH oxidized/min/g protein.
[c] Eight subjects.
Source: Lane et al. (1983), reproduced with permission.

TABLE 4. Selenium Levels in Erythrocytes and Plasma and GSHPx Activities in Erythrocytes of Free-Living Elderly[a]

Measures	$\bar{x}$	SD	Range
Erythrocytes Se (μg/ml)	.20	.06	.09−.32
Males	.17[b]	.04	
Females	.20	.06	
Plasma Se (μg/ml)	.10	.03	.06−.13
Males	.09	.02	
Females	.10	.02	
Erythrocyte GSHPx activity (units[c]/g of protein)	27.5	5.0	19.8−35.1
Males	25.9[b]	4.0	
Females	29.0	3.6	

[a] Thirty-six subjects.
[b] $P < .001$ as measured by Student's t test.
[c] Units are expressed as NADPH oxidized/min.
Source: Lane et al. (1983), reproduced with permission.

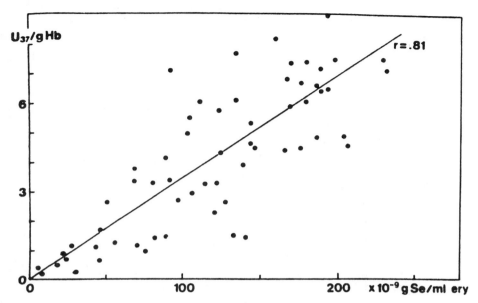

FIG. 11.   Erythrocyte glutathione peroxidase activity vs selenium content of erythrocytes.
Source: Lombeck et al. (1981), reproduced with permission.

idation of membrane lipids exposed to oxidizing conditions has been reported both *in vivo* and *in vitro*. In chicks with exudative diathesis, dietary selenium and α-tocopherol protected against peroxidation (Noguchi *et al.* 1973; Combs *et al.* 1975). In rat erythrocytes, the same protection was afforded by dietary selenium (Rotruck *et al.* 1972). Similar results were obtained by utilizing the measurement of pentane production as an index of lipid peroxidation (Hafeman and Hoekstra 1977; Dillard *et al.* 1978). Use of other antioxidants, such as ascorbic acid, reduced both selenium requirement and mortality rate significantly in α-tocopherol-depleted chicks (Combs and Scott 1974). Recently, Forman *et al.* (1983) obtained good protection in rats exposed to hypoxia by feeding them a high selenium diet. The same results were obtained with sulfur amino acids which are known to be the precursors of the selenium enzyme GSHPx. The efficacy of other selenoproteins in electron-transfer reactions has been described by several investigators (Schum and Murphy 1972; Turner and Stadtman 1973; Whanger *et al.* 1973).

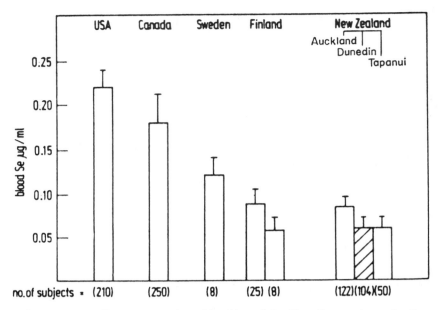

FIG. 12.   Blood Se concentration of healthy adults. Results are shown for the United States, Canada, Sweden, Finland, and New Zealand.

CONCLUSION

Trace metals in general, and selenium by virtue of its antioxidant property in particular, have been implicated in the aging processes. In animal studies, the literature is not very clear as to whether selenium increases or decreases with age. Data supporting both conditions have been reported. It is likely, however, that a better explanation for the increase can be made on the basis of adaptation response of the animal to the aging process. Since in advancing years, markers of aging such as lipid peroxidation and lipofuscin pigmentation are likely to increase, availability of selenium in the environment enhances the activity of GSHPx, hence the slowing down of the aging process. In the absence of adaptation or selenium availability in the environment, aging may be accelerated and body reserve of selenium would thus diminish.

In humans the data are very scanty, and what is available is quite controversial. Large-scale cross-sectional studies of selenium concentration of various tissues from necropsy materials of a wide range of age groups, similar to those done for zinc and copper, can help answer some of these questions. With the advent of modern technology and the availability of sensitive methods for selenium determination in tissues, it is not unlikely that such a task would be easily carried out in the near future.

With regard to the mechanism of action of selenium in peroxidation as a nonspecific antioxidant, three lines of evidence could be presented in favor of the biological antioxidant hypothesis: Rotruck's discovery that selenium is a component of GSHPx; the repeated findings by several investigators that selenium protects cellular membrane of selenium-deficient animals *in vivo* and *in vitro;* and the demonstration that at the subcellular level selenium protects the microsomes and mitochondria of liver cells from peroxidative damage. However, further clarification of the role of selenium in aging is needed to define precisely its relationship to other antioxidants, particularly vitamin E and other selenoproteins.

Finally, it is fitting to mention that the elderlies in this day and age enjoy a better nutritional status than their predecessors in view of the availability of multivitamin preparations which include selenium and vitamin E. Although the recommended dietary allowances of selenium and vitamin E are not well established, their inclusion in the multivitamin preparation would certainly ensure adequate intake by our elderly population.

ACKNOWLEDGMENTS

This work was supported by the Medical Research Service of the Veterans Administration. The secretarial assistance of Ms. Lois Sullivan and Mrs. Pat Greene of the Veterans Administration Pathology Department is gratefully acknowledged.

REFERENCES

Alfin-Slater, R. B. 1979. Nutrition and aging. Introduction. Fed. Proc., Fed. Am. Soc. Exp. Biol. *38,* 1993.

Ames, B. N. 1983. Dietary carcinogens and anticarcinogens: Oxygen radicals and degenerative diseases. Science *221,* 1256–1264.

Burch, R. E., Sullivan, J. E., Jetton, M. M., and Hahn, H. K. J. 1979. The effect of aging on trace element content of various rat tissues. I. Early stages of aging. Age *2,* 103–107.

Chen, L. H. 1973. Effect of vitamin E and selenium on tissue antioxidant status of rats. J. Nutr. *103,* 503–508.

Chio, K. S., and Tappel, A. L. 1969. Synthesis and characterization of the fluorescent products derived from malondialdehyde and amino acids. Biochemistry *8,* 2821–2827.

Christopherson, B. O. 1969. Reduction of linolenic acid hydroperoxide by a glutathione peroxidase. Biochim. Biophys. Acta *176,* 463–470.

Combs, G. F., Jr., and Scott, M. L.   1974.   Antioxidant effects on selenium and vitamin E function in the chick. J. Nutr. *104*, 1297–1303.

Combs, G. F., Jr., Noguchi, T., and Scott, M. L.   1975.   Mechanisms of action of selenium and vitamin E in protection of biological membranes. Fed. Proc., Fed. Am. Soc. Exp. Biol. *34*, 2090–2095.

Csallany, A. S., Zaspel, B. J., and Ayaz, K. L.   1981.   Selenium and aging. *In* Selenium in Biology and Medicine. J. E. Spallholz, J. L. Martin, and H. E. Ganther (Editors), pp. 119–131. AVI Publishing Co., Westport, CT.

Dillard, C. J., Litov, R. E., and Tappel, A. L.   1978.   Effects of dietary vitamin E, selenium and polyunsaturated fat on *in vivo* lipid peroxidation in the rat as measured by pentane production. Lipids *13*, 396–402.

Diplock, A. T.   1981.   The role of vitamin E and selenium in the prevention of oxygen-induced tissue damage. *In* Selenium in Biology and Medicine. J. E. Spallholz, J. L. Martin, and H. E. Ganther (Editors), pp. 303–316. AVI Publishing Co., Westport, CT.

Diplock, A. T., and Lucy, J. A.   1973.   The biochemical modes of action of vitamin E and selenium: A hypothesis. FEBS Lett. *29*, 205–210.

Forman, H. J., Rotman, E. I., and Fisher, A. B.   1983.   Roles of selenium and sulfur-containing amino acids in protection against oxygen toxicity. Lab. Invest. *49*, 148–153.

Fridovich, I.   1978.   The biology of oxygen radicals. Science *201*, 875–880.

Hafeman, D. G., and Hoekstra, W. G.   1977.   Lipid peroxidation *in vivo* during vitamin E and selenium deficiency in the rat as monitored by ethane evolution. J. Nutr. *107*, 666–672.

Harman, D.   1965.   The free radical theory of aging: Effect of age on serum copper levels. J. Gerontol. *20*, 151–153.

Harman, D., and Eddy, D. E.   1979.   Free radical theory of aging: Beneficial effect of adding antioxidants to the maternal mouse diet on life span of offspring; possible explanation of the sex difference in longevity. Age *2*, 109–122.

Hill, C. H.   1975.   Interrelationships of selenium with other trace elements. Fed. Proc., Fed. Am. Soc. Exp. Biol. *34*, 2096–2101.

Hsu, J. M.   1979.   Current knowledge on zinc, copper, and chromium in aging. World Rev. Nutr. Diet. *33*, 42–69.

Lane, H. W., Warren, D. C., Taylor, B. J., and Stool, E.   1983.   Blood selenium and glutathione peroxidase levels and dietary selenium of free living and institutionalized elderly subjects. Proc. Soc. Exp. Biol. Med. *173*, 87–95.

Liebovitz, B. E., and Siegel, B. V.   1980.   Aspects of free radical reactions in biological systems. Aging. J. Gerontol. *35*, 45–56.

Little, C., and O'Brien, P. J.   1968.   An intracellular GSH-peroxidase with a lipid peroxide substrate. Biochem. Biophys. Res. Commun. *31*, 145–150.

Lombeck, I., Kasperek, K., Feinendegen, L. E., and Bremer, H. J.,   1981.   Low selenium state in children. *In* Selenium in Biology and Medicine. J. E. Spallholz, J. L. Martin, and H. E. Ganther (Editors), pp. 269–282. AVI Publishing Co., Westport, CT.

Lucy, J. A.   1972.   Functional and structural aspects of biological membranes: A suggested structural role for vitamin E in the control of membrane permeability and stability. Ann. N.Y. Acad. Sci. *203*, 4–11.

McCay, P. B., Bigson, D. D., Fong, K. L., and Hornbrook, K. R.   1976.   Effect of glutathione peroxidase activity on lipid peroxidation in biological membranes. Biochim. Biophys. Acta *431*, 459–468.

Moment, G. B.   1982.   Introduction. *In* Nutritional Approaches to Aging Research. G.

B. Moment, R. C. Adelman, and G. S. Roth (Editors), pp. 2–11. CRC Press, Inc., Boca Raton, FL.

Noguchi, T., Cantor, A. H., and Scott, M. L. 1973. Mode of action of selenium and vitamin E in prevention of exudative diathesis in chicks. J. Nutr. *103*, 1502–1511.

Perry, H. M., Jr., and Erlanger, M. W. 1977. Effect of a second metal on cadmium-induced hypertension. *In:* Trace Substances in Environmental Health–XI. D. D. Hemphill (Editor), pp. 280–288. Univ. of Missouri Press, Columbia.

Pinto, R. E., and Bartley, W. 1969. The effect of age and sex on glutathione reductase and glutathione peroxidase activities and on aerobic glutathione oxidation in rat liver homogenates. Biochem. J. *112*, 109–115.

Porta, E. A., and Hartroft, W. S. 1969. Lipid pigments in relation to aging and dietary factors (lipofuscins), *In* Pigments in Pathology. M. Wolman (Editor). Academic Press, New York.

Reichel, W. 1968. Lipofuscin pigment accumulation and distribution in five rat organs as a function of age. J. Gerontol. *23*, 145–153.

Robinson, M. E., and Thomson, C. D. 1981. Selenium levels in humans vs environmental sources. *In* Selenium in Biology and Medicine. J. E. Spallholz, J. L. Martin, and H. E. Ganther (Editors), pp. 283–302. AVI Publishing Co., Westport, CT.

Rotruck, J. T., Pope, A. L., Ganther, H. E., and Hoekstra, W. G. 1972. Prevention of oxidative damage to rat erythrocytes by dietary selenium. J. Nutr. *102*, 689–696.

Schum, A. C., and Murphy, J. C. 1972. Effects of selenium compounds on formate metabolism and coincidence of selenium-75 incorporation and formic dehydrogenase activity in cell-free preparations of *Escherichia coli.* J. Bacteriol. *110*, 447–449.

Smith, J. C., and Hsu, J. M. 1982. Trace elements in aging research: Emphasis on zinc, copper, chromium and selenium. *In* Nutritional Approaches to Aging Research. G. B. Moment, R. C. Adelman, and G. S. Roth (Editors), pp. 120–134. CRC Press, Inc., Boca Raton, FL.

Sun, A.-Y., and Sun, G-Y., 1982. Dietary antioxidants and aging on membrane functions. *In* Nutritional Approaches to Aging Research G. B. Moment, R. C. Adelman, and G. S. Roth (Editors), pp. 136–155. CRC Press, Inc., Boca Raton, FL.

Tappel, A. L. 1965. Free radical lipid peroxidation and its inhibition by vitamin E and selenium. Fed. Proc., Fed. Am. Soc. Exp. Biol. *24*, 73–78.

Tappel, A. L. 1975. Lipid peroxidation and fluorescent molecular damage to membranes. *In* Pathobiology of Cell Membranes. B. F. Trump and A. V. Arstila (Editors), Vol. 1. Academic Press, New York.

Tappel, A. L. 1980. Vitamin E and selenium protection from *in vivo* lipid peroxidation. Ann. N.Y. Acad. Sci. *355*, 18–30.

Tappel, A., Fletcher, B., and Dreamer, D. 1973. Effect of antioxidants and nutrients on lipid peroxidation fluorescent products and aging parameters in the mouse. J. Gerontol. *28*, 415–424.

Thomson, C. D., and Robinson, M. F. 1980. Selenium in human health and disease with emphasis on those aspects peculiar to New Zealand. Am. J. Nutr. *33*, 303–323.

Turner, D. C., and Stadtman, T. C. 1973. Purification of protein components of the clostridial glycine reductase system and characterization of protein A as a selenoprotein. Arch. Biochem. Biophys. *154*, 366–381.

Westermarck, T., and Sandholm, M. 1977. Decreased erythrocyte glutathione peroxidase activity in neuronal ceroid lipofuscinosis (NCL)—corrected with selenium supplementation. Acta Pharmacol. Toxicol. *40*, 70–74.

Westermarck, T., Raunu, P., Kirjarinta, M., and Lappalainen, L. 1977. Selenium content of whole blood and serum in adults and children of different ages from different parts of Finland. Acta Pharmacol. Toxicol. *40*, 465–475.

Whanger, P. D.   1981.   Selenium and heavy metal toxicity. *In* Selenium in Biology and Medicine. J. E. Spallholz, J. L. Martin, and H. E. Ganther (Editors), pp. 230–255. AVI Publishing Co., Westport, CT.

Whanger, P. D., Pederson, N. D., and Weswig, P. H.   1973.   Selenium proteins in ovine tissues. II. Spectral properties of a 10,000 molecular weight selenium protein. Biochem. Biophys. Res. Commun. *53,* 1031–1035.

Yunice, A. A., and Hsu, J. M.   1984.   Homeostasis of trace elements in the aged. *In* Metabolism of Trace Metals in Man: Developmental Biology and Genetic Implication. O. M. Rennert and W. Y. Chan (Editors), Vol. 1, pp. 99–127. CRC Press, Inc., Boca Raton, FL.

Yunice, A. A., Lindeman, R. D., Czerwinski, A. W., and Clark, M.   1976.   Effect of age and sex on serum copper and ceruloplasmin levels. *In* The Biochemical Role of Trace Elements in Aging. J. M. Hsu (Editor), Eckerd College Gerontology Center, St. Petersburg, FL.

69

Effects on Human Health of Exposure to Selenium in Drinking Water

Jane L. Valentine
Lesley S. Reisbord
Han K. Kang
Mark D. Schluchter

Human health associated with exposure to selenium through drinking water has been poorly documented. In the 1930s, Smith and co-workers (1) surveyed rural families residing in highly seleniferous areas of South Dakota and Nebraska. Of the water samples, 23% showed the presence of selenium at concentrations of 50–330 µg/liter. They observed an increased prevalence of bad teeth, icteroid skin, dermatitis, arthritis, gastrointestinal disturbances, and diseased nails. However, the frequency of symptoms did not show a direct correlation with urinary output of selenium despite a wide range of values from trace to 1330 µg/liter.

The same subjects were divided into two groups representing high and low selenium excretion, with high representing equal to or greater than 200 µg/liter. Subjects with high levels were more often associated with pathological disturbances of the nails, with gastrointestinal disorders, and with icteroid skin. No differences between the two groups were determined for arthritis and dermatitis.

In a follow-up study using a much larger area in the same location, Smith and Westfall (2) reported similar symptoms. They felt, however, that only the high prevalence of gastrointestinal disturbances could be regarded as specific to selenium poisoning. Unfortunately, neither study used a control area for comparison.

Lemley and Merryman in 1941 (*3*) associated selenium poisoning to contaminated water and foods. Subjects were drawn from geographic locations in North and South Dakota, Montana, Wyoming, and Nebraska. Urine specimens (24 hr) in ranges of 124–600 μg/liter were associated with a wide variety of symptoms, including neurologic and emotional disorders, skin lesions, and gastrointestinal disturbances. Upon 5-day administrations of bromobenzene to these individuals, selenium concentration in the urine fell to normal levels (≤ 100 μg/liter), and the symptoms generally disappeared.

Jaffe and co-workers (*4*) reported a clinical and biochemical study of Venezuelan children, comparing a small group living in a seleniferous area with a control group. Blood and urinary selenium were much higher in the seleniferous area; however, liver function tests were normal in both groups. Symptoms of nausea and vomiting, skin depigmentation, hair loss, and depressed hemoglobin were more frequent in the exposed children, but the authors felt this may been due to a greater incidence of intestinal parasites in the exposed group. They concluded that the level of selenium intake did not pose a severe health hazard in the population studied.

One additional study was that of Rosenfeld and Beath (*5*). They reported an isolated case of selenium poisoning by well water containing 9000 μg/liter selenium. The family experienced alopecia, abnormal nails, and lassitude. These symptoms disappeared when selenium exposure through drinking was eliminated.

We have previously reported (*6,7*) the findings of elevated selenium concentrations in urine, blood, and hair in populations exposed to selenium through their drinking water. Concentrations in urine equaled or exceeded those reported in the earlier studies cited above. The present study assesses the health significance of such exposure and body burden concentrations through the evaluation of responses to a health history questionnaire.

METHODS

Selection of Participants

The study participants were selected from geographic locations having elevated selenium concentration in the local water supply. Levels were considered excessive if greater than the United States drinking water standard of 10 μg/liter. Three communities, Red Butte and Jade Hills of Wyoming, and Grants, New Mexico had selenium in water at average concentrations of 494, 194, and 327 μg/liter, respectively. Two communities, Sun Valley, Nevada and Casper, Wyoming, were selected as controls. People residing in the control communities received water containing selenium at average concentrations of 3 and 2 μg/liter, respectively.

Participants were obtained from randomly selected households. In cases where the communities were too small for such random selection, i.e., Grants, New Mexico, those willing to participate were taken. No participant was included in the study unless he had used his present water supply for at least 1 year.

The overall general income of the selenium-exposed communities and their educational background was used as an initial screen for matching. The exposed participants were then group matched to the control subjects on the basis of age. Sample size for the various study groups was based on the ability to determine differences in mean levels of selenium found in blood, urine, and hair.

Data Collection

A 5-g hair sample, a 20-ml blood sample, and 24-hr urine specimens were collected from each individual where possible. A sample of tap water was obtained from each household.

The questionnaire developed for this study solicited responses pertaining to length of residence, occupational history, dietary habits, water consumption, and possible exposure to other chemicals in addition to the symptom and disease history. Based on previous studies, participants were questioned regarding symptoms of gastrointestinal disturbances, depression, lassitude, dizziness, headaches, abnormal loss of hair and nails, skin rashes, and irregular menses. Also, each respondent was asked if a doctor had diagnosed liver, kidney, or heart disease, emotional illness, cancer, arthritis, anemia, ulcers, or diseases of the nerves within the past 2 years. The questionnaire, which was administered by trained interviewers, was pretested, but no attempt at validation was made. A copy of the questionnaire is available upon request.

Laboratory Methods

Prior treatment of samples before hydride generation was necessary. Sample preparation of hair consisted of a thorough washing with nonionic detergent, ethyl alcohol, and distilled, deionized water rinse. The entire hair sample was used for analysis, since a steady state was assumed with respect to intake and excretion for the element due to the requirement that participants consume the water at least 1 year prior to the data collection.

Blood (5–10 ml), water (10–100 ml), and approximately 1 g of hair were digested with equal volumes of (5 ml each) nitric and perchloric acid on a hot plate. Digestion was continued until the organic matter was destroyed, and the solution evaporated to almost dryness.

To each of the digested samples 15 ml of 6 N HCl was added for selenium measurement. Generation of hydrogen selenide in the acidic solution was accomplished by the addition of a sodium borohydride pellet to the acidic solutions and the gas flushed with nitrogen into the burner of the atomic absorption spectrophotometer, Perkin–Elmer model 603 or 303. Standards including a reagent blank were analyzed using the same procedure.

Duplicate samples of trace metal reference standards prepared by the U.S. Environmental Protection Agency (EPA) were analyzed with each batch of biological sample. In all cases, accuracy was within ±5% of the EPA value for the element. When known amounts of selenium were added to distilled, deionized water and carried through the entire procedure (acid digestion, hydride generation), recoveries of added selenium were within ±10% of added amounts.

Statistical Methods

Associations of mean selenium concentrations in blood, hair, and urine of the exposed vs control populations were compared using Student's t test. The relationship between urinary selenium concentration and concentrations of selenium in water was examined by linear regression analaysis after taking the log transformation of both the independent and dependent variables.

The proportions reporting each symptom in exposed vs nonexposed communities were compared using Fisher's exact test. Symptom rates were also compared across groups formed according to concentration of selenium in urine and blood, again using Fisher's exact test.

RESULTS

Body Burden Concentrations of Selenium

Selenium concentrations in the various body specimens and in drinking water samples are given in Table 1. The control populations had selenium in their water at concentrations of less than 4 μg/liter.

TABLE 1. Selenium Concentrations in Human Blood, Urine, Hair, and Water for Selenium-Exposed and Control Communities

| Sample | Community | No. of samples | Present study | |
			Mean ± SD	Range
Blood (μg/ 100 ml)	Jade Hills	19	12.19 ± 3.25	10.90–16.50
	Red Butte	29	13.79 ± 2.17	9.60–20.0
	Casper	68	7.15 ± 2.13	2.99–12.0
	Grants	33	17.12 ± 2.53	13.30–24.8
	Sun Valley	84	12.09 ± 3.46	3.54–20.63
Hair (μg/g)	Jade Hills	18	0.38 ± 0.20	0.01–0.96
	Red Butte	24	0.50 ± 0.29	0.01–0.94
	Casper	57	0.34 ± 0.29	0.05–2.03
	Grants	30	0.46 ± 0.44	0.02–1.98
	Sun Valley	—[a]	—	—
Urine (μg/liter)	Jade Hills	32	100.31 ± 79.56	7–356.22
	Red Butte	36	299.59 ± 201.88	18–907.2
	Casper	66	33.52 ± 16.15	.96–70.21
	Grants	28	79.21 ± 48.12	14.4–337.5
	Sun Valley	70	24.33 ± 13.47	6.0–65.73
Water (μg/liter)	Jade Hills	19	194.42 ± 7.88	178.0–202.0
	Red Butte	31	494.39 ± 59.38	363.0–560.0
	Casper	67	1.71 ± 0.38	0.60–2.90
	Grants	19	327.46 ± 549.1	26–1800
	Sun Valley	3[b]	2.9 ± .46	2.4–3.3

[a] Hair selenium concentrations not determined.
[b] 69 measured unconcentrated and in all selenium concentrations were less than 1 μg/liter.

The exposed populations had as much as 1800 μg/liter of selenium in their drinking water. Water in one exposed community (Grants, NM) came from individual wells. Thus, those participants had the most variation in elevated exposures.

Participants from both of the control communities had very low selenium concentrations in their blood and urine. Their values were similar to those reported by other researchers for persons with no exposure to selenium (8–12). Generally elevated levels of selenium were found in the biological samples from the exposed populations. Participants from the Red Butte and Jade Hills communities had average blood and urine selenium concentrations that were determined to be significantly different from control values ($P < .001$). For Grants, selenium concentrations in blood and urine also reflect the increased exposure of that population and were significantly greater than those of the controls ($P < .001$). An example of the varying exposures resulting in increased body burden is shown in Fig. 1 where urinary selenium is compared with selenium concentrations in drinking water for the Grants participants.

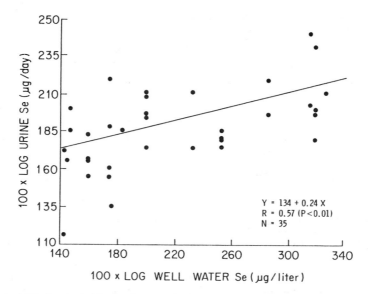

FIG. 1. Individual urinary selenium concentration plotted as a function of selenium concentration in well water for the Grants participants.
Source: Valentine et al. (6).

Hair samples were not analyzed for control participants from Sun Valley. Values for hair measurements of selenium done for the control community of Casper were found to be lower than those reported by Schroeder (*13*). Among the exposed communities only for Red Butte participants were average mean hair selenium values significantly higher than those for controls (*P* < .05).

Symptom Reporting

Tables 2 and 3 show the number and percentage of participants in each community responding positively to each of the symptoms and disease questions. These data are for the age group 18–55 and only for participants who were users of tap water.

Low symptom reporting was observed in all of the survey communities. No statistically significant differences in symptom and disease reporting were found for comparisons between selenium-exposed communities or for comparisons of each selenium-exposed community with each of the control groups.

Tables 4 and 5 give the number and percentages of symptoms reported by individuals according to various body burden concentrations.

TABLE 2. Response to Symptoms Queries Relating to Selenium Exposure (for Ages 18–55)

Symptoms	Communities				
	Grants	Red Butte	Jade Hills	Sun Valley	Casper
Gastrointestinal					
Nausea or vomiting	1/13 (7.7)[a]	1/25 (4.0)	1/12 (8.3)	3/52 (5.8)	6/47 (12.8)
Stomach pains or cramps	4/13 (30.8)	3/25 (12.0)	2/12 (16.7)	6/52 (11.5)	7/47 (14.9)
Diarrhea	4/13 (30.8)	7/25 (28.0)	2/12 (16.7)	9/52 (17.3)	7/47 (14.9)
Loss of appetite	0/13	1/25 (4.0)	3/12 (25.0)	4/52 (7.7)	3/47 (6.4)
Weight loss	0/13	0/25	2/12 (16.7)	2/52 (3.8)	2/47 (4.3)
Other					
Depression	2/13 (15.4)	3/25 (12.0)	4/12 (33.3)	6/52 (11.5)	5/47 (10.0)
Dizziness	5/13 (38.5)	3/25 (12.0)	3/12 (25.0)	7/52 (13.5)	3/47 (6.4)
Lassitude	1/13 (7.7)	6/25 (24.0)	5/12 (41.7)	11/52 (21.2)	9/47 (19.1)
Skin rashes	3/13 (23.1)	1/25 (4.0)	2/12 (16.7)	7/52 (13.5)	3/47 (6.4)
Irregular menses					
Abnormal loss of nail	0/13	0/25	0/12	0/52	0/47
Brittle nails	1/13 (7.7)	3/25 (12.0)	2/12 (16.7)	8/52 (15.4)	6/47 (12.8)
Pain in muscle and joints	2/13 (15.4)	3/25 (12.0)	2/12 (16.7)	5/52 (9.6)	3/47 (6.4)
Severe or frequent headaches	2/13 (15.4)	8/25 (32.0)	2/12 (16.7)	9/52 (17.3)	7/47 (14.9)
Abnormal loss of hair	0/13	0/25	0/12	0/52	0/47

[a] Numbers in parentheses represent percentage of response.

TABLE 3. Response to Disease Queries Relating to Selenium Exposure (for Ages 18–55)

Disease	Communities				
	Grants	Red Butte	Jade Hills	Sun Valley	Casper
Liver disease	0/13	1/25 (4.0)	0/12	0/52	0/47
Kidney disease	1/13 (7.7)[a]	1/25 (4.0)	0/12	1/52 (1.9)	2/47 (4.3)
Diseases of nerves, paralysis or numbness, etc.	2/13 (15.4)	1/25 (4.0)	0/12	1/52 (1.9)	0/47
Emotional illness	1/13 (7.7)	0/25	0/12	4/52 (7.7)	0/47
Heart disease	2/13 (15.4)	0/25	0/12	1/52 (1.9)	1/47 (2.1)
Cancer	0/13	0/25	0/12	0/52	0/47
Arthritis	2/13 (15.4)	1/25 (4.0)	1/12 (8.3)	8/52 (15.4)	5/47 (10.6)
Ulcers	0/13	0/25	0/12	3/52 (5.8)	0/47
Anemia	0/13	0/25	1/12 (8.3)	4/52 (7.7)	2/47 (4.3)

[a] Numbers in parentheses represent percentage of response.

TABLE 4. Response to Symptoms Queries Relating to Selenium Exposure (for Ages 18–55)

Symptoms	Urine Se (μg/day)		Blood (μg/100 ml)	
	≦ 70 μg/day	< 70 μg/day	≦ 12 μg/100 ml	> 12 μg/100 ml
Gastrointestinal				
Nausea or vomiting	6/60 (10.0)[a]	3/37 (8.1)	7/56 (12.5)	2/41 (4.9)
Stomach pains or cramps	11/60 (18.3)	5/37 (13.5)	7/56 (12.5)	9/41 (22.0)
Diarrhea	11/60 (18.3)	9/37 (24.3)	9/56 (16.1)	11/41 (26.8)
Loss of appetite	4/60 (6.7)	3/37 (8.1)	6/56 (10.7)	1/41 (2.4)
Weight loss	1/60 (1.7)	3/37 (8.1)	3/56 (5.4)	1/41 (2.4)
Other				
Depression	8/60 (13.3)	6/37 (16.2)	6/56 (10.7)	8/41 (19.5)
Dizziness	7/60 (11.7)	7/37 (18.9)	4/56 (7.1)	10/41 (24.4)
Lassitude	11/60 (18.3)	10/37 (27.0)	11/56 (19.6)	10/41 (24.4)
Skin rashes	8/60 (13.3)	1/37 (2.7)	4/56 (7.1)	5/41 (12.2)
Irregular menses	—	—	—	—
Abnormal loss of nail	0/60	0/37	0/56	0/41
Brittle nails	11/60 (18.3)	1/37 (2.7)	8/56 (14.3)	4/41 (9.8)
Pain in muscle and joints	5/60 (8.3)	5/37 (13.5)	4/56 (7.1)	6/41 (14.6)
Severe or frequent headaches	11/60 (18.3)	8/37 (21.6)	10/56 (17.1)	9/41 (22.0)
Abnormal loss of hair	0/60	0/37	0/56	0/41

[a] Numbers in parentheses represent percentage of response.

TABLE 5. Response to Disease Queries Relating to Selenium Exposure (for Ages 18–55)

Disease	Urine Se (µg/day)		Blood (µg/100 ml)	
	≤ 70 µg/day	< 70 µg/day	≤ 12 µg/100 ml	> 12 µg/100 ml
Liver disease	0/60	1/37 (2.7)	0/56	1/41 (2.4)
Kidney disease	3/60 (5.0)[a]	1/37 (2.7)	2/56 (3.6)	2/41 (4.9)
Diseases of nerves, paralysis or numbness, etc.	1/60 (1.7)	2/37 (5.4)	1/56 (1.8)	2/41 (4.9)
Emotional illness	1/60 (1.7)	0/37	0/56	1/41 (2.4)
Heart disease	3/60 (5.0)	0/37	1/56 (1.8)	2/41 (4.9)
Cancer	0/60	0/37	0/56	0/41
Arthritis	6/60 (10.0)	3/37 (8.1)	5/56 (8.9)	4/41 (9.8)
Ulcers	0/60	0/37	0/56	0/41
Anemia	2/60 (3.3)	1/37 (2.7)	2/56 (3.6)	1/41 (2.4)

[a] Numbers in parentheses represent percentage of response.

Cut points for blood, hair, and urine were done on the basis of the upper range of normal as determined in the control group of Casper. No statistically significant differences were found between the groups.

DISCUSSION

We have clearly shown that selenium in hair, blood, and urine will increase with elevated exposure. This was dramatically shown in the Grants population where individuals were exposed to varying amounts of selenium in their drinking water. However, levels of selenium in the hair did not always reflect environmental exposure in the other two exposed communities.

The good correlation between urine selenium and intake through drinking water is also in agreement with several animal studies. Smith and co-workers [14] observed that urinary excretion of selenium by rats was directly related to dietary intake, with 80% being eliminated through this route. In addition, Burk and co-workers [15] showed that the amount of ^{75}Se appearing in the urine was related to the level of selenium in the diet fed to rats.

We were unable to relate the elevated levels of selenium in the body to any disease state, even though blood, hair, and urine are often used for such relationships. For example, similar body samples have been used to link selenium levels to the disease states of cancer, hypertension, uremia, and protein–calorie malnutrition [16–20]. Our present findings differ from those of Smith and Westfall [1,2] as well as Lemley and Merryman [3] where they suggested an association between selenium exposure via drinking water and various illness symptoms. Whether their correlations were statistically significant is not known.

Reasons for the finding of no significance between body burden concentration and symptom reporting may be many. One possible cause is that of sample size. In the present study sample size for each group was calculated on the basis of providing differences in body burden and as such may have been too small to detect differences between groups for the more common illnesses and symptoms. It was also too small to be able to pick up the rare illnesses such as cancer.

A second cause of no significance found in the present study may relate to the higher quality of water now received by United States communities compared to that received in the 1930s when the studies of Smith and co-workers and of Lemley and Merryman were conducted. The high incidence of gastrointestinal disturbances noted in previous studies could have been related to microbial contaminants and/or generally lower quality of water.

We have presented data suggesting that an exposure to selenium in water representing a mean daily intake of as much as 988 μg does not produce demonstrable ill effects as determined by questionnaire response. The lack of correlation between increased oral intake of selenium and subsequent elevated body burdens as noted in this study is by no means definitive. It does suggest, however, that at the levels of exposure documented no frank deleterious effect on human health is presented.

REFERENCES

1. Smith, M. L., Frank, K. W., and Westfall, B. B.   1936.   The selenium problem in relation to public health. Public Health Rep. *51*, 1496–1505.
2. Smith, M. I., and Westfall, B. B.   1937.   Further studies on the selenium problem in relation to public health. Public Health Rep. *52*, 1375–1384.
3. Lemley, R. E., and Merryman, M. P.   1941.   Selenium poisoning in the human. Lancet *61*, 435–438.
4. Jaffe, W. G., Raphael, M. D., Mondragon, M. C., and Cueras, M. A.   1972.   Clinical and biochemical study in children from a seleniferous zone. Arch. Latinoam. Nutr. *22*, 595–611.
5. Rosenfeld, I., and Beath, O. A.   1964.   Selenium: Geobotany, Biochemistry, Toxicity and Nutrition. Academic Press, New York.
6. Valentine, J. L., Kang, H. K., and Spivey, G. H.   1978.   Selenium levels in human blood, urine, and hair in response to exposure via drinking water. Environ. Res. *17*, 347–355.
7. Valentine, J. L., Kang, H. K., Dang, P.-M., and Spivey, G.   1981.   Selenium levels in humans as a result of drinking water exposure. *In* Selenium in Biology and Medicine. J. E. Spallholz, J. L. Martin, and H. E. Ganther (Editors), pp. 354–357. AVI Publishing Co., Westport, CT.
8. Dickinson, R. C., and Tomlinson, R. H.   1967.   Selenium in blood and human tissues. Clin. Chim. Acta *16*, 311–321.
9. Allaway, W. H., Kubota, J., Losee, F., and Roth, M.   1968.   Selenium, molybdenum, and vanadium in human blood. Arch. Environ. Health *16*, 342–348.
10. Shamberger, R. J., and Willis, C. E.   1971.   Selenium distribution and human cancer mortality. CRC Crit. Rev. Clin. Lab. Sci. *2*, 211–221.
11. Glover, J. R.   1967.   Selenium in human urines: A tentative maximum allowable concentration for industrial and rural populations. Ann. Occup. Hyg. *10*, 3–14.
12. Sterner, J. H., and Lindfelt, V.   1941.   The selenium content of "normal" urine. J. Pharmacol. Exp. Ther. *73*, 205–211.
13. Schroeder, H. A., Frost, D. V., and Balassa, J. J.   1970.   Essential trace metals in man: Selenium. J. Chronic Dis. *23*, 227–243.
14. Smith, M. I., Westfall, B. B., and Stohlman, E. F.   1938.   Studies on the fate of selenium in the organism. Public Health Rep. *53*, 1199–1216.
15. Burk, R. F., Brown, D. G., Seely, R. J., and Scaief, C. C.   1972.   Influence of dietary and injected selenium on whole body retention, route of excretion, and tissue retention of $^{75}Se\ O_3^{2-}$ in the rat. J. Nutr. *102*, 1049–1056.

16. McConnell, K. P., Broghamer, W. L., Blotcky, A. J., and Hart, O. J. 1975. Selenium levels in human blood and tissue in health and in disease. J. Nutr. *105*, 1026–1031.
17. Shamberger, R. J., Rukovena, E., Longfield, A. K., Tytko, S. A., Deodhar, S., and Willis, C. E.   1973.   Antioxidants and cancer. I. Selenium in the blood of normal and cancer patients. J. Natl. Cancer Inst. (U.S.) *50*, 863–870.
18. Wester, P. O.   1973.   Trace elements in serum and urine from hypertensive patients before and during treatment with chlorthalidone. Acta Med. Scand. *194*, 505–512.
19. Brune, D., Samsahl, K., and Wester, P. O.   1966.   A comparison between the amounts of As, Au, Br, Cu, Fe, Mo, Se, and Zn in normal and uraemic human whole blood by means of neutron activation analysis. Clin. Chim. Acta *13*, 285–291.
20. Levine, R. J., and Olson, R. E.   1970.   Blood selenium in Thai children with protein–calorie malnutrition. Proc. Soc. Exp. Biol. Med. *234*, 1030–1034.

Dietary Selenium Requirements of Pregnant Women and Their Infants

Judy A. Butler
Philip D. Whanger

Blood Se levels vary widely throughout the United States. A range of 0.1–0.34 ppm Se with a mean value of 0.21 ppm was reported in a survey of healthy donors in 19 different areas of the country (Allaway *et al.* 1968).

The Se levels of the soils and plants in the Pacific Northwest region are among the lowest in the United States. Kubota *et al.* (1967) reported that plants growing in this area (which includes the state of Oregon) contained an average of less than 0.05 ppm Se. This may be one of the reasons that Oregon residents have been found to have lower blood Se levels than the national average (values as low as 0.12 ppm have been reported).

Some investigators have suggested that the estimated daily dietary Se intake by Americans of between 134 (Watkinson 1974) and 168 µg (Schrauzer *et al.* 1977) should be doubled for optimal health benefits (Schrauzer and White 1978). The blood Se concentrations resulting from such a dietary intake have been estimated to be between 0.25 and 0.30 ppm (Schrauzer 1978).

Certain nutrient requirements are known to be greater during pregnancy and lactation. If the Se intake of Americans is marginal, then a decrease in blood Se levels might be expected to occur during pregnancy. Various studies of Se nutrition during pregnancy have been reported with rats and humans. Serum Se levels were found to drop significantly (Behne *et al.* 1976, 1978a) and plasma glutathione peroxidase (GSHPx) levels decreased while erythrocyte red blood cell (RBC)

GSHPx activity remained unchanged (Behne *et al.* 1978b) during pregnancy in rats. In a comparison of pregnant and nonpregnant women, Rudolph and Wong (1978) reported lower RBC and plasma GSHPx activities and plasma Se levels (but not RBC Se levels) in pregnant women as compared to nonpregnant controls. Behne and Wolters (1979) found that the plasma Se content and GSHPx activity were significantly lower in women in their third trimester of pregnancy than in nonpregnant controls. Decreased plasma GSHPx activity in pregnant women was also noted in a stable isotope study which demonstrated that pregnant women excreted less Se in their urine than nonpregnant women and that conservation of Se was most pronounced in late pregnancy (Swanson *et al.* 1983). Most investigations of Se requirements during pregnancy have involved cross-sectional studies of pregnant and nonpregnant women. In 1978 our group conducted a longitudinal investigation of the influence of pregnancy on blood Se levels and GSHPx activities in a small group of women and found that whole blood and plasma Se levels decreased, whereas RBC and plasma GSHPx activities increased as pregnancy progressed (Butler *et al.* 1982). These somewhat unexpected results provided the basis for the present study of the influence of Se supplements on metabolism and health in women during the course of pregnancy and lactation and in their infants through 2 years of age.

EXPERIMENTAL PROCEDURE

Sixty-three normal healthy women between the ages of 20 and 38 were recruited over a 9-month period for this study. All women were in the early stages (months 2–4) of their pregnancies. Sixty-one women continued to participate in the study until the births of their infants and 30 women are still in the study at the end of the first postpartum year. As the women joined the study, they were assigned to one of three groups and instructed to take either yeast tablets containing no additional Se (controls) or yeast tablets containing 100 or 200 µg Se on a daily basis. Supplementation ceased at parturition. All women followed normal dietary recommendations of their physicians and the majority consumed one of the standard commercial prenatal vitamin and mineral supplements (containing no Se) routinely prescribed for obstetrics patients in this community.

The women were asked to keep daily records of their health during pregnancy through the first postpartum checkup (generally at 6 weeks) and of their infants through 2 years of age. Five women in each of the three groups were asked to keep a diet record one day per month during pregnancy and to save small samples of all foods eaten that day for an estimation of daily dietary Se intake. Analysis of dietary intakes will be reported at the conclusion of the study, although early results show dietary intakes to be in the range of 70–85 μg per day.

Blood samples were collected monthly from participants (at the time of their routine obstetric appointments) into Se-free heparinized vacutainer tubes at one of two physician-affiliated medical laboratories. A maternal sample was also collected at parturition and at 1.5, 3, 6, 12, and 24 months postpartum. A sample of fetomaternal cord blood was collected at the birth of the baby and small, heparinized blood samples were also collected from some of the infants (via heel prick or later by using pediatric vacutainers) at 3, 6, 12, and 24 months of age. One aliquot of whole blood was removed and refrigerated at 4°C and another aliquot was frozen within 30 min of drawing for later analysis of GSHPx activity. The remainder of the blood sample was centrifuged for 15 min at 1000 g. The plasma portion was removed and both plasma and RBC fractions were frozen immediately after separation. Hematocrit determinations were done on the refrigerated samples as soon as possible after collection and then the whole blood sample was frozen along with the other aliquots. Plasma, RBC, and whole blood samples were later thawed and analyzed within 30 min for GSHPx activity at 30°C using a coupled enzyme procedure (Paglia and Valentine 1967) with modifications (Whanger et al. 1977). Although both hydrogen peroxide and t-butyl hydroperoxide were used as substrates in this assay, only the activities with hydrogen peroxide are represented in the graphs at this time. In general, the GSHPx activities with t-butyl hydroperoxide followed similar trends to the activities with hydrogen peroxide. Whole blood and plasma Se concentrations were determined by the semiautomated fluorimetric procedure of Brown and Watkinson (1977). All work followed guidelines established by the Committee for the Protection of Human Subjects, Oregon State University.

RESULTS

Although this study is not complete, certain trends are evident from the data collected to date. The maternal whole blood (Fig. 1) and plasma (Fig. 2) Se levels gradually decreased during pregnancy in the control group of women, reaching the lowest point just before parturition. At the time of delivery the whole blood and plasma values were approximately 83 and 80%, respectively, of those found during the first trimester. The most rapid decline appears to occur during the first

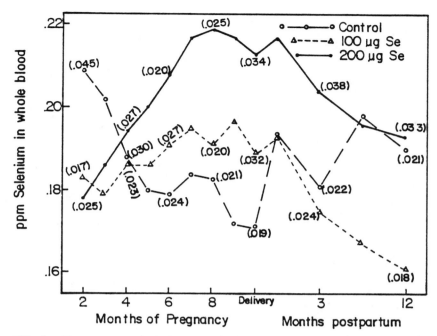

FIG. 1.   Mean maternal whole blood Se levels (ppm) during pregnancy and postpartum. Numbers in parentheses are standard deviations from the means. (○), Control; (△), 100 mg Se; (●), 200 mg Se.

5–6 months of pregnancy, then again between the eighth and ninth months in whole blood and between the ninth month and delivery in plasma. These results are similar to those obtained in the pilot study (Butler *et al.* 1982), although the percentage of decline is more modest in the present investigation. In contrast to the control group, whole blood Se levels in the two supplemented groups gradually increased during pregnancy, although there was a slight drop in both groups just before parturition (parturition values were 4 and 2% lower than ninth-month values in the 100- and 200-μg groups, respectively). The increase noted in the 100-μg group was less than that found in the 200-μg Se-supplemented group (8 and 17% increases, respectively, from the first trimester to ninth-month values). In all three groups there was a subsequent increase in values between parturition and the next sampling at about 6 weeks postpartum. This increase was greatest in the control group. Since all women terminated their supplements at the time of delivery, a gradual decline in Se values was to be

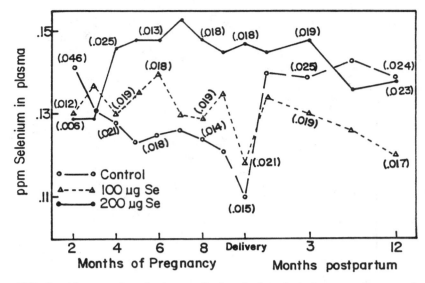

FIG. 2. Mean maternal plasma Se levels (ppm) during pregnancy and postpartum. Numbers in parentheses are standard deviations from the means. See legend to Fig. 1 for symbols.

expected in both supplementation groups during the postpartum period.

The supplementation profile was somewhat different for the plasma Se. Values for the 200-µg group increased 16% from the first trimester through month 7 of pregnancy, then declined very slightly (4%) at parturition. In contrast, the 100-µg group showed an increase of 4% during the first 6 months of pregnancy, then a slight decrease (4%) during the last trimester and a marked decline (13%), similar to the control group, between the ninth month and parturition. A subsequent increase was noted between the parturition and 6 weeks postpartum samples. A slow, gradual decline in plasma Se was noted for the rest of the first postpartum year.

The GSHPx activity profile (using hydrogen peroxide as substrate) of the control group during the course of pregnancy was in contrast to the Se pattern. The whole blood GSHPx activities (Fig. 3) in the control group showed a 36% increase from the first trimester through the ninth month of pregnancy, followed by a 13% drop in activity between the ninth month and parturition. There appeared to be a slight increase in activity between the 6 weeks and 3 months postpartum samples, although the general trend was for a gradual decline in activity during the entire postpartum period. A similar trend was noted in the

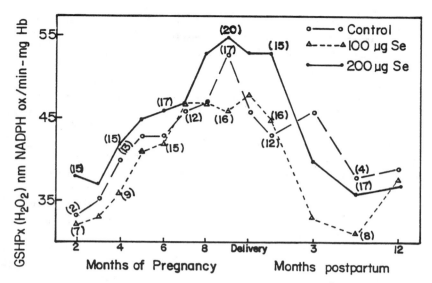

FIG. 3. Mean maternal whole blood GSHPx activities during pregnancy and postpartum. Activity is expressed in nanomoles NADPH oxidized per minute per milligram of hemoglobin at 30°C using hydrogen peroxide as substrate. Numbers in parentheses are standard deviations from the means. See legend to Fig. 1 for symbols.

two supplemented groups, although the prenatal percentage increases were slightly less in both groups than in the control (30 and 31% for the 100- and 200-µg groups, respectively). Only a very slight (3%) decrease in activity was noted between the ninth month and parturition and in the 200-µg group, and a 4% increase in activity was found during this period in the 100-µg group. Values for both supplementation groups declined throughout at least the sixth postpartum month.

Similarly to the whole blood GSHPx profile, the RBC GSHPx activities (data not shown) increased from the first trimester through the eighth month of pregnancy in all three groups (percentage increases were 30, 35, and 43, respectively, for the control, 100-µg, and 200-µg Se groups), and values at parturition were similar to those at the eighth-month sampling period. Values in all groups declined through at least the 3-month postpartum sampling. The plasma GSHPx activity profile (data not shown) appeared to differ from that in the pilot study in that all three groups showed an increase in activity during the first 5–6 months of pregnancy followed by a decrease in the last trimester and a subsequent increase in activity just before delivery. Mean delivery values for all three groups were greater than first trimester values. Plasma activities should be normalized to plasma protein content in the final analysis of the data to determine the possible effect of the hemodilution of pregnancy on GSHPx activity.

In a matched set of maternal parturition and fetomaternal cord blood samples, mean GSHPx activities in the whole blood, plasma, and RBCs, and plasma Se levels were found to be lower than maternal values for all three supplementation groups. For plasma Se (Fig. 4), the cord values were 73, 81, and 80%, respectively, of the maternal parturition values for the control, 100-μg, and 200-μg groups. GSHPx activities for whole blood cord samples (Fig. 5) were 64, 78, and 75%, respectively, of maternal activities, while RBC cord GSHPx activities (data not shown) were 72, 92, and 75%, respectively, of maternal parturition values for the three groups. Cord plasma GSHPx activities (data not shown) were 79, 76, and 59%, respectively, of the maternal

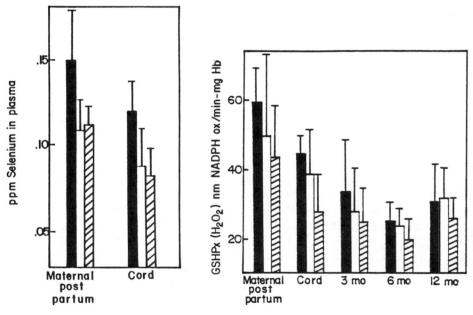

FIG. 4. Matched mean maternal parturition and fetomaternal cord blood plasma Se levels (ppm). Error bars represent standard deviations from the means. (▨), Control; (☐), 100 μg Se; (■), 200 μg Se.

FIG. 5. Matched mean maternal parturition and infant whole blood GSHPx activities from birth (fetomaternal cord sample) through 12 months of age. Activity is expressed in nanomoles NADPH oxidized per minute per milligram of hemoglobin at 30°C using hydrogen peroxide as substrate. Error bars represent standard deviations from the means. See legend to Fig. 4 for symbols.

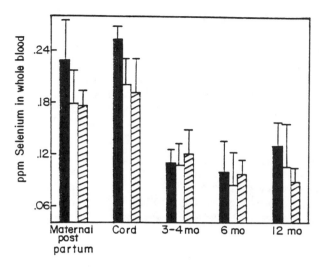

FIG. 6.   Matched mean maternal parturition and infant whole blood Se levels (ppm) from birth (fetomaternal cord sample) through 12 months of age. Error bars represent standard deviations from the means. See legend to Fig. 4 for symbols.

parturition values. In contrast to the general trend for cord values to be lower than matched maternal parturition values, the cord whole blood Se levels (Fig. 6) were slightly though consistently greater than maternal values for all treatment groups (8, 11, and 9%, respectively, above maternal parturition values for the control, 100-μg, and 200-μg Se groups). For all parameters measured, except plasma GSHPx, cord values for the 200-μg Se group tended to be higher than for the control group.

In all three experimental groups, a slow progressive drop in infant RBC (data not shown), whole blood (Fig. 5) GSHPx, and whole blood Se (Fig. 6) values were noted from birth (cord blood sample) through the sixth-month sampling followed by a slight increase at the next sampling at age 1 year. The trend for plasma GSHPx (data not shown) was similar, although the increase may have started to occur slightly earlier, between the third and sixth month of age. There was little difference in mean values between the three experimental groups for any of the parameters from age 3 months through 1 year, suggesting that there was little carry-over to this period from maternal prenatal supplementation. The primary nutritional source during the first 5–6 months for nearly all of the infants involved in the postnatal sampling was breast milk, with only a few receiving commercial formula, except as a supplement to breast milk. Most were gradually started on solid foods at about 4–6 months of age.

In nearly all cases, there was significant intragroup variation in both maternal and infant values. Comparison of blood parameters to such variables as maternal diet and health during the pregnancy and lactation periods, and the sex, health, and nutriture of their infants will be used in the final analysis to investigate these variations.

DISCUSSION

The preliminary results obtained in the present investigation from the nonsupplemented control group tend to confirm the results of the pilot study. A drop in whole blood and plasma Se levels and an increase in RBC GSHPx activities occurred as pregnancy progressed. The plasma GSHPx activity profile appeared to be somewhat different, although mean values at delivery for all three groups were higher than first trimester values. Interpretation of these results must await further investigation. Other researchers have reported RBC GSHPx activities to be no different (Behne and Wolters 1979) and slightly but not significantly lower in pregnant women than in nonpregnant controls (Perona et al. 1979), while Rudolph and Wong (1978) reported significantly higher activities in pregnant than in nonpregnant women. RBC GSHPx activity was reported to be significantly increased in women taking oral contraceptives for more than 6 months (Capel et al. 1981), and it has also been noted that nonpregnant women taking multiple vitamin and mineral (no Se) supplements tend to have higher RBC and plasma GSHPx activities and Se levels than women not taking supplements (Pleban et al. 1982). These results are of interest because oral contraceptives mimic the hormonal conditions of pregnancy, and almost all of the women involved in the present study were taking prenatal vitamin–mineral supplements during pregnancy and lactation.

Many studies of plasma GSHPx activity and Se levels during pregnancy have reported that both parameters are lower in pregnant than in nonpregnant women (Behne and Wolters 1979; Rudolph and Wong 1978; Verlinden et al. 1983). Plasma Se levels found in the present investigation are in agreement with these studies, but the plasma GSHPx activity profile is less clear. The possible effects of the alteration in protein/water ratio due to increased blood volume that occurs during pregnancy need to be evaluated, although Behne and Wolters (1979) still found a decrease in plasma Se levels and GSHPx activity when calculations were corrected for a dry weight basis.

Cord RBC GSHPx activity was found to be significantly lower than adult controls (Emerson *et al.* 1972) and maternal parturition samples (Perona *et al.* 1979). However, Lombeck *et al.* (1977) found similar activities of the enzyme between cord and adult samples. Plasma Se was lower in cord samples than in matched maternal samples (Verlinden *et al.* 1983), and Rudolph and Wong (1978) found that cord plasma and RBC Se levels and GSHPx activities were all lower than maternal samples. Results from the present study were similar to those noted above, except for the observation that whole blood Se values tended to be higher than maternal values at parturition. The cord samples had higher hematocrits than maternal samples, however, and since plasma cord values were lower than maternal values, it would seem that cord RBC Se levels would have to be higher per milliliter of blood for the whole blood Se levels to appear higher. For a valid comparison, all whole blood Se levels should be normalized to hematocrit values.

The drop in infant Se and GSHPx values found in the present study from birth through the early months of life has also been noted by others (Lombeck *et al.* 1977, 1978; Emerson *et al.* 1972; Gross 1979; McKenzie *et al.* 1978). Emerson *et al.* (1972) suggested that RBC GSHPx activity may be under partial hormonal control, since values for children through the age of 12 years were still less than those of adults. The small differences found in the present study between the three treatment groups after birth (cord samples) suggest that there is little long-term carry-over into the lactation period from maternal prenatal supplementation if it is terminated at birth. It has been reported that commercial infant formulas provide less Se per day than human breast milk (Zabel *et al.* 1978; Lombeck *et al.* 1977: Smith *et al.* 1982). Smith *et al.* (1982) found that the serum Se content of 3-month-old infants that had been nursing was significantly higher than in infants receiving commercial formulas. Since the majority of the infants in the present study were breast fed during the early months of life, no definite conclusions can be made about the differences, if any, between the effects of the two forms of nutriture.

Longitudinal studies of breast milk have provided varied results. No significant decrease in mature milk Se content was found between 2 weeks and 3 months (Smith *et al.* 1982) and 1 and 6 months (Levander *et al.* 1981) in two United States studies, while a Finnish study showed a significant decrease in Se content between 1 and 3 months of lactation (Kumpulainen *et al.* 1983). Higashi *et al.* (1983) reported that colostrum had the highest Se levels, followed by a significant decline in Se during the first 1–4 weeks of lactation, after which a plateau was reached that continued through month 5. They found no significant correlation between maternal serum Se and milk Se levels. Although milk samples were not collected in the present study, a slow gradual decline was noted in whole blood and plasma Se levels in maternal blood after supplementation ceased at delivery. The amount of this decline, however, did not appear to correlate very well with the drop in infant values during the first 6 months of life.

The daily Se intake of the women in the present study appears to be about 70–85 μg, as determined by preliminary dietary sample analyses. The drop in whole blood selenium levels noted in the control group seems to be adequately prevented in women taking 100 μg Se per day. Although less, this pattern was similar to that found in women taking 200 μg Se per day. Using blood Se levels as the criterion, it is hypothesized that the selenium requirements of pregnant women are not more than 170–185 μg per day.

ACKNOWLEDGMENTS

This study was supported by U.S. Department of Agriculture grant number 8000566. We wish to express our appreciation to Nutrition 21, San Diego, CA, for supplying the yeast tablets, and to the laboratory staffs of the Corvallis Clinic and Good Samaritan Hospital for drawing and preparing the blood samples. We also thank Dr. Thomas Hart, Jr., and Dorothea Crisp of the Corvallis Clinic for their assistance and support during the recruitment phase of the study.

REFERENCES

Allaway, W. H., Kubota, J., Losee, F., and Roth, M.   1968.   Selenium, molybdenum, and vanadium in human blood. Arch. Environ. Health 16, 342–348.

Behne, D., and Wolters, W. 1979. Selenium content and glutathione peroxidase activity in the plasma and erythrocytes of nonpregnant and pregnant women. J. Clin. Chem. Clin. Biochem. 17, 133–135.

Behne, D., Elger, W., Schmelzer, W., and Witte, M.   1976.   Effects of sex hormones and of pregnancy on the selenium metabolism. Bioinorg. Chem. 5, 199–202.

Behne, D., von Berswordt-Wallrabe, R., Elger, W., Hube, G., and Wolters, W.   1978a.   Effects of pregnancy and lactation on the serum selenium content of rats. Experientia 34, 270–271.

Behne, D., von Berswordt-Wallrabe, R., Elger, W., and Wolters, W. 1978b. Glutathione peroxidase in erythrocytes and plasma of rats during pregnancy and lactation. Experientia 34, 986–987.

Brown, M. W., and Watkinson, J. H.   1977.   An automated fluorimetric method for the determination of nanogram quantities of selenium. Anal. Chim. Acta 89, 29–35.

Butler, J. A., Whanger, P. D., and Tripp, M. J.   1982.   Blood selenium and glutathione peroxidase activity in pregnant women: Comparative assays in primates and other animals. Am. J. Clin. Nutr. 36, 15–23.

Capel, I. D., Jenner, M., Williams, D. C., Donaldson, D., and Nath, A. 1981. The effect of prolonged oral contraceptive steroid use on erythrocyte glutathione peroxidase activity. J. Steroid Biochem. *14*, 729–732.

Emerson, P. M., Mason, D. Y., and Cuthbert, J. E. 1972. Erythrocyte glutathione peroxidase content and serum tocopherol levels in newborn infants. Br. J. Haematol. *22*, 667–680.

Gross, S. 1979. Antioxidant relationship between selenium-dependent glutathione peroxidase and tocopherol. Am. J. Pediatr. Hematol. Oncol. *1*, 61–69.

Higashi, A., Tamari, H., Kuroki, Y., and Matsuda, I. 1983. Longitudinal changes in selenium content of breast milk. Acta Paediatr. Scand. *72*, 433–436.

Kubota, J., Allaway, W. Y., Carter, D. L., Carey, E. E., and Lazar, V. A. 1967. Selenium in crops in the United States in relation to selenium-responsive diseases of animals. J. Agric. Food Chem. *15*, 448–453.

Kumpulainen, J., Vuori, E., Kuitunen, P., Makinen, S., and Kara, R. 1983. Longitudinal study on the dietary selenium intake of exclusively breast-fed infants and their mothers in Finland. Int. J. Vitam. Nutr. Res. *53*, 420–426.

Levander, O. A., Morris, V. C., and Moser, P. B. 1981. Dietary selenium intake and selenium content of breast milk and plasma in lactating and nonlactating women. Fed. Proc., Fed. Am. Soc. Exp. Biol. *40*, 890.

Lombeck, I., Kasperek, K., Harbisch, H. D., Feinendegen, L. E., and Bremer, H. J. 1977. The selenium state of healthy children. I. Serum selenium concentrations at different ages; activity of glutathione peroxidase of erythrocytes at different ages; selenium content of food of infants. Eur. J. Pediatr. *125*, 81–88.

Lombeck, I., Kasperek, K., Harbisch, H. D., Becker, K. *et al.* 1978. The selenium state of children. II. Selenium content of serum, whole blood, hair and the activity of erythrocyte glutathione peroxidase in dietetically treated patients with phenylketonuria and maple syrup urine disease. Eur. J. Pediatr. *128*, 213–223.

McKenzie, R. L., Rea, H. M., Thomson, C. D., and Robinson, M. F. 1978. Selenium concentration and glutathione peroxidase activity in blood of New Zealand infants and children. Am. J. Clin. Nutr. *31*, 1413–1418.

Paglia, D. E., and Valentine, W. N. 1967. Studies on the quantitative and qualitative characterization of erythrocyte glutathione peroxidase. J. Lab. Clin. Med. *70*, 158–169.

Perona, G., Guidi, G. C., Piga, A., Cellerino, R. *et al.* 1979. Neonatal erythrocyte glutathione peroxidase deficiency as a consequence of selenium imbalance during pregnancy. Br. J. Haematol. *42*, 567–574.

Pleban, P. A., Munyani, A., and Beachum, J. 1982. Determination of selenium concentration and glutathione peroxidase activity in plasma and erythrocytes. Clin. Chem. (Winston-Salem, NC) *28*, 311–316.

Rudolph, N., and Wong, S. L. 1978. Selenium and glutathione peroxidase activity in maternal and cord plasma and red cells. Pediatr. Res. *12*, 789–792.

Schrauzer, G. N. 1978. Trace elements, nutrition and cancer: Perspectives of prevention. *In* Inorganic and Nutritional Aspects of Cancer, pp. 323–344. Plenum Press, New York.

Schrauzer, G. N., and White, D. A. 1978. Selenium in human nutrition: Dietary intakes and effects of supplementation. Bioinorg. Chem. *8*, 303–318.

Schrauzer, G. N., White, D. A., and Schneider, C. J. 1977. Cancer mortality correlation studies. III. Statistical associations with dietary selenium intakes. Bioinorg. Chem. *7*, 23–34.

Smith, A. M., Picciano, M. F., and Milner, J. A. 1982. Selenium intakes and status of human milk and formula-fed infants. Am. J. Clin. Nutr. *35*, 521–526.

Swanson, C. A., Reamer, D. C., Veillon, C., King, J. C., and Levander, O. A. 1983. Quantitative and qualitative aspects of selenium utilization in pregnant and non-pregnant women: An application of stable isotope methodology. Am. J. Clin. Nutr. *38*, 169–180.

Verlinden, M., van Sprundel, M., Van der Auwera, J. C., and Eylenbosch, W. J. 1983. The selenium status of Belgian population groups. II. Newborns, children and the aged. Biol. Trace Elem. Res. *5*, 103–113.

Watkinson, J. H.   1974.   The selenium status of New Zealanders. N. Z. J. Med. *80*, 202–205.

Whanger, P. D., Weswig, P. H., Schmitz, J. A., and Oldfield, J. E.   1977.   Effects of selenium and vitamin E deficiencies on reproduction, growth, blood components, and tissue lesions in sheep fed purified diets. J. Nutr. *107*, 1288–1296.

Zabel, N. L., Harland, J., Gormican, A. T., and Ganther, H. E.   1978.   Selenium content of commercial formula diets. Am. J. Clin. Nutr. *31*, 850–858.

Vitamin E and Selenium Supplementation in Geriatric Patients: A Double-Blind Clinical Trial

Matti Tolonen
Markku Halme
Seppo Sarna

Aging often brings with it a variety of illnesses and complaints. Some 12% of people over the age of 65 suffer from mild dementia accompanied by depression, loss of memory, confusion, and untidiness. Older people are also less capable of looking after themselves. Milder symptoms of aging are even more common: regression in emotional life, diminishing interest in the environment, declining independent activity, slowing down of movements, and difficulty in adapting to anything new.

The biological basis of aging is not yet fully understood. One reason is probably lipid peroxidation in cell membranes (2,7). The supplementary intake of antioxidants may inhibit lipid peroxidation and accumulation of lipofuscin, thus slowing down the aging process (1,3,4,9). We tested this hypothesis by administering large doses of two antioxidants, selenium and vitamin E, to residents of an old people's home for 1 year.

SUBJECTS AND METHODS

Subjects

In the autumn of 1982, there were altogether 47 residents in Hartola old people's home. Thirty of them were included in this study, 15 propositae and 15 controls. The remaining 17 patients were either too severely ill, too deeply demented, or mentally ill with a paranoid attitude toward this 1-year trial.

At the start of the study in the autumn of 1982, the mean age of the verum group was 76.8 years (range 58–90) and that of the control group 76.2 years (range 50–92). The verum group contained 4 men and 11 women, and the control group 3 men and 12 women. Allocation into groups was random. Initially (see Results), the groups were very much alike in regard to the parameters studied. There were no significant differences regarding diseases, medication, diet, standard of education, or other similar factors. Any prior medication was continued throughout the study as far as possible.

Four subjects dropped out. One 90-year-old member of the group of propositae died of a cardiac infarction toward the end of the study. One 75-year-old male in the verum group moved back home and neither assessments nor blood tests were therefore carried out on him. One female in the control group discontinued the supplementation as well as other medication because of stomach cancer. One person in the control group died of myocardial infarction 1 week after the start of the study.

Antioxidant Therapy

The therapy group received 8 mg of sodium selenate (corresponding to 1720 μg of pure selenium) per day, divided into two doses. In addition, they received one 50-μg capsule of organic selenium (Vita-Hiven) per day, and 400 mg of D-α-tocopherol (Ido-E) divided into two daily doses. Subjects in the control group received corresponding placebos. All the tablets and capsules were dispensed in coded packages, and neither the personnel of the old people's home nor the assessors or the doctor managing the study (M.H.) knew which subjects received the active ingredient and which the placebo. Throughout the intervention, the code was in the possession of the head of the study (M.T.) and was not broken until the end of the study in September, 1983.

Assessment of Variables

Two nurses working in the old people's home assessed each subject at the beginning of the study and during it at 2-month intervals using the Sandoz Clinical Assessment Geriatric (SCAG) scale (*6,8,10,11*). This method has been accepted by the U.S. Food and Drug Administration (FDA) for geriatric pharmaceutical studies. In Finland, it has been used at the Halikko, Salo, and Kiikala-Kisko-Suomusjärvi old people's homes (5). SCAG measures changes in mental well-being and is useful in assessing the effects of drugs on symptoms of senility.

Before starting the study, both nurses and the doctor in charge of the old people's home (M.H.) practiced assessment and discussed inter-individual differences. Despite this there was a small systematic difference in the results: One of the nurses kept to a stricter scale than the other. This has been taken into account in the statistical analysis. The arithmetic mean values obtained are presented graphically.

Laboratory Studies

The blood and urinary selenium (B-Se and U-Se, μmol/liter), blood glutathione peroxidase (GSHPx) activity, plasma α-tocopherol and retinol, hematocrit and serum cholesterol, triglyceride, calcium, and free thyroxine were measured. Blood selenium levels were measured at the beginning of the study, after 1, 2, and 3 weeks, and after 3, 6, 9, and 12 months. Plasma α-tocopherol was measured at the beginning of the study, at 2 and 4 days, and at 3 and 12 months. GSHPx activity was measured only once, at the end of the study.

Statistical Methods

Covariance analysis using BMDP software was used to correct for the differences between statistical significances and between the assessors. Statistical tests were performed two-sidedly. Correlations were calculated with Pearson's product moment formula, since the distributions were sufficiently normal.

RESULTS

SCAG Assessment

Among the propositae (19 subjects), a statistically significant improvement (compared with the control group as well as with the initial status) was seen in a total of 12 variables:

Depression $(P < .001)$
Self-care $(P < .01)$
Anxiety $(P < .01)$
Mental alertness $(P < .001)$
Emotional lability $(P < .001)$
Motivation and initiative $(P < .025)$

Hostility $(P < .025)$
Interest in the environment $(P < .05)$
Fatigue $(P < .001)$
Anorexia $(P < .025)$
General impression $(P < .01)$
Sum of variables $(P < .01)$

The group receiving antioxidants showed an improvement after only 2 months. The assessors considered the changes so evident that they spontaneously claimed to be able to guess which of the subjects were receiving active therapy and which were controls. At the request of the head of the study (M.T.), the nurses wrote down their assessments of which of the patients whose condition had improved belonged to the group receiving antioxidants. At the end of the study, comparison with the code showed that the nurses had been right in 80% of the cases. The differences between the groups in regard to the variables mentioned above increased throughout the study. The treatment had no side effects.

Results of Laboratory Tests

The mean blood selenium contents (B-Se) for the two groups were initially the same, 1.57 μmol/liter (range 1.0–2.5 μmol/liter). The corresponding urinary selenium content (U-Se) was 0.31 μmol/liter (range 0.1–0.5 μmol/liter). The blood selenium level fell slightly in the control group during the year, the mean value at the end of the study being 1.34 μmol/liter. The changes in blood and urinary selenium in the verum group are shown in Table 1.

GSHPx Activity

The mean blood GSHPx activity at the end of the study was 1.55 mU/mg Hb in the control group and 2.22 mU/mg Hb in the group of propositae. The activity of the enzyme was thus some 40% higher in the therapy group than in the control group and correlated with blood selenium content.

Plasma α-Tocopherol and Retinol

The mean plasma α-tocopherol content was initially 20.3 μmol/liter in the therapy group (range 11.6–46.4 μmol/liter) and 19.9 μmol/liter in the control group (range 11.6–34.8 μmol/liter). In the therapy group, the level increased within 2 days to an average of 55 μmol/liter,

TABLE 1. Average Blood and Urinary Selenium Contents (μmol/liter) in the Group Receiving Active Therapy ($n = 15$) at Various Stages of the Study

Time	Blood selenium		Urinary selenium	
	x	SD	x	SD
Beginning	1.57 ±	0.27	0.31 ±	0.12
1 week	1.97 ±	0.28	11.31 ±	6.26
2 weeks	2.15 ±	0.24	11.58 ±	4.52
3 weeks	1.87 ±	0.24	10.34 ±	4.57
3 months	2.30 ±	0.35	13.76 ±	4.13
6 months	2.55 ±	0.33	24.42 ±	10.81
9 months	2.54 ±	0.35	20.72 ±	6.74
12 months	2.21 ±	0.31	14.26 ±	8.29

where it remained until the end of the study. The α-tocopherol content correlated with that of serum triglyceride in the verum group ($r = .87$, $P < .001$). There was no such correlation in the control group. The level of α-tocopherol also correlated with that of serum cholesterol ($r = .76$, $P < .001$ in the therapy group and $r = .50$, $P < .10$ in the control group).

The mean plasma retinol contents at the beginning of the study were the same in both groups and correlated with serum triglyceride ($r = .62$, $P < .01$ in the therapy group, $r = .62$, $P < .02$ in the control group). There were no significant differences or changes in the rest of the laboratory results.

DISCUSSION

Selenium and vitamin E supplementation markedly increased the levels of plasma α-tocopherol and blood selenium. It was especially noteworthy that GSHPx activity increased in this group by 40% compared to the control group. The activity of this enzyme correlated linearly with blood selenium. It is generally believed that inorganic selenium does not increase the activity of this enzyme. However, it now seems that the administration of a sufficiently large amount of selenate has an effect of this kind.

Vitamin and trace element supplementation has repeatedly been criticized for its side effects. In this study no side effects were observed within 1 year, even though the dosage of vitamin E was 100 times the recommended daily allowance (RDA) and the daily dosage of selenium about 60 times the current daily intake in Finland. This observation is especially noteworthy in a population of elderly patients in poor condition.

In the present study, clinically and statistically significant improvements in several typical aging symptoms were observed in the group receiving vitamin E and selenium. Improvements were seen after only 2 months of supplementation, and the differences in a number of variables between verum and control groups increased steadily throughout the follow-up period.

Our study was designed to conform to the requirements set for pharmaceutical clinical trials. However, the number of subjects was limited, with only 15 patients receiving active therapy and 15 controls. In spite of these small groups, the results are valid within the study. However, to prove their external validity, studies in larger series of geriatric patients would be required.

SUMMARY

We used a double-blind approach in studying the effects of selenium and vitamin E supplementation on the geriatric symptoms and conditions of residents in an old people's home. Over a period of 1 year we gave 15 older people a daily dose of vitamin E 100 times larger than the recommended daily allowance and a daily dose of selenium 60 times as high as the current dietary intake in Finland. Statistically significant improvements were observed in the therapy group compared with the placebo group in the following variables: anxiety, depression, fatigue, hostility, anorexia, mental alertness, self-care, emotional lability, motivation, and initiative and interest in the environment.

The plasma α-tocopherol content tripled and blood GSHPx activity increased by about 40% in the group receiving vitamin E and selenium. Their blood selenium content increased from 1.6 μmol/liter to 2.5 μmol/liter. A linear correlation was observable between GSHPx activity and blood selenium content.

The effects of this antioxidant therapy proved beneficial in this study group. Thus, the results are internally valid, but to determine their external validity a carefully controlled study with a larger numbers of patients is necessary.

REFERENCES

1. Barber, A., and Bernheim, F.   1967.   Lipid peroxidation, its measurement, occurrence, and significance in animal tissues. Adv. Gerontol. Res. 2, 77–120.

2. Dowson, J. H. 1982. Neuronal lipofuscin accumulation in aging and Alzheimer dementia: A pathogenic mechanism? Br. J. Psychiatry pp. 140–148.
3. Harman, D. 1956. Aging: A theory based on free radical and radiation chemistry. J. Gerontol. pp. 298–300.
4. Hoekstra, W. 1975. Biochemical function of selenium and its relation to vitamin E. Fed. Proc., Fed. Am. Soc. Exp. Biol. *34*, 11.
5. Junnila, S., Lyra, H., and Wigren, T. 1982. SCAG suomeksi—arvioita käyttökelpoisuudesta. Sandoz Rep. *13* (6), 9–12.
6. Maurer, W., Ferner, U., Patin, J., and Hamot, H. 1982. Sandoz Clinical Assessment Geriatric Scale (SCAG): Eine transkulturelle faktorenanalytische Studie. Z. Gerontol. *15*, 26–30.
7. Pohjolainen, P., and Hervonen, H. 1983. Gerontologian perusteet. Lääketieteellinen oppimateriaalikustantamo, Tampere.
8. Shader, R., Harmatz, J., and Salzman, C. 1974. A new scale for clinical assessment in geriatric populations: Sandoz Clinical Assessment—Geriatric (SCAG). J. Am. Geriatr. Soc. *22*, 107–113.
9. Tappel, A. 1972. Vitamin E and free radical peroxidation of lipids. Ann. N.Y. Acad. Sci. *203*, 12–28.
10. Venn, R. 1983. The Sandoz Clinical Assessment Geriatric (SCAG) Scale. A general-purpose psychiatric rating scale. Gerontology *29*, 185–198.
11. Yesavage, J., Adey, M., and Werner, T. 1981. Development of a geriatric behavioural self-assessment scale. J. Am. Geriatr. Soc. *29*, 285–288.

Endemic Selenosis and Fluorosis

Bian-Sheng Liu
Shen-Si Li

At the Health Institute of the Chinese Academy of Medical Sciences we had confirmed in 1966 that there were some cases of endemic selenosis in humans and animals in Enshi County, Hebei Province.

In other areas of Enshi County more serious cases of endemic fluorosis were found in 1977. The source of selenium and fluorine causing these two diseases was bone coal which abounded in these areas. Part of the coal Se and F got into the soil, and their (Se and F) content in corn and vegetables was rich. Through these foods, considerable amounts of Se and F entered the human body and caused intoxication. Much research revealed that there were some patients of fluorosis in the areas of endemic selenosis. But in areas of high fluorine the residents did not have any symptoms of endemic selenosis despite their higher level of blood and hair Se. We have studied the relationship of both diseases by epidemiological and clinical methods and measured Se and F of many samples.

MATERIALS AND METHODS

In the area of fluorosis, we have made on-the-spot investigations, carried out physical examinations, and made X-ray and epidemiological studies in 91 cases. In the area of selenosis, 22 cases were examined by clinical, X-ray, and epidemiological methods.

The analysis of selenium was accomplished by 2,3-diaminonaphthalene fluoroscopy. The analysis of fluorine was determined spectrophotometrically.

RESULTS

Geography

Both areas are part of the Wudang Mountains. The area is 200 km^2. The elevation is 1000 m above sea level. The structure of the earth's crust in these areas is sandstone and limestone. These areas abound in bone coal, which is rich in selenium and fluorine. The residents use bone coal as fuel. The area of fluorosis also abounds in sulfur mines. The bone coal slag is used to fertilize the soil.

Contents of Selenium and Fluorine

The contents of selenium and fluorine in bone coal is illustrated in Table 1. These results suggest that the content of selenium in coal in areas of selenosis is higher than in areas of fluorosis, but the content of fluorine in areas of fluorosis is higher than in the areas of selenosis.

Content of Selenium and Fluorine in Corn

The analytical result shows that the F content of corn in the area of fluorosis is obviously higher than in areas of selenosis ($P < .01$). The Se content of corn is not different in these two regions ($P > .05$) (see Table 2).

Physical Examinations

Selenosis Region. In 22 cases, residents possessed "brown spotted teeth." Twelve cases (55%) had fluorotic deformed elbow joints. Through X-ray examination, 13 cases (59%) were confirmed to have fluorotic osteopathy. The changes are all scleroskeletal. In this region, many selenosis patients were found between 1961 and 1965. Three

TABLE 1. Content of Se and F in Bone Coal (ppm)

	Area of selenosis	Area of fluorosis
Fluorine	442	815
	(252–657)[a]	(433–1128)
Selenium	540	212
	(210–1332)	(166–272)

[a] Range.

TABLE 2. Content of Se and F in Corn (ppm)

	Area of selenosis	Area of fluorosis	P
F	15.20 ± 4.56[a] (n = 12)	40.94 ± 8.03 (n = 12)	<.01
Se	4.44 ± 2.33 (n = 5)	4.52 ± 0.20 (n = 5)	>.05

[a] Mean ± SD.

cases were diagnosed as chronic selenosis, manifested as deformed and thick fingernails and toenails.

Case Example.   In 1965, a male patient, aged 64 yr, had eaten corn rich in selenium, resulting in complete loss of scalp hair. After therapy, the patient's hair grew again, but his nails remained deformed and thick. Since 1972, the patient complained of an uncomfortable elbow joint, which disturbed its activity. The patient was given an X ray and had evidence of fluorine osteopathy, stage II.

Fluorosis Region.   We examined 91 people living in a fluorosis region. The results showed that all residents had brown spotted teeth, 81 cases (89%) had fluorotic deformed elbows, and 11 cases (12%) had fluorotic hunchback. The bone changes were of two types: the scleroskeleton type and the porous skeleton type. All the 91 cases showed no symptoms of selenosis, which included trichomadesis, loss of hair and nails, despite the fact that their hair and blood Se levels were high (see Table 3). In a retrospective study of this disease, there was also evidence of selenosis.

TABLE 3. Content of Se and F in Blood, Hair, and Urine (ppm)

	Area of selenosis	Area of fluorosis	P
Blood Se	3.48 ± 1.32[a] (n = 22)	2.98 ± 0.36 (n = 21)	>.05
Hair Se	21.58 ± 8.74 (n = 22)	29.10 ± 13.17 (n = 21)	<.05
Urine F	7.36 ± 4.57 (n = 22)	13.85 ± 6.62 (n = 114)	<.01

[a] Mean ± SD.

TABLE 4. Comparison of Epidemiological Features of Two Areas

Feature	Area of selenosis	Area of fluorosis
Source of Se and F	Bone coal	Bone coal
Sulfur mine	A few	Majority
Content of Se in bone coal	Higher	High
Content of F in bone coal	High	Higher
Bone coal slag	Fertilize soil	Fertilize soil
Fuel	Bone coal and firewood	All bone coal
Content of Se in corn	High	Same as in selenosis area
Content of F in corn	High	Higher
Human selenosis	High incidence, currently only a little chronic selenosis	No cases
Animal selenosis	More cases of alkali disease	No cases
Human fluorosis	Some cases	Many cases

Epidemiological Investigation

Our study suggests that bone coal is rich in Se and F in both areas. Epidemiological features of two regions are given in Table 4.

CONCLUSIONS

This chapter reports the relationship between endemic selenosis and fluorosis. The investigation suggests that the source of selenium and fluorine is bone coal. The principal food of residents is corn, which is rich in both selenium and fluorine. The cause of poisoning by selenium and fluorine is consumption of this corn.

Our study also shows that there are some fluorosis patients in the area of selenosis, but the residents who live in areas of high fluorine do not have any symptoms of selenosis, despite the fact that their blood and hair selenium is at a high level.

This study suggests two possibilities: (1) High fluorine may inhibit selenium in the human body and accelerate the excretion of selenium through hair, urine, etc.; and (2) an excess of sulfur content in the environment and in food may affect the utilization of selenium and selectively excrete selenium from the body.

Section VII

Selenium in Animals

Pathology of Selenium and Vitamin E Deficiency in Animals

John F. Van Vleet

INTRODUCTION

A wide variety of diseases may be produced in animals with vitamin E–selenium deficiency (VESD), vitamin E deficiency alone, or selenium deficiency only. The spectrum of lesions produced, species affected, and the various terms that have been applied to these diseases are listed in Table 1. The onset of many of these deficiency diseases may be potentiated by exposure to diets containing large amounts of polyunsaturated fats or trace elements that act as Se antagonists.

The pathology of VESD in animals has been the subject of many excellent reviews (Hadlow 1962, 1973; Jubb and Kennedy 1970; Telford 1971; Mason and Horwitt 1972; Mason 1973; Bradley 1975; Lannek and Lindberg 1976; Underwood 1977; Burk 1978; Nelson 1980; Bradley and Fell 1981; Jones and Hunt 1983; Shamberger 1983; Subcommittee on Selenium 1983).

DEFICIENCY DISEASES

Skeletal Muscle

Involvement of skeletal muscle is the most frequent lesion observed in the many species in which the pathologic alterations of VESD have been described. Many terms have been used to describe the muscle lesions (Table 1). Affected animals have muscular weakness. Grossly,

TABLE 1. Diseases Associated with Selenium and Vitamin E Deficiency in Animals

Tissue	Disease	Other terms	Species affected	
			Spontaneous	Experimental
Skeletal muscle	Nutritional myodegeneration	Nutritional muscular dystrophy, nutritional myopathy, "white muscle disease," "stiff lamb disease," "late-lactation paralysis," enzootic muscular dystrophy, paralytic myoglobinuria	Cow, sheep, horse, pig, goat, chicken, turkey, duck, rabbit, mink, dog, cat, kangaroo, wallaby, nyala	Rat, mouse, hamster, guinea pig, monkey, deer, salmon
Cardiac muscle	Nutritional cardiomyopathy	"Mulberry heart disease," dietetic microangiopathy, white muscle disease	Pig, cow, sheep, turkey, duck, goat, horse, mink	Rat, mouse, rabbit, guinea pig, wallaby, monkey
Smooth muscle				
1. Gizzard, intestine	Nutritional myodegeneration	Gizzard myopathy	Turkey, duck	Quail
2. Intestine, uterus	Lipofuscinosis	"Brown dog gut," ceroid deposition	—	Dog, rat
Liver	Nutritional hepatic necrosis	Hepatosis dietetica, toxic liver dystrophy, hemorrhagic necrotic hepatitis	Pig	Rat, mouse
Pancreas	Nutritional pancreatic necrosis	Pancreatic fibrosis, nutritional pancreatic dystrophy, nutritional pancreatic atrophy	Chicken	—
Stomach	Esophagogastric ulceration	—	Pig	Rat

Tissue	Lesion			
Adipose tissue—skin, body cavities	Nutritional steatitis	Nutritional panniculitis, "yellow fat disease"	Cat, mink, horse	Rat, pig
Skin				
1. Capillaries in subcutis and skeletal muscles	Exudative diathesis	—	Chicken, turkey, duck, pig	Quail
2. Hair and feather follicles	Alopecia	—	—	Rat, monkey, quail
Nervous tissues				
1. Cerebellum	Nutritional encephalomalacia	—	Chicken	—
2. Spinal cord and nerves	Localized axonal dystrophy	—	—	Rat, dog
Testicle	Testicular degeneration	—	—	Rat, mouse, hamster, guinea pig, rabbit, dog, monkey, chicken
Embryo	Embryonic death and resorption	—	—	Rat, mouse, pig, hamster, guinea pig
Bone marrow and blood	Nutritional anemia	—	—	Monkey, rat, pig
Eye-lens	Cataract	—	—	Rat
Lung	Pulmonary hemorrhage	—	—	Rat
Kidney	Nephrosis	—	—	Rat, mouse
Teeth	Incisor depigmentation	—	—	Rat, hamster

the affected muscles appear pale and dry. The lesions generally are seen as scattered white streaks, but may produce diffuse involvement and tend to be bilaterally symmetrical.

Many microscopic and ultrastructural studies have established the sequential alterations in damaged muscle fibers (Howes *et al.* 1964; Van Vleet *et al.* 1967, 1968, 1976; Bergmann 1970, 1972, 1979; Oksanen and Poukka 1972; Sweeny *et al.* 1972; Van Vleet and Ferrans 1976, 1977a; Dahlin *et al.* 1978; Bergmann and Kursa 1979; Lin and Chen 1982). Fibers with early alterations have mitochondrial swelling and myofibrillar lysis. Subsequently, affected fibers develop hyaline or granular necrosis. In hyaline necrosis, the fibers have hypereosinophilic and swollen sarcoplasm, and cross striations are obliterated (Fig. 1). In granular necrosis, the fibers have sarcoplasmic disruption into scattered basophilic granules. Ultrastructurally, the necrotic muscle fibers have masses of dense calcified mitochondria lying between dense clumps of lysed and disrupted contractile material (Fig. 2). Muscle nuclei have pyknosis and karyorrhexis. The external lamina of necrotic fibers persists, but has scattered focal disruptions through which macrophages enter and engulf the sarcoplasmic debris (Fig. 3).

Affected muscles, with the various degradative alterations described above, will usually have concurrent efforts at muscle cell regeneration. In necrotic fibers, thin satellite cells lying against the external lamina of the sarcolemma become activated to form myoblasts (Fig. 4). Myoblasts fuse with each other to form elongated multinucleated myotubes. The cytoplasm of myoblasts and myotubes contains abundant polysomes that synthesize contractile proteins for myofilament formation and incorporation into newly formed sarcomeres. Myotubes mature into fully differentiated muscle fibers. Regeneration of necrotic fibers may be nearly complete in the affected muscles of animals with VESD.

Cardiac Muscle

Myocardial lesions in Se–E-deficient animals are seen most frequently in calves, lambs, pigs, turkey poults, and ducklings (Vawter and Records 1947; Obel 1953; Grant 1961; Scott *et al.* 1967; Michel *et al.* 1969; Nafstad and Tollersrud 1970; Van Vleet *et al.* 1970; Bradley 1975; Van Vleet 1982a,b). Affected calves have extensive pale areas of necrosis and calcification in the left ventricular free wall and ventricular septum while in lambs the pale lesions are present in the subendocardial myocardium of the right ventricle (Jubb and Kennedy 1970; Hadlow 1973). Histologically, areas of myocardial damage have

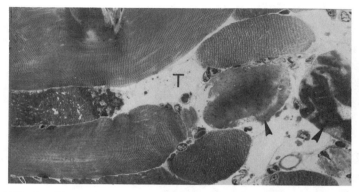

FIG. 1.   Photomicrograph of skeletal myodegeneration in a pig with VESD. Necrotic fibers have disrupted sarcoplasm (arrowheads). Tube (T) of sarcolemma persists in disrupted fiber. Plastic embedded; Toluidine Blue stain (×320).

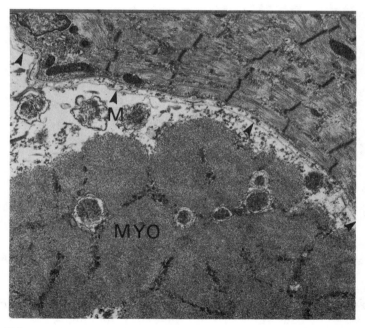

FIG. 2.   Electron micrograph of necrotic skeletal muscle fibers in a duckling with VESD. The disrupted fiber at the bottom has lysing myofibrils (MYO), disrupted mitochondria (M), and persistence of the external lamina (arrowheads). Intact, but partially lysed fiber is at top (×9600).

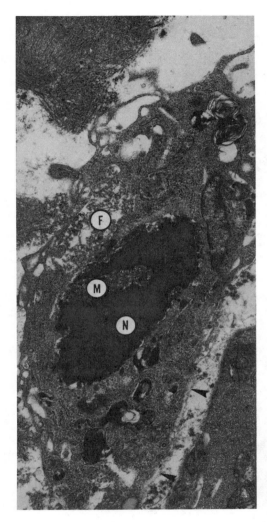

FIG. 3. Electron micrograph of a necrotic skeletal muscle fiber in a pig with VESD. The disrupted fiber is invaded by a macrophage that has engulfed a pyknotic nucleus (N), fragmented mitochondria (M), and disrupted fibrils (F) from the necrotic muscle fiber(×11,600).

hyaline necrosis with or without accompanying calcification, subsequent macrophagic invasion, and eventual formation of areas of stromal collapse and fibrosis.

Growing pigs, usually 2 to 4 months old, may have the cardiac form of VESD (Van Vleet *et al.* 1970). At necropsy, abundant serous transudates are generally present in the body cavities, and the lungs have severe congestion and edema. The heart may have scattered pale streaks in the ventricular myocardium (Fig. 5), but the most striking alterations are widespread epicardial and myocardial hemorrhages

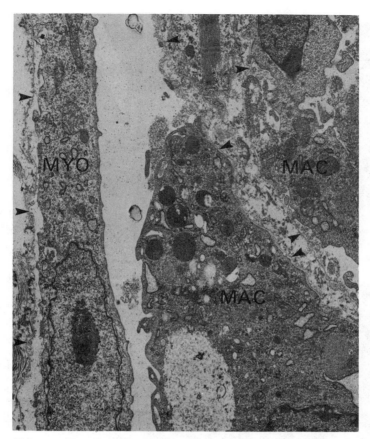

FIG. 4. Electron micrograph of regenerating skeletal muscle fibers in a duckling with VESD. A myoblast (MYO) and several macrophages (MAC) lie within the external lamina (arrowheads) of disrupted muscle fibers (×9600).

that have resulted in the term "mulberry heart disease" for this lesion. Histologically, the hearts have both vascular and myocyte lesions. Vascular changes include fibrinoid necrosis in intramyocardial arteries and arterioles and numerous fibrin microthrombi in myocardial capillaries (Fig. 6). Myocardial hemorrhage and edema accompany the vascular lesions. Multifocal hyaline necrosis and calcification is followed by macrophagic invasion and myocardial fibrosis (Fig. 7) in some pigs with prolonged survival, but the majority of animals have only the acute vascular and myocyte lesions. The myocardial lesions are present in the walls of all 4 chambers, but tend to be most severe in

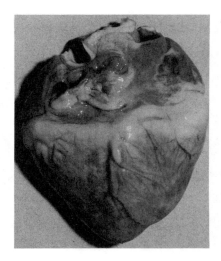

FIG. 5. Pale streaks of myocardial necrosis are present in the ventricles of a pig with VESD.

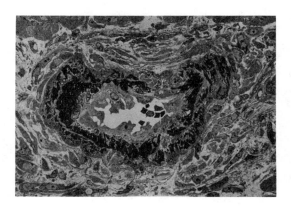

FIG. 6. Photomicrograph of a myocardial arteriole with fibrinoid necrosis in a VESD pig with mulberry heart disease. Dense masses of fibrin are present in the muscular wall of the vessel. Plastic-embedded tissue; Toluidine Blue stain (×340).

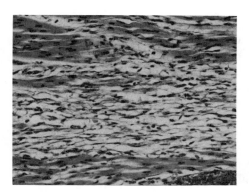

FIG. 7. Photomicrograph of an area of resolved myocardial necrosis in a pig with VESD. Only stromal tissue remains in the area of necrosis. H&E stain (×145).

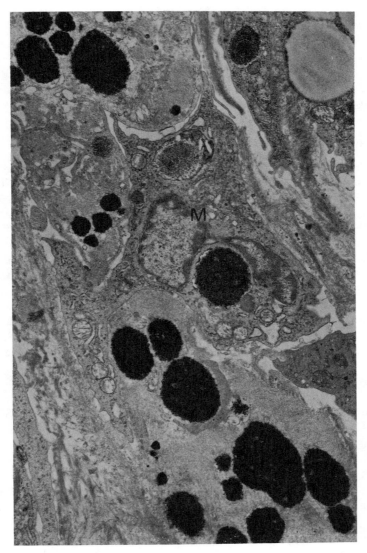

FIG. 8.   Electron micrograph of the necrotic myocardium of a duckling with VESD. A necrotic myocyte contains numerous prominent dense mineralized mitochondria. A macrophage (M) has invaded the necrotic fiber and has engulfed several mineralized mitochondria (×11200).

the atria. Ultrastructural study of these hearts (Sweeny and Brown 1972; Van Vleet *et al.* 1977a,b; Van Vleet and Ferrans 1977b, 1982) has demonstrated myocyte alterations that included mitochondrial swelling and mineralization, and myofibrillar lysis and necrosis with contraction bands (Fig. 8). Endothelial cell damage and necrosis with fibrin accumulation in the walls and lumens was observed in affected vessels.

Smooth Muscle

Necrosis of smooth muscle of the gizzard and intestine may be a prominent lesion of VESD in turkey poults, ducklings, and quail (Van Vleet 1982a). Grossly, the affected gizzard has scattered white to yellowish white areas that are most apparent upon incising the muscular wall (Fig. 9). The intestinal lesion appears as multiple scattered white circular rings of necrosis (Fig. 10). Microscopically, the affected areas show hyaline necrosis and calcification with distention of the interstitial tissues by edema and infiltrated heterophils and macrophages (Fig. 11). Resolving lesions have fibrosis. Ultrastructural examination showed damaged smooth muscle cells to have swollen and calcified mitochondria and distended elements of sarcoplasmic reticulum (Yarrington and Whitehair 1975; Van Vleet and Ferrans 1977c). Necrotic fibers had myofibrillar lysis and macrophagic invasion.

A second lesion of smooth muscle occurs in the intestine of the dog and in the uterus of the rat fed diets deficient in vitamin E and rich in polyunsaturated fats (Cordes and Mosher 1966; Mason 1973; Van Vleet 1975). Lipofuscinosis is not associated with clinical disease, but produces a striking brown discoloration of the affected muscle. Microscopically, the muscle fibers have perinuclear accumulation of small brown granules. By electron microscopy, the granules appear as residual bodies composed of dense granular and membranous debris (Gedigk and Wessel 1964).

Liver

Liver lesions occur in pigs, rats, and mice with VESD (Obel 1953; Trapp *et al.* 1970; Van Vleet *et al.* 1970; Mason and Horwitt 1972). In pigs, the acute lesions appear as multiple scattered red swollen lobules and edema of the gallbladder wall (Fig. 12). Chronic lesions are seen as scattered collapsed lobules that give the hepatic surface a rough granular appearance. In rats and mice, the liver lesions appear as large areas of pallor and swelling that may involve several lobes or the entire liver.

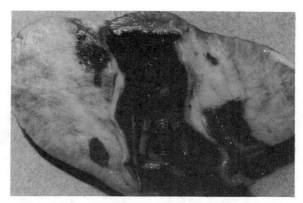

FIG. 9. Cut surface of the gizzard from a duckling with VESD has extensive pale areas of myonecrosis and calcification.

FIG. 10. The loop of duodenum from a duckling with VESD has multiple white circular rings of myonecrosis.

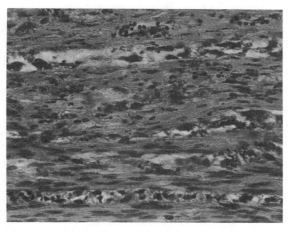

FIG. 11. Photomicrograph of the necrotic smooth muscle in the gizzard of a duckling with VESD. Hyalinized necrotic fibers are present at top. H&E stain (×320).

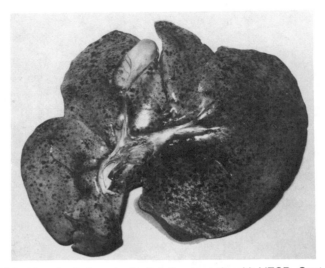

FIG. 12.   Acute hepatosis dietetica in a pig with VESD. Scattered dark lobules have hemorrhagic necrosis.

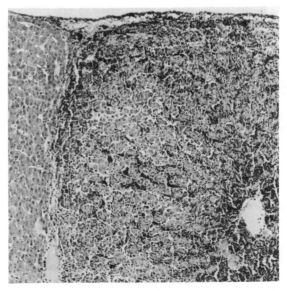

FIG. 13.   Photomicrograph of acute massive hemorrhagic necrosis in the liver of a VESD pig with acute hepatosis dietetica. H&E stain (×80).

Microscopically, massive necrosis with or without hemorrhage is present in the acute phase (Fig. 13). The necrotic lobules are invaded by leukocytes during resolution, and in the chronic stage the necrotic lobules have collapsed stroma and fibrosis. Ultrastructural study of the early alterations in rat livers revealed the initial damage to be disruption of hepatocyte plasma membranes at the sinusoidal border (Svoboda and Higginson 1963; Porta *et al.* 1968).

Pancreas

Pancreatic acinar lesions are produced in chicks with Se deficiency or VESD (Gries and Scott 1972; Rebar and Van Vleet 1977). Grossly, the earliest alteration is a granular appearance of the capsular surface. Chronic lesions appear as marked atrophy of the pancreas. Microscopically, the initial alteration is cytoplasmic vacuolation followed by patchy necrosis of acinar tissue with selective preservation of the acini that surround the islets. Late lesions include atrophy with loss of acini and proliferation of ductal elements and fibrous connective tissue. Ultrastructurally, focal dilatation of elements of endoplasmic reticulum and formation of autophagic vacuoles are early findings (Rebar and Van Vleet 1977). Necrosis and disruption of acinar cells follow. Ductal elements persist and are surrounded by proliferated fibroblasts and increased collagen deposition.

Stomach

Gastric ulceration may accompany liver lesions in pigs and rats with VESD (Lannek and Lindberg 1976). The squamous portion of the gastric mucosa that surrounds the esophageal entrance is the constant site of the ulceration. In affected pigs, the carcass may be pale from hemorrhagic anemia associated with intraluminal hemorrhage. The stomach and intestine may be filled with dark masses of clotted blood. The site of ulceration appears as a deep reddish brown excavated area. Microscopically, the squamous mucosa is lacking and the ulcerated area becomes filled by proliferating granulation tissue.

Adipose Tissue

Steatitis is associated with feeding of diets deficient in vitamin E and high in unsaturated fats. The disease occurs in cats, mink, horses, rats, and pigs (Gershoff and Norkin 1962; Stowe and Whitehair 1963; Lannek and Lindberg 1976). Grossly, the subcutaneous and abdominal fat depots have multiple yellow, firm lesions. Microscopically, the altered lipocytes contain eosinophilic deposits of ceroid with surrounding areas of neutrophilic infiltration.

Skin

Exudative diathesis is a frequent lesion of VESD in chicks (Cheville 1966) and also occurs in VESD of turkey poults, ducklings, quail, and pigs. Grossly, the disease is characterized by extensive edema, with or without accompanying hemorrhage, in the subcutis and extending into underlying skeletal muscles. Affected areas may appear blue to green and often are present over the distal limbs, at the thoracic inlet, and along the inner thigh. Microscopically, the subcutis has severe edema, and small vessels are congested and may contain thrombi. Ultrastructural study showed affected vessels had endothelial damage with necrosis and adhering masses of fibrin thrombi (Van Vleet and Ferrans 1976).

Alopecia is reported in rats, monkeys, and quail associated with pure Se deficiency (Shamberger 1983).

Nervous Tissues

Encephalomalacia is an important lesion of vitamin E deficiency in chicks. Grossly, the cerebellum appears red, swollen, and soft. Microscopically, hemorrhage and necrosis is present in the white matter (Wolf and Pappenheimer 1931; Jungherr 1949). Ultrastructurally, the initial site of damage was in the endothelium of capillaries (Bergmann 1970; Young et al. 1973; Yu et al. 1974; Nelson 1980). Edema and necrosis occurred in the surrounding white matter.

Localized axonal dystrophy was described in rats and dogs with chronic vitamin E deficiency (Mason and Horwitt 1972; Nelson 1980). Affected animals had kyphoscoliosis and bilaterally symmetrical atrophy of skeletal muscles. Microscopic and ultrastructural study revealed degeneration in the posterior columns of the spinal cord and axonal swelling.

Testicle

Testicular degeneration has been described in a wide variety of animal species with vitamin E deficiency, including the rat, mouse, hamster, guinea pig, rabbit, dog, monkey, and chicken (Mason and Horwitt 1972; Nelson 1980). Grossly, the affected testes are atrophic. Microscopically, the germinal epithelium lining the seminiferous tubules is degenerated.

Embryo

Embryonic death with subsequent resorption was an early lesion described with vitamin E deficiency. Species affected include the rat, mouse, pig, hamster, and guinea pig (Mason and Horwitt 1972; Nelson 1980). The affected embryo and placenta may have multiple associated developmental abnormalities.

Bone Marrow and Blood

Anemia has been described in the monkey, rat, and pig with chronic vitamin E deficiency (Mason and Horwitt 1972; Nelson 1980). Histopathologic examination of marrow revealed alterations in erythroid precursors.

Eye

Cataracts have been reported in rats with VESD (Whanger and Weswig 1975). The affected lens was apparent grossly by a white opaque appearance.

Lung

Alveolar hemorrhage has been seen in rats with VESD (Mason and Horwitt 1972).

Kidney

Nephrosis with necrosis of the proximal tubular epithelium was described in rats and mice with VESD (Mason and Horwitt 1972).

Teeth

Depigmentation of the maxillary incisors has been seen in rats and hamsters with VESD (Mason and Horwitt 1972). Microscopically, the enamel organ of affected teeth was atrophic.

REFERENCES

Bergmann, V. 1970. Studies on avian encephalomalacia. 2. Electron microscopic findings in the cerebellar medulla in spontaneous cases. Arch. Exp. Veterinaermed. *24*, 935–949.

Bergmann, V. 1972. Electron-microscopic findings in the skeletal musculature of sheep with enzootic muscular dystrophy. Arch. Exp. Veterinaermed. *26*, 646–660.

Bergmann, V. 1979. Electron-microscopic findings recorded from skeletal muscles of broilers with nutritional encephalomalacia. Arch. Exp. Veterinaermed. *33*, 13–20.

Bergmann, V., and Kursa, J. 1979. Electron-microscopic investigation of enzootic myodystrophy of cattle. Arch. Exp. Veterinaermed. *33*, 1–12.

Bradley, R. 1975. Selenium deficiency and bovine myopathy. *In* The Veterinary Annual. C. S. G. Grunsell and F. W. G. Hill (Editors), 15th issue, pp. 27–36. Wright-Scientechnica, Bristol, UK.

Bradley, R., and Fell, B. F. 1981. Myopathies in animals. *In* Disorders of Voluntary Muscle. J. Walton (Editor), 4th Edition, pp. 824–872. Churchill-Livingstone, London and New York.

Burk, R. F. 1978. Selenium in nutrition. World Rev. Nutr. Diet. *30*, 88–106.

Cordes, D. O., and Mosher, A. H. 1966. Brown pigmentation (lipofuscinosis) of canine intestinal muscularis. J. Pathol. Bacteriol. *92*, 197–206.

Dahlin, K. J., Chen, A. C., Benson, E. S., and Hegarty, P. V. J. 1978. Rehabilitating effect of vitamin E therapy on the ultrastructural changes in skeletal muscles of vitmain E-deficient rabbits. Am. J. Clin. Nutr. *31*, 94–99.

Gedigk, P., and Wessel 1964. Electron-microscopic investigation of vitamin E-deficient pigments in myometrium of the rat. Virchows Arch. A: Pathol. Anat. *337*, 367–382.

Gershoff, S. N., and Norkin, S. A. 1962. Vitamin E deficiency in cats. J. Nutr. *77*, 303–308.

Grant, C. A. 1961. Morphological and etiological studies of dietetic microangiopathy in pigs ("mulberry heart disease"). Acta Vet. Scand. *2*, Suppl. 2, 1–107.

Gries, C. L., and Scott, M. L. 1972. Pathology of selenium deficiency in the chick. J. Nutr. *102*, 1287–1296.

Hadlow, W. J. 1962. Diseases of skeletal muscle. *In* Comparative Neuropathology. J. R. M. Innes and L. Z. Saunders (Editors), pp. 147–243. Academic Press, New York.

Hadlow, W. J. 1973. Myopathies of animals. *In* The Striated Muscle, Int. Acad. Pathol. Monogr. No. 12, pp. 364–409. Williams & Wilkins Co., Baltimore, MD.

Jones, T. C., and Hunt, R. D. 1983. Hypovitaminosis E. *In* Veterinary Pathology, 5th Edition, pp. 1044–1050. Lea & Febiger, Philadelphia, PA.

Jubb, K. V. F., and Kennedy, P. C. 1970. The nutritional myopathies. *In* Pathology of Domestic Animals, 2nd Edition, Vol. 2, pp. 482–494. Academic Press, New York.

Jungherr, E. L. 1949. Ten-year incidence of field encephalomalacia in chicks and observations on its pathology. Ann. N.Y. Acad. Sci. *52*, 104–112.

Lannek, N., and Lindberg, P. 1976. Vitamin E and selenium deficiencies (VESD) of domestic animals. Adv. Vet. Sci. Comp. Med. *19*, 127–164.

Lin, C.-T., and Chen, L.-H. 1982. Ultrastructural and lysosome enzyme studies of skeletal muscle and myocardium in rats with long-term vitamin E deficiency. Pathology *14*, 375–382.

Mason K. E. 1973. Effects of nutritional deficiency on muscle. *In* The Structure and Function of Muscle. G. H. Bourne (Editor), 2nd Edition, Vol. 4, pp. 155–206. Academic Press, New York.

Mason, K. E., and Horwitt, M. K. 1972. Effects of deficiency in animals. *In* The Vitamins: Chemistry, Physiology, Pathology, Methods. W. H. Sebrell, Jr. and R. S. Harris (Editors), Vol. 5, pp. 272–292. Academic Press, New York.

Michel, R. L., Whitehair, C. K., and Keahey, K. K. 1969. Dietary hepatic necrosis associated with selenium–vitamin E deficiencies in swine. J. Am. Vet. Med. Assoc. *155*, 50–59.

Nafstad, I., and Tollersrud, S. 1970. The vitamin E deficiency syndrome in pigs. I. Pathological changes. Acta Vet. Scand. *11,* 452–480.

Nelson, J. S. 1980. Pathology of vitamin E deficiency. *In* Vitamin E: A Comprehensive Treatise. L. J. Machlin (Editor), pp. 397–428. Marcel Dekker, New York.

Obel, A. L. 1953. Studies on the morphology and etiology of so-called toxic liver dystrophy (hepatosis dietetica) in swine. Acta Pathol. Microbiol. Scand. Suppl. *94,* 1–118.

Oksanen, A., and Poukka, R. 1972. An electron microscopical study of nutritional muscular degeneration (NMD) of myocardium and skeletal muscle in calves. Acta Pathol. Microbiol. Scand., Sect. A *80A,* 440–448.

Porta, E. A., de la Iglesia, F. A., and Hartroft, W. S. 1968. Studies on dietary hepatic necrosis. Lab. Invest. *18,* 283–297.

Rebar, A. H., and Van Vleet, J. F. 1977. Ultrastructural changes in the pancreas of selenium–vitamin E-deficient chicks. Vet. Pathol. *14,* 629–642.

Scott, M. L., Olson, G., Krook, L., and Brown, W. R. 1967. Selenium-responsive myopathies of myocardium and of smooth muscle in the young poult. J. Nutr. *91,* 573–583.

Shamberger, R. J. 1983. Selenium deficiency diseases in animals. *In* Biochemistry of Selenium, pp. 31–58. Plenum Press, New York.

Stowe, H. D., and Whitehair, C. K. 1963. Gross and microscopic pathology of tocopherol-deficient mink. J. Nutr. *81,* 287–300.

Subcommittee on Selenium, Committee on Animal Nutrition, Board on Agriculture, National Research Council 1983. Selenium in Nutrition, Revised Edition, pp. 77–106. National Academy Press, Washington, DC.

Svoboda, D. J., and Higginson, J. 1963. Ultrastructural changes in rats on a necrogenic diet. Am. J. Pathol. *43,* 477–495.

Sweeny, P. R., and Brown, R. G. 1972. Ultrastructural changes in muscular dystrophy. I. Cardiac tissue of piglets deprived of vitamin E and selenium. Am. J. Pathol. *68,* 479–492.

Sweeny, P. R., Buchanan-Smith, J. G., deMille, F., Pettit, J. R., and Moran, E. T. 1972. Ultrastructure of muscular dystrophy. II. A comparative study in lambs and chickens. Am. J. Pathol. *68,* 493–510.

Telford, I. R. 1971. Experimental Muscular Dystrophies in Animals: A Comparative Study, pp. 3–243. Charles C. Thomas, Springfield, IL.

Trapp, A. L., Keahey, K. K., Whitenack, D. L., and Whitehair, C. K. 1970. Vitamin E–selenium deficiency in swine: Differential diagnosis and nature of the field problem. J. Am. Vet. Med. Assoc. *157,* 289–300.

Underwood, E. J. 1977. Selenium. *In* Trace Elements in Human and Animal Nutrition, 4th Edition pp. 302–346. Academic Press, New York.

Van Vleet, J. F. 1975. Experimentally induced vitamin E–selenium deficiency in the growing dog. J. Am. Vet. Med. Assoc. *166,* 769–744.

Van Vleet, J. F. 1982a. Amounts of 12 elements required to induce selenium–vitamin E deficiency in ducklings. Am. J. Vet. Res. *43,* 851–857.

Van Vleet, J. F. 1982b. Comparative efficacy of five supplementation procedures to control selenium–vitamin E deficiency in swine. Am. J. Vet. Res. *43,* 1180–1189.

Van Vleet, J. F., and Ferrans, V. J. 1976. Ultrastructural changes in skeletal muscle of selenium–vitamin E-deficient chicks. Am. J. Vet. Res. *37,* 1081–1089.

Van Vleet, J. F., and Ferrans, V. J. 1977a. Ultrastructural alterations in skeletal muscle of ducklings fed selenium–vitamin E-deficient diet. Am. J. Vet Res. *38,* 1399–1405.

Van Vleet, J. F., and Ferrans, V. J. 1977b. Ultrastructure of hyaline microthrombi in myocardial capillaries of pigs with spontaneous mulberry heart disease. Am. J. Vet. Res. *38*, 2077–2080.

Van Vleet, J. F., and Ferrans, V. J. 1977c. Ultrastructural alterations in gizzard smooth muscle of selenium–vitamin E-deficient ducklings. Avian Dis. *21*, 531–542.

Van Vleet, J. F., and Ferrans, V. J. 1982. Myocardial ultrastructural alteration in ducklings fed tellurium. Am. J. Vet. Res. *43*, 2000–2009.

Van Vleet, J. F., Hall, B. V., and Simon, J. 1967. Vitamin E deficiency: A sequential study by means of light and electron microscopy of the alterations occurring in regeneration of skeletal muscle of affected weanling rabbits. Am. J. Pathol. *51*, 815–830.

Van Vleet, J. F., Hall, B. V., and Simon, J. 1968. Vitamin E deficiency: A sequential light and electron microscopic study of skeletal muscle degeneration in weanling rabbits. Am. J. Pathol. *52*, 1067–1079.

Van Vleet, J. F., Carlton, W., and Olander, H. J. 1970. Hepatosis dietetica and mulberry heart disease associated with selenium deficiency in Indiana swine. J. Am. Vet. Med. Assoc. *157*, 1208–1219.

Van Vleet, J. F., Ruth, G. R., and Ferrans, V. J. 1976. Ultrastructural alterations in skeletal muscle of pigs with selenium–vitamin E deficiency. Am. J. Vet. Res. *37*, 911–922.

Van Vleet, J. F., Ferrans, V. J., and Ruth, G. R. 1977a. Ultrastructure alterations in nutritional cardiomyopathy of selenium–vitamin E-deficient swine. I. Fiber lesions. Lab. Invest. *37*, 188–200.

Van Vleet, J. F., Ferrans, V. J., and Ruth, G. R. 1977b. Ultrastructure alterations in nutritional cardiomyopathy of selenium–vitamin E-deficient swine. II. Vascular lesions. Lab. Invest. *37*, 201–211.

Vawter, L. R., and Records, E. 1947. Muscular dystrophy (white muscle disease) in young calves. J. Am. Vet. Med. Assoc. *110*, 152–157.

Whanger, P. D., and Weswig, P. H. 1975. Effects of selenium, chromium, and antioxidants on growth, eye cataracts, plasma cholesterol, and blood glucose in selenium deficient, vitamin E-supplemented rats. Nutr. Rep. Int. *12*, 345–358.

Wolf, A., and Pappenheimer, A. M. 1931. The histopathology of nutritional encephalomalacia of chicks. J. Exp. Med. *54*, 399–406.

Yarrington, J. T., and Whitehair, C. K. 1975. Ultrastructure of gastrointestinal smooth muscle in ducks with a vitamin E–selenium deficiency. J. Nutr. *105*, 782–790.

Young, P. A., Taylor, J. J., Yu, W.-H., Yu, M. C., and Tureen, L. L. 1973. Ultrastructural changes in chick cerebellum induced by vitamin E deficiency. Acta Neuropathol. *25*, 149–160.

Yu, W.-H., Yu, M. C., and Young, P. A. 1974. Ultrastructural changes in the cerebrovascular endothelium induced by a diet high in linoleic acid and deficient in vitamin E. Exp. Mol. Pathol. *21*, 289–299.

Selenium and Hormones in the Male Reproductive System

D. Behne
T. Höfer-Bosse
W. Elger

The starting point of our studies on the relations between selenium and hormones in the male reproductive system was a series of experiments in which we investigated the distribution and retention of selenium in tissues of rats that had been fed a selenium-deficient diet (1,2). The values for the liver, erythrocytes, and testis obtained from the results of these experiments are shown in Fig. 1.

After a feeding period of 70 days there was a considerable decrease in the selenium content in most of the tissues of the depleted animals compared with the values of a control group which had been fed the same diet with added selenium. As shown in Fig. 1A, the change was particularly marked in the liver, which is the main pool for glutathione peroxidase (GSHPx). In the erythrocytes, the second largest GSHPx pool, the selenium content also decreased considerably. In the testis, however, no significant change in the selenium level was observed during the feeding period. This cannot be explained solely by a very long biological half-life of the element in the male gonads, as in this case the selenium level would already have been reduced as a result of the pubertal growth of the testes during the depletion period.

Great differences in the behavior of selenium in the individual tissues were also observed when the animals were given a small amount of radioactively labeled selenium (Fig. 1B). Thirty-five days after administration, the highest value for the ^{75}Se activity was found in the testis, although the selenium content had remained more or less unchanged in this organ and most of the other tissues had a low se-

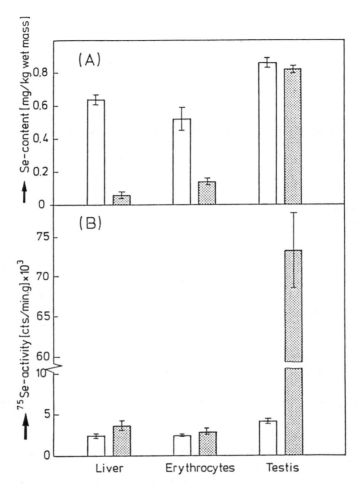

FIG. 1. Effects of a selenium-deficient diet on the distribution of selenium in male rats. (A) Selenium content (mean ± SD, *n* = 4) in animals fed on a low selenium and low vitamin E diet or the same diet with added selenite for 70 days. (B) 75Se activity (mean ± SD) in animals fed on the selenium-sufficient (*n* = 5) or the selenium-deficient diet (*n* = 4) for 95 days, 35 days after the injection of selenite labeled with 75Se. (□), 0.25 mg Se/kg diet; (▦), <0.015 mg Se/kg diet.

lenium status. In comparison with the control group, the depleted animals retained 17 times more selenium in the testes, whereas in the case of the GSHPx pools, liver and erythrocytes, retention in the deficient animals was only slightly increased.

The results of this study show that in nutritional selenium deficiency the GSHPx activity in the body of rats decreases more rapidly than the selenium content. They also show that regulation mechanisms exist which are responsible for priority supply of this element to certain tissues, such as the testes, and that the maintenance of the GSHPx status of the animals appears to be of secondary importance compared with the maintenance of the testis selenium content.

EFFECTS OF HORMONES ON SELENIUM METABOLISM

In order to obtain more information about these regulation mechanisms and to find out whether the hormones which control the male reproductive processes also influence the supply of selenium to the testes, experiments were carried out on rats. By removing the pituitary gland the production of the two gonadotropic hormones, the luteinizing hormones (LH) and the follicle-stimulating hormone (FSH), was interrupted. The testicular function was subsequently regenerated, either by administering pregnant mare's serum gonadotropin (PMS), which has the same effect as both LH and FSH, or testosterone.

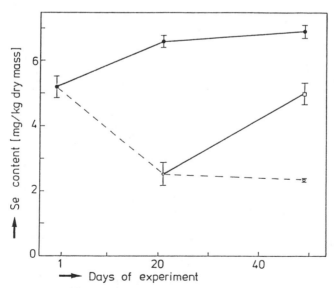

FIG. 2.   Effects of hypophysectomy and subsequent treatment with pregnant mare's serum gonadotropin (PMS) on the testis selenium content in rats (mean ± SD, n = 4). (●), Intact; (○), hypophysectomized; (□), hypophysectomized + PMS.

During these hormonal changes the testis selenium content was determined.

From Fig. 2 it can be seen that the selenium content of the male gonads changed considerably under these conditions. After hypophysectomy the testis became atrophic and the selenium level in the remaining testicular tissue decreased to values of about 2.5 mg Se/kg dry mass, whereas in the intact controls the values were between 6.6 and 7 mg Se/kg. After 28 days of treatment with PMS, the testis selenium content increased again to 5 mg Se/kg. This increase after the administration of PMS to hypophysectomized animals was also found in rats that had been fed a selenium-deficient diet (1). This means that selenium is transported to the testes from other already selenium-deficient tissues and shows again the priority of the selenium supply to the male reproductive organs.

Figure 3 shows the results of a second experiment in which the findings of the study described above, namely, the decrease in the testis selenium content after hypophysectomy and the rise again after

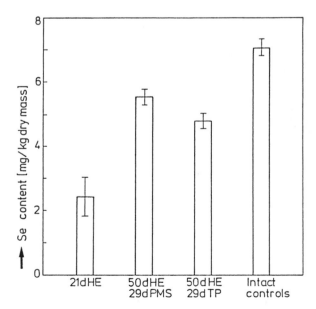

FIG. 3.   Effects of hypophysectomy (HE) and subsequent treatment with pregnant mare's serum gonadotropin (PMS) or testosterone proprionate (TP) on the testis selenium content in rats (mean ± SD, $n =$ 5).

administration of PMS, were confirmed. In addition, in this experiment the effect of testosterone in the absence of FSH was investigated. The selenium level in the male gonads of the hypophysectomized animals also rose after application of this hormone, but the rise was slightly less than after the administration of PMS.

The results all indicate that the testis selenium content is regulated by the hormones responsible for spermatogenesis. The regulation most probably serves mainly to ensure the supply of sufficient amounts of selenium to the spermatozoa, which, with values of about 15 mg Se/kg dry mass, had the highest selenium content of all the tissues and body fluids in the rat. In the sperm, selenium is located in the outer membrane of the mitochondria where it is present as a specific selenoprotein and most probably has structural functions (3).

EFFECTS OF SELENIUM ON
HORMONE METABOLISM

In addition to contributing to the formation of spermatozoa, selenium also seems to be involved in other processes in the testes. In the course of the experiments described above, we measured the serum testosterone level in rats that had been fed a selenium-deficient diet for over 5 months. In these animals the serum testosterone concentration was slightly lower than in control animals that had received sufficient amounts of the element. As this was an indication that selenium could be involved in the steroidogenesis of the testis Leydig cells, we carried out experiments to check this possibility.

The first step in this study was to administer to rats the luteinizing hormone-releasing hormone (LHRH) which stimulates the pituitary secretion of LH and FSH. LH controls the Leydig cells and thus the secretion of testosterone.

As far as the increase in the serum LH concentration was concerned, no difference was found between the selenium-deficient animals and the control group. A difference did exist, however, in the secretion of testosterone. As can be seen in Fig. 4A, the rise in the serum testosterone level 2 hr after the administration of LHRH was lower in the animals that had been fed a selenium-deficient diet for 170 days than the value for the control group, which had been fed sufficient amounts of the element.

In order to localize this selenium effect, testosterone secretion was studied in a further experiment under direct stimulation with LH in the form of human chorionic gonadotropin (HCG). Here we also found a similar, statistically significant difference in the rise in the serum

D. BEHNE *et al.*

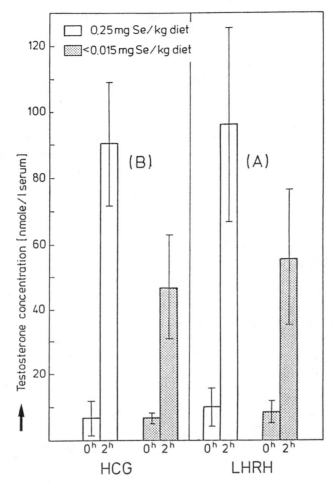

FIG. 4. Rise in the serum testosterone concentration
(mean ± SD, *n* = 6) following stimulation with luteinizing
hormone-releasing hormone (LHRH) or human chorionic
gonadotropin (HCG) in male rats fed on a low selenium
diet or the same diet with added selenite for 170 days.

testosterone level between the selenium-deficient animals and the
control group (Fig. 4B).

The findings of both experiments indicate that selenium deficiency
has an effect on the Leydig cells and suggest a biological function of
the element in this steroidogenic organ.

CONCLUSIONS

Studies on rats have shown that the testis selenium content is controlled by regulation mechanisms and that the supply of sufficient amounts of selenium to the testis has priority over the supply to other tissues and also over the maintenance of the GSHPx status of the organism. The testis selenium level is regulated by the hormones responsible for spermatogenesis. The regulation most probably serves mainly to ensure the supply of sufficient amounts of the element to the spermatozoa. In the testis, in addition to contributing to the formation of spermatozoa, selenium seems to have a further biological function in the Leydig cells.

REFERENCES

1. Behne, D., Höfer, T., von Berswordt-Wallrabe, R., and Elger, W. 1982. Selenium in the testis of the rat: Studies on its regulation and its importance for the organism. J. Nutr. 112, 1682–1687.
2. Behne, D., and Höfer-Bosse, T. 1984. Effects of a low selenium status on the distribution and retention of selenium in the rat. J. Nutr. 114, 1289–1296.
3. Calvin, H. I., Cooper, G. W., and Wallace, E. 1981. Evidence that selenium in rat sperm is associated with a cysteine-rich structural protein of the mitochondrial capsules. Gamete Res. 4, 139-149.

Importance of Selenium Quantity in Soil and Fodder in Regard to the Occurrence of Some Diseases in Cattle, Pigs, Sheep, and Poultry in Yugoslavia

Branko Gavrilović
Dubravka Matešić

We have been carrying on investigations concerning the quantity of selenium in soil and fodder in relation to the occurrence of some diseases in domestic animals in the Pozega Valley in Yugoslavia. This region is a fertile plateau, surrounded by mountains, having an altitude of about 200 m and a surface area of 1255 km².

The reasons prompting us to investigate the levels of selenium in soil and plants relate to the fact that we often encountered degeneration of skeleton and cardiac musculature in cattle, pigs, and sheep, causing serious economic losses. We were able to find a large number of reports in the literature indicating the connection between vitamin E and selenium and the occurrence of these diseases.

Our successes in the prevention and therapy of these diseases with vitamin E and selenium as well as data from elsewhere that these diseases occur more frequently in regions poor in selenium indicated that our region might be selenium deficient. Therefore, we decided to investigate the main agricultural products and soil whereon they were produced to determine what kind of region we are in with regard to selenium quantity. The results of such investigations will enable us to better realize how to prevent the aforementioned diseases and to reduce the production losses.

In our studies, we have been investigating in the same samples of crops besides selenium also the quantity of vitamin E and protein. We investigated vitamin E due to its metabolic interrelationships with selenium and the quantity of proteins, since selenium in the plant material is predominantly protein-bound.

MATERIAL AND METHODS

The investigation of the selenium quantity in soil and plants of Pozega Valley has been going on for 2 years, 1980 and 1981. We used soil samples of lands under cultivation by the Agricultural and Processing Plant (PPK), Kutjevo, and wheat, maize, and soybeans produced on these lands. Every year seven samples of wheat (Dukat, NS Rana-2, NS-7000, Zlatna dolina, Super Zlatna, Osjecanka, Zlatoklasa), four samples of maize hybrids (Edo, KH 343, Forla, Mutin), and samples of soil whereon these cultures have been grown were subjected to analysis for selenium. The quantity of proteins and vitamin E was also investigated in the samples of wheat, maize, and soybeans. In 1981, we investigated the quantity of selenium, vitamin E, and proteins in representative samples of maize and soybean forage stored in silos.

Methods of Selenium, Vitamin E, and Protein Determination

Selenium was determined by hydride generation flame atomic absorption spectroscopy. For preparation of the samples, we used the method and apparatus described by Matesic *et al.* (1981). The quantity of selenium is expressed in ppm in relation to the dry matter of the sample.

Vitamin E was determined by high-pressure liquid chromatography (HPLC) in accordance with the method "Determination of Tokoferol" (JUS E, A.1.048), adapted in the Center of Poultry Science, Zagreb. The quantity of vitamin E was expressed in international units (IU). Crude protein was determined by the Kjeldahl macromethod. The quantity of protein was expressed as percentages of air-dry matter of the sample.

Clinical Investigations

We have been investigating the etiologies of animal diseases connected with selenium in the region of the Pozega Valley since 1975. In order to determine the occurrence of these diseases, we used our own

FIG. 1.   Yugoslavia in Europe. (▦), Pozega Valley; (▨), dystrophogenic area for sheep.

production practice, the practice of other veterinarians working in the region of Pozega Valley, and also laboratory results of the Veterinary Faculty in Zagreb and Ljubljana, Veterinary Institute in Zagreb, to which the material for researching has been supplied.

The clinical investigations were carried out with the following species and categories of animals: (1) Calves and young cattle of the Simmental breed under conditions of intensive fattening (from 120 to 500 kg) in villages and socially owned farms; (2) pigs (piglets, hogs, breeding class) of Swedish Landrace and crosses of this breed, under conditions of socially organized pig breeding in villages of the Pozega Valley; (3) sheep (crosses of Sjenica and Wurttemberg sheep) under conditions of cooperative sheep raising managed by the Veterinary Station in Slavonska Pozega; and (4), poultry during raising of brood flocks.

During clinical investigations the following examinations and methods have been undertaken: (1) clinical examinations, establishment of diagnosis, and therapeutic and prophylactic treatments; (2) pathoanatomic and pathohistologic examinations; (3) bacteriological tests; and (4) biochemical analyses of tissues and fodder.

RESULTS

From Laboratory Analyses

The results of laboratory analyses of soil, wheat, maize, and soybeans are presented in Tables 1–3. The highest level of selenium (0.018 ppm) occurred in wheat of the 1980 harvest, cultivated on soil with the highest level of selenium among all the investigated samples of soil. Assuming that feed has to contain 0.1 ppm of selenium in order to satisfy nutritive needs of the animals, the quantities of 0.018 ppm found in our samples are far below those which satisfy the nutritive needs of domestic animals. All the investigated samples of wheat from the 1981 crop and of maize from the 1980 and 1981 crops show selenium rate below 0.01 ppm, i.e., only about 10% of the nutritive need.

The results indicate that the selenium content of wheat depends more on the level of selenium in the soil than on the protein level in the plant. We found similar relationships from the quantity of selenium and protein in soybeans. For example, we had in the year 1980 an average level of 0.014 ppm selenium in all the wheat samples, with an average protein content of 11.21%. This wheat was grown on soil

TABLE 1. Wheat Samples Analyses (Yield 1980 and 1981)

Wheat sorts	Selenium (ppm)		Proteins (%)		Vitamin E (IU/kg)	
	1980	1981	1980	1981	1980	1981
Sort 1	0.015	<0.01	10.61	11.35	10.53	7.9
Sort 2	0.011	<0.01	10.52	12.24	20.15	10.3
Sort 3	0.013	<0.01	11.21	12.62	11.97	10.1
Sort 4	0.016	<0.01	10.08	11.44	83.42	9.4
Sort 5	0.013	<0.01	10.99	11.05	8.62	11.0
Sort 6	0.012	<0.01	12.64	13.69	1.40	7.9
Sort 7	0.018	<0.01	12.42	14.90	2.82	10.2
Soil sample	0.048	0.02	—	—	—	—

TABLE 2. Maize Sample Analyses (Yield 1980 and 1981)

Maize hybrid	Selenium (ppm)		Proteins (%)		Vitamin E (IU/kg)	
	1980	1981	1980	1981	1980	1981
Hybrid 1	<0.01	<0.01	11.83	10.83	1.95	15.2
Hybrid 2	<0.01	<0.01	10.14	8.92	6.40	16.8
Hybrid 3	<0.01	<0.01	10.88	10.11	4.18	15.6
Hybrid 4	<0.01	<0.01	10.64	9.17	12.22	12.2
Soil sample	0.042	0.043	—	—	—	—

containing 0.048 ppm selenium. All wheat samples in the year 1981 had less than 0.01 ppm selenium, but this wheat had on the average 12.47% protein and had been cultivated on soil containing 0.020 ppm of selenium.

The vitamin E levels in samples of wheat and maize, taken for each year individually, show a great variability among different samples within the same year (wheat, 1980, from 1.40 to 83.42 IU; wheat, 1981, from 7.9 to 11.00 IU; maize, 1980, from 1.95 to 12.22 IU; maize, 1981, from 12.20 to 16.80 IU). Likewise, there is a great variability for vitamin E between the years for the same plants (wheat, 1980, average 19.72 IU and 1981 average 9.54 IU; maize, 1980 average 6.18 IU and 1981 14.95 IU).

From Production Practice

We present the results from production practice as descriptions of clinical cases of the diseases in cattle, pigs, and sheep on which we used with more or less success preventive or therapeutical application of medicinal preparations of Na selenite and vitamin E Tokoselen product of Pliva Chemopharmaceutical Products, Zagreb, Yugoslavia, and Evitaselen product of Galenika, Belgrade, Yugoslavia.

Cattle. Our first experiences of the efficiency of vitamin E and selenium in the prevention of large losses in young fattening cattle date to the years 1975 and 1977. At that time on the socially owned

TABLE 3. Maize and Soya Average Silo Sample Analyses (Yield 1981)

Samples	Selenium (ppm)	Proteins (%)	Vitamin E (IU/kg)
Soya	<0.01	24.84	17.9
Maize	<0.01	9.22	8.7

farm in Pozega Valley, young cattle were fed a diet containing 80% maize conserved with propionic acid and 20% of a protein–vitamin–mineral supplement, to which a premix had been added, which enriched the diet with 10 mg/kg of vitamin E. The young cattle were healthy, in very good condition, and of satisfactory export quality. When the animals reached 300 kg of weight, they were transported to Italy, but during the transport, taking 4 days, some deaths occurred. The total death rate during transport amounted to 30%. The remaining animals arrived at the destination in very poor condition—lying depressed and not interested in feed. Laboratory findings from the carcasses of the animals determined that death was caused by degeneration of cardiac and skeletal musculature. Live, but very sick animals were treated with selenium and vitamin E, and most of them recovered. After these experiences, we treated the next shipment of young cattle 1 or 2 days before the transport with 5 mg intramuscularly with 5 mg sodium selenite and 600 mg vitamin E. These animals, raised under identical conditions as the former ones and transported under identical conditions as before, but receiving injection of vitamin E and selenium before the transport, arrived at the destination without losses, in good condition, and left the freight cars running and leaping.

We had further experiences in the use of vitamin E and selenium during the initial stage of fattening the calves. In this period, we often encountered some calves lying, afebrile, and not interested in feed. Good therapeutic effect was obtained by intramuscular administration of 10 ml of Tokoselen or Evitaselen, and generally after 24–48 hr, many of the calves recovered and normal fattening could be continued. The preventive application of the same treatment at the time of the entry into fattening proved also very efficient, because there have not been further occurrences of the problems.

Pigs. Our perceptions concerning death of piglets after weaning and pigs in fattening date back to 1976. These sudden death cases were reducing our planned production each year and influencing the economy of production. Pathoanatomical and pathohistological diagnosis identified the so-called mulberry heart disease (MHD) in the death losses, and application of vitamin E and selenium resulted in reduced losses (Gavrilović 1981). In addition to MHD occurrence, we often noted death of piglets aged 2–5 days after preventive treatments with iron preparations. Such piglets seemed overly sensitive to iron preparations, and in order to prevent this problem, we gave the piglets simultaneously iron, vitamin E, and selenium, but no positive results followed. We did obtain positive results in the prevention of oversensitivity of piglets against iron preparations when 1 month before farrowing, their mothers were treated with vitamin E and selenium. We noted a favorable effect of vitamin E and selenium on gilts at first breeding. A greater number of gilts treated with vitamin E and selenium attained normal fertile estrus than those not treated with these preparations. We also often encountered and diagnosed, by necropsy, dystrophy of the liver in weaned piglets, another disease whose etiology is connected with the lack of vitamin E and selenium.

Sheep. The occurrences of mass deaths of lambs in the Pozega Valley in 1981, when 40 lambs aged 2–7 days died in a group of 50, caused disastrous losses to sheep producers. Muscular dystrophy was diagnosed in the carcasses of these lambs. After the diagnosis, the remaining lambs and newcomers were subjected to intramuscular administration of vitamin E and selenium (Evitaselen and Tokoselen in accordance with the manufacturers' instructions), and the occurrences of the disease, as well as further deaths, disappeared completely.

Poultry. In 1982 a flock of 2000 chickens from Ross hybrid parents was raised for the production of hatching eggs for broiler production. The well-known disease exudative diathesis occurred in the third week and affected 30% of the flock, endangering further rearing. We attacked the problem by adding selenium and vitamin E to the drinking water (0.1 mg of sodium selenite and 30 mg of tocopherol acetate per liter). Upon completed therapy, the flock started to recover by the third day and continued to grow normally until the age of egg production, when more than 80% hatchability of very good, healthy broiler chickens was obtained.

DISCUSSION

With laboratory investigations concerning the quantity of selenium in soil and crops on the one hand, and laboratory-confirmed occurrences of selenium- and vitamin E-responsive myopathies and other diseases on the other, we believe that we have contributed to the evidence that the Pozega Valley in Yugoslavia belongs among the so-called selenium-poor regions. As mentioned in the work published by Kubota et al. (1967), the regions of the United States with crops containing from 0.1 to 0.03 ppm are considered poor in selenium, and in these regions, often diseases responding to selenium and vitamin E occur. According to our investigations, the quantity of selenium in main crops of the Pozega Valley range from 0.01 to 0.018 ppm.

We did not find increased quantities of selenium in samples of wheat, maize, and soybeans containing high quantities of protein, but we had the highest rates of selenium in wheat cultivated on the area where the soil contained the highest quantity of selenium. Our results confirm that the selenium level in crops depends first on its quantity in soil. Accordingly, in relation to selenium in crops, the locality of their cultivation is of primary importance. The quantities of vitamin E in the samples of wheat and maize show a great variability between sorts (or hybrids) in 1 year as well as between the years 1980 and 1981. These results conform with data concerning the quantity and variation of vitamin E contents in wheat and maize (Bunnell *et al.* 1968, as cited by Ullrey 1981). As Kivimae and Carpena (1973) have published, the quantity of vitamin E in plants depends upon genetic variety, agroecological and agrotechnical conditions, and on further technological processes, such as drying, heating, grinding, and storing.

We can conclude from our investigations that the quantity of selenium in maize samples from our region does not satisfy the nutritive needs of animals, and that it ranged below these needs in both years of the study. However, the quantity of vitamin E varied considerably. Accepting the explanation of Rotruck *et al.* (1973) that the biochemical result of vitamin E and selenium activity in the organism is the same, we can explain why there are more occurrences of diseases in some years than in others.

Mass deaths in young cattle during shipping, laboratory diagnosis of the cause of death, and beneficial effects of selenium and vitamin E in the prevention of these losses are in conformity with data published by Forenbacher *et al.* (1975), Karlovic *et al.* (1977), and Van Vleet (1980). The so-called mulberry heart disease, whose occurrence has been described by us in detail in a separate paper (Gavrilović 1981) is the most noticeable, selenium- and vitamin E-responsive disease of pigs. Zintzen and Belčić (1975) wrote that "in Norway it is besides the diseases of the digestive tract the most frequent disease of pigs." Lindberg (1968) has noted that a normal pig heart contains 1.05 ± 0.01 ppm selenium, and the diseased pig heart only about 10% of that amount. Our application of vitamin E and selenium at the time of first breeding of gilts and the beneficial working of these preparations agree with data published by Underwood (1977) and Piatkowski *et al.* (1979).

In our opinion, it is very important to point out the danger of selenium poisoning by adding excessive quantities of selenium either to the feed, water, or parenterally. Selenium is biochemically very active, and it has been mentioned by Stadtman (1980) that it should be in an optimal range to be able to perform its normal function.

We have determined the existence of direct and indirect material damages caused by selenium and vitamin E deficiency. Ullrey (1980) states that lack of selenium in the feed for pigs and some types of poultry has caused annual losses of $82 million in the United States and similar losses of $545 million in cattle breeding and sheep raising. We cannot immediately provide such estimates of the losses in our region; however, we suggest that the supplementation of selenium to deficient foods, now allowed by government action in the United States, should be considered for our area also.

CONCLUSIONS

On the basis of completed investigations concerning the selenium in Yugoslavian soil and fodder in relation to the occurrence of some diseases in cattle, pigs, sheep, and poultry, we conclude the following:

Quantities of selenium in soil under cultivation in Pozega Valley, from 0.020 to 0.048 ppm, in wheat (all samples less than 0.018 ppm), and in maize (all samples less than 0.01 ppm) grown on these lands show that this region is poor in selenium.

Mass occurrences of selenium-responsive diseases and considerable direct and indirect losses in cattle, pigs, sheep, and poultry indicate the great importance of selenium in soil and fodder in relation to health and productivity of animals, as well as to the economy of animal husbandry of the region.

Constant low levels of selenium in soil and crops and great variations within the years of the crop vitamin E level suggest why during some years there are more frequent occurrences of the disease in whose etiology vitamin E and selenium play some role.

In the regions poor in selenium, due to its great importance, we recommend that fodder for animals contains quantities of at least 0.01 ppm selenium.

ACKNOWLEDGMENTS

We wish to express our gratitude to PPK Kutjevo for the financial help. We are most obliged to Professors Herak, Sviben, and Forenbacher from the Veterinary Faculty in Zagreb for assistance during these investigations. We also owe thanks to Mrs. Eva Rezo for her trouble in typing and Mrs. Zora Farbak for translating this work.

REFERENCES

Forenbacher, S., Herceg, M., and Feldhofer, S. 1975. Systemic myopathy of fattening cattle caused by lack of vitamin E.

Gavrilović, B. 1981. On mulberry heart disease in pigs of Požega Valley. Prax. Vet. (3–4).

Karlović, M., Blagović, S., Lerman, S., and Hečej, Z. 1977. Systemic myopathy in fattening young cattle and the possibility of its treatment. Prax. Vet. (3).

Kivimae, A., and Carpena, C. 1973. The level of vitamin E content in some conventional feeding stuffs and the effects of genetic variety, harvesting, processing, and storage. Acta Agric. Scand., Suppl. 19.

Kubota, J., Allaway, W. H., Carter, D. R., Cary, E. E., and Lazar, A. V. 1967. Selenium in crops in the United States in relation to selenium-responsive diseases of animals. J. Agric. Food Chem. 15 (3)

Lindberg, P. 1968. Selenium determination in plant and animal material and in water. Acta Vet. Scand., Suppl. 23.

Matešić, D., Kos, K., and Strašek, A. 1981. The quantity of selenium in some forages and poultry feed from Croatia. Vet. Arch. 51 (2), 79–82.

Piatkowski, T. L., Mahan, D. C., Cantor, A. H., Moxon, A. L., Cline, J. H., and Grifo, A. P. 1979. Selenium and vitamin E in semipurified diets for gravid and nongravid gilts. J. Anim. Sci. 48.

Rotruck, J. T., Pope, A. L., Ganther, H. E., and Hoekstra, W. G. 1973. Selenium's biochemical role as a component of glutathione peroxidase. Science 179, 588.

Stadtman, T. C. 1980. Biological functions of selenium. Trends Biochem. Sci. 5, August, pp. 203–206.

Ullrey, E. D. 1980. Regulation of essential nutrient additions to animal diets (selenium—a model case). J. Anim. Sci. 51, 645–651.

Ullrey, E. D. 1981. Vitamin E for swine. J. Anim. Sci. 54.

Underwood, E. J. 1977. Trace Elements in Human and Animal Nutrition. 4th Edition. Academic Press, New York.

Van Vleet, J. F. 1980. Current knowledge of selenium-vitamin E deficiency in domestic animals. J. Am. Vet. Med. Assoc. 174 (4).

Zintzen, H., and Belčić, I. 1975. Aspekti vitamina E i selena u svinja. Published report of lecture read on April 25, 1975, during SVIND Symposium, Zagreb.

Some Activity Changes of Enzymes in Peroxidative Pathways in Mice Fed Se-Deficient Cereals from a Keshan Disease Endemic Area of China

Guang-Lu Xu   *Wu-Hong Tan*
Ying-Dou Niu   *Jian-Ye Li*
Yue-Ai Han   *Lan-Hua Zhou*
Wen-Lan Xue

Previous studies (*1,2*) in this laboratory have indicated that the agricultural populations in Keshan disease (KD) endemic areas have an Se-poor status; the selenium contents of locally grown staple cereals and daily diets in the endemic areas also are lower than those in the nonendemic areas, and the disease has been shown to be effectively prevented by the administration of a prophylactic oral dose of 1–4 mg sodium selenite every 10 days. The results showed that KD is an Se-responsive disease of human beings, the endemic areas of KD seem to be an Se-deficient region, and Se deficiency might be a primary factor in the pathogenesis of the disease.

Se has been identified as an integral component of Se-dependent glutathione peroxidase (GSHPx), an enzyme which is capable of catalyzing the reduction of hydrogen peroxide and a wide range of lipid hydroperoxides (*3*). Its activity in various tissues from several species is dependent on dietary Se intake (*4,5*). The defense mechanism that functions *in vivo* to limit or prevent free radical-initiated peroxidative damage also includes a number of other enzyme systems and factors. Superoxide dismutase (SOD) is a group of metalloenzymes containing

Cu/Zn or Mn at their active centers which catalyze the disproportionation of the superoxide free radical anion (6). Catalase, an iron-containing enzyme, catalyzes the decomposition of hydrogen peroxide (7). Glutathione S-transferase (GST) catalyzes the conjugation to glutathione of a large group of xenobiotics; however, it is also capable of acting as a peroxidase to catalyze the reduction of organic hydroperoxide, but not of hydrogen peroxide, and it is sometimes referred to as "Se-independent GSHPx" (8). Vitamin E (VE) functions in vivo to limit proliferation of free radical damage by scavenging lipid peroxy radicals (9), and may also play a structural role in stabilizing biological membranes (10). Hence, all these protective agents may be involved in various parts of peroxidative pathways to protect the cell from damage by unwanted peroxidation. Xanthine oxidase (XOD), containing Mo in its active center, catalyzes the one-electron reduction of dioxygen to superoxide anion (11).

The purpose of this study was to investigate the activity changes of enzymes, most of them being metalloenzymes, involved in peroxidative pathways in tissues of mice fed Se-deficient cerals from a KD endemic area. Since the intravenous injection of large doses of ascorbic acid has been dramatically effective in treating the KD patient with cardiogenic shock (12), the effect of ascorbic acid on the above parameters was also investigated.

MATERIALS AND METHODS

Animals and Diet

Weanling mice were divided randomly into 5 groups, which were given the designated diet and distilled water ad libitum during a 2- to 3-week experimental period. Group 1 was given the KD endemic area cereal diet (KD diet), which was composed of corn flour (76%), wheat flour (10%), soybean flour (12%), salt (1%), and cod-liver oil (vitamin A, 1,500 IU/g and vitamin D, 150 IU/g) (1%); group 2 received the KD nonendemic area cereal diet (non-KD diet) with the same composition as the KD diet; group 3 was fed the KD diet supplemented with a low dose of sodium selenite (KD-low Se diet); group 4 was reared using the KD diet to which was added a higher dose of sodium selenite (KD-high Se diet); and group 5 was fed the stock diet from the animal house in Xi'an Medical College (XD), which consisted of corn flour (20%), wheat flour (30%), soybean flour (30%), bran (10%), fish meal (5%), bone meal (1%), salt (1%), cod-liver oil (1%), yeast (1%), mineral mix (0.5%), and vitamin mix (0.5%).

TABLE 1. Elemental Contents of the Diets[a]

Diet	Se	Zn	Cu	Mn	Fe
KD	0.012	26.2	4.0	12.7	125
Non-KD	0.043	31.0	4.3	12.6	438
KD + low Se	0.053	24.0	4.0	11.3	182
KD + high Se	0.222	24.0	3.9	11.4	195
XD	0.111	45.0	10.8	29.0	500

[a] ppm, mean of three determinations.

Diets were analyzed for Se by a fluorometric procedure (*13*) using 2,3-diaminonaphthalene with minor modification, and for Cu, Zn, Fe, and Mn by conventional atomic absorption spectrophotometry. The values are presented in Table 1.

Sampling and Homogenization of Tissues

Six to eight mice in each diet group were exsanguinated by scission of the ophthalmic artery into a heparinized test tube 2 to 3 weeks after beginning the experiments. For investigating the effect of ascorbic acid, 5 mice in group 1 or 4 were exsanguinated 2 hr after an intravenous injection of *injectio acidi ascorbici* (10%, 0.2 ml/10 g body weight). Plasma was prepared from whole blood by centrifugation at 1000 g for 10 min. Livers were perfused with cold 0.9% NaCl and samples of each liver and of each heart ventricle were excised, trimmed, washed in cold 0.9% NaCl, blotted dry, and immediately held on ice. For the enzyme assays, portions of tissues were weighed and homogenized with 10 volumes of ice-cold 0.25 M sucrose using a motor-driven glass-glass homogenizer at 4°C. For catalase assay, the crude homogenates were centrifuged at 700 g for 10 min at 5°. Ethanol was added to the supernatant fraction to a final concentration of 0.17 M. The mixture was placed on ice for 30 min, then 100 μl 0.05 M phosphate buffer, pH 7.0, containing 10% Triton X-100, was added to 1 ml of the mixture (*14*). For CSHPx, GST, SOD, and XOD activities, the crude homogenates were centrifuged at 12,000 g for 15 min at 5° and the supernatant fractions were recovered. For lipid peroxidation determination, portions of tissues were weighed and homogenized with 9 volumes of ice-cold 0.05 M phosphate buffer, pH 7.4.

Analytical Methods

GSHPx activity was measured by the glutathione reductase-coupled assay of Paglia and Valentine (*15*), using 0.25 mM H_2O_2 as the substrate for the Se-dependent enzyme and 1.5 mM cumene hydroperoxide to determine total GSHPx activity. Se-independent GSHPx activity was estimated as the difference between total GSHPx activity and the activity of the Se-dependent GSHPx. One enzyme unit represented 1 μmol of NADPH oxidized per min at 25°.

GST was determined by measuring the conjugation of 1-chloro-2,4-dinitrobenzene with GSH as described by Habig *et al.* (*16*) One unit of enzyme activity represented 1 μmol of CDNB utilized per min at 30°C. SOD activity was assayed by the method of Misra and Fridovich (*17*). The amount of SOD required to inhibit the rate of epinephrine autooxidation by 50% was considered to contain one unit of enzyme activity. Catalase activity was measured by the method of Aebi (*7*). One enzyme unit represented 1 μmol H_2O_2 reduced/min. Xanthine oxidase was determined by the nitrotetrazolium blue chloride reduction method (*18*). The results were expressed as Δ Abs-min-mg protein. Plasma VE concentrations were determined by the fluorometric method of Hansen (*19*). The thiobarbituric acid (TBA) test for lipid peroxide formation was performed according to the method of Wilbur *et al.* (*20*). Protein was measured by the method of Lowry *et al.* (*21*), with bovine serum albumin as standard.

Statistics

All the determinations were made on parts of tissues from individual animals. The results obtained from each group were expressed as mean value ± standard error. The experiment was repeated once. The significance of the differences was assayed using Student's t test, with $P = 0.05$ as the limit of significance.

RESULTS

The activities of Se-dependent GSHPx in liver were significantly lower in mice fed the KD diet than in those fed the non-KD diet (Table 2). Se supplementation resulted in significant elevations of Se-dependent GSHPx in the liver. The total GSHPx activity of liver measured was consistently higher than the Se-dependent GSHPx activity. The Se-dependent GSHPx amounted to 78% of the total GSHPx activity in Se-supplemented group 4 mice. The activity of Se-dependent GSHPx in group 1 was 55, 61, 17, and 42% of the values in groups 2–5, respectively, but total GSHPx was 72, 73, 31, and 60%, respectively. There were no significant differences in the Se-independent GSHPx activities between group 1 and other groups.

TABLE 2. Effect of Diet on Hepatic GSHPx Activity in Mice

| | GSHPx activity (mEU/mg protein) | | |
Diet	Total	Se dependent	Se independent
KD	150 ± 17	64 ± 9	86 ± 9
KD + VC[a]	144 ± 12	66 ± 13	78 ± 5
Non-KD	207 ± 19[b]	116 ± 9[c]	92 ± 12
KD + low Se	206 ± 5[c]	104 ± 6[c]	102 ± 9
KD + high Se	491 ± 45[d]	384 ± 48[d]	107 ± 7
KD + high Se + VC	438 ± 51	354 ± 58	85 ± 4
XD	249 ± 17[c]	148 ± 5[c]	101 ± 10

[a] VC, Vitamin C.
[b] $P < .05$; [c] $P < .01$; [d] $P < .001$, as compared with KD diet.

The cardiac Se-dependent GSHPx in group 1 mice was not signicantly lower than that in groups 2 and 3, but was significantly lower than that in groups 4 and 5. In each of the 5 dietary groups, no significant difference was apparent between the Se-dependent and the total GSHPx activities (Table 3).

No significant differences between group 1 and the other groups were detected in hepatic and cardiac GST activities, mice had extremely low activity of GST in heart (Table 4).

The KD diet had no significant effect on hepatic and cardiac SOD activities when compared to the non-KD and Se-supplemented KD diets; however, the activities in liver were significantly higher in group 5 than in group 1 (Table 4).

Catalase activities of liver showed no significant difference between group 1 and 2, 3, and 5, respectively, but the activities were higher in group 4 than in group 1 (Table 5).

TABLE 3. Effect of Diet on Cardiac GSHPx in Mice

Diet	Total GSHPx (mEU/mg protein)	Se-dependent GSHPx (mEU/mg protein)
KD	40 ± 4	37 ± 3
KD + VC[a]	47 ± 11	47 ± 9
Non-KD	45 ± 4	42 ± 4
KD + low Se	49 ± 9	47 ± 7
KD + high Se	65 ± 8[b]	59 ± 6[b]
KD + high Se + VC	68 ± 6	65 ± 4
XD	53 ± 4[c]	50 ± 4[c]

[a] VC, Vitamin C.
[b] $P < .01$ as compared to KD diet.
[c] $P < .05$ as compared to KD diet.

TABLE 4. Effect of Diet on GST and SOD in Mice Tissues

Diet	GST (EU/mg protein)		SOD (EU/mg protein)	
	Liver	Heart	Liver	Heart
KD	1.34 ± 0.14	0.097 ± 0.025	89.6 ± 5.7	52.2 ± 5.4
KD + VC[a]	0.94 ± 0.11	0.098 ± 0.007	104.9 ± 7.6	65.0 ± 19.3
Non-KD	1.07 ± 0.09	0.074 ± 0.011	103.4 ± 3.5	40.7 ± 4.9
KD + low Se	1.34 ± 0.09	0.097 ± 0.011	104.6 ± 12.2	38.7 ± 3.9
KD + high Se	1.53 ± 0.20	0.098 ± 0.006	98.5 ± 3.0	44.7 ± 3.7
KD + high Se + VC	1.58 ± 0.03	0.117 ± 0.018	108.5 ± 5.4	47.9 ± 2.4
XD	1.01 ± 0.08	0.073 ± 0.006	125.7 ± 12.2[b]	51.4 ± 3.4

[a] VC, Vitamin C.
[b] $P < .05$, as compared to KD diet.

TABLE 5. Effect of Diet on Hepatic Catalase and Plasma Vitamin E in Mice

Diet	Catalase (EU/mg protein)	Vitamin E (μg/ml)
KD	11.7 ± 1.5	1.13 ± 0.11
KD + VC[a]	14.9 ± 1.4	1.15 ± 0.16
Non-KD	11.8 ± 1.4	0.97 ± 0.08
KD + low Se	13.3 ± 1.8	0.80 ± 0.08
KD + high Se	15.5 ± 0.5[b]	0.83 ± 0.08
KD + high Se + VC	14.5 ± 2.1	1.18 ± 0.13
XD	15.6 ± 3.6	2.88 ± 0.30[c]

[a] VC, Vitamin C.
[b] $P < .05$ as compared to KD diet.
[c] $P < .01$ as compared to KD diet.

No appreciable differences of the cardiac and hepatic XOD activities were noted between group 1 and other groups (Table 6).

The plasma VE concentrations were comparable in groups 1–4, although they were higher in group 5 (Table 5).

TBA values in liver homogenates were insignificantly increased in the KD diet group as compared with non-KD diet group, but were significantly higher than in other groups (Table 7).

The activities of GSHPx, GST, SOD, catalase, and XOD in tissues studied were not influenced 2 hr after an intravenous injection of *injectio acidi ascorbici* in mice fed the KD diet or the KD diet with a high level of Se (Tables 2–6). In mice fed the KD diet, the TBA values of liver homogenates were not significantly lower in the ascorbic acid-treated mice than in control mice (Table 7).

TABLE 6. Effect of Diet on XOD in Mice

Diet	XOD (Δ λ/min/mg protein)	
	Liver	Heart
KD	0.238 ± 0.007	0. ·9 ± 0.022
KD + VC[a]	0.215 ± 0.019	0.125 ± 0.040
Non-KD	0.232 ± 0.006	0.178 ± 0.033
KD + low Se	0.226 ± 0.011	0.154 ± 0.021
KD + high Se	0.217 ± 0.007	0.135 ± 0.013
KD + high Se + VC	0.163 ± 0.004	0.103 ± 0.014
XD	0.210 ± 0.016	0.113 ± 0.011

[a] VC, Vitamin C.

TABLE 7. Effect of Diet on Hepatic TBA Value in Mice

Diet	TBA (OD_{535}/mg protein)
KD	0.208 ± 0.021
KD + VC[a]	0.162 ± 0.012
Non-KD	0.173 ± 0.009
KD + low Se	0.151 ± 0.010[b]
KD + high Se	0.139 ± 0.007[c]
KD + high Se + VC	0.135 ± 0.019
XD	0.135 ± 0.006[c]

[a] VC, Vitamin C.
[c] $P < .05$ as compared to KD diet.
[c] $P < .01$ as compared to KD diet.

DISCUSSION

The Se contents determined show that the KD diet was indeed an Se-deficient one. For the other elements measured, the contents of Zn, Cu, Mn, and Fe in the stock diet from the animal house in Xi'an Medical College and that of Fe in the non-KD diet were higher as compared to other diets used.

The results of the GSHPx measurements indicate that the Se-deficient KD diet resulted in a significant fall in the activity of this enzyme in liver when compared to the non-KD diet; the KD diet supplemented with sodium selenite produced a significant increase of Se-dependent GSHPx activity, and in liver of Se-supplemented mice (group 4, dietary Se 0.222 ppm), most of the activity was due to the Se-dependent enzyme. It has been reported by Lawrence and Burk (22) and Lawrence et al. (23) that, in rats, Se deficiency is accompanied by a compensatory rise in Se-independent GSHPx activity. Unlike rats, no change in the activity of Se-independent GSHPx was observed in mice, indicating that mice lack a mechanism for compensating for acute Se deficiency, similar to the situation in ducklings fed a torula yeast-based Se-deficient diet (24). This result is also borne out by the GST measurements in mouse liver.

In these experiments, the KD diet produced a small but not significant fall in the activity of GSHPx in mouse heart as compared to the non-KD diet and the KD diet containing 0.053 ppm Se, although the KD diet containing 0.222 ppm Se caused a significant increase in GSHPx activity. This might be due to the fact that heart tissue in mice has a low content of GSHPx and is less sensitive to changes in dietary Se than liver. The fact that no significant difference was apparent between Se-dependent and total GSHPx activities suggests the absence of significant amounts of Se-independent GSHPx in mouse heart, in agreement with the results in rats and mice fed a torula yeast-based Se-deficient diet (*22,25*). This is also consistent with the very low GST activity in mouse heart.

Hepatic catalase activity was not influenced by dietary Fe content in this study, although the Fe contents of the non-KD diet and the XD diet were higher than that in the KD diet. Presumably, the Fe levels in the diets determined may have been sufficient to meet the dietary requirement for Fe. Mouse hepatic catalase activity was not altered by the KD diet as compared to the non-KD diet and the KD diet with a low level of Se. The reason that the hepatic catalase activity was higher in mice fed the KD diet containing 0.222 ppm Se than in mice fed the Se-deficient KD diet is in need of further study, since it has been shown that rat hepatic catalase activity is unchanged during Se deficiency (*26*).

In addition, there was no change in the activity of SOD in tissues from mice fed KD diets with or without Se supplementation, or a non-KD diet, in agreement with other workers who showed that changes in Se status are reflected in changes in the activity of GSHPx, but not of SOD (*24*). The higher level of SOD activities in liver of the XD diet group is probably a reflection of the higher Cu and Zn contents in the diet of that group.

Xanthine oxidase activities were not influenced by the Se-deficient KD diet when compared with other diets, similar to the report that the activity of liver xanthine oxidase is not affected by Se deficiency in rats fed a 20% casein-based diet (*27*).

Lipid peroxidation (expressed as TBA value) in liver homogenates showed increase that was not statistically significant in mice fed the KD diet during a 2- to 3-week period when compared with the non-KD diet, although the activity of GSHPx was significantly decreased in the KD diet group mouse liver. KD diets with Se levels higher than those in the KD diet caused a significant depression of the TBA values observed.

An intravenous injection of *injectio acidi ascorbici* had no significant effect on the parameters investigated. Further studies are needed to investigate the influence of ascorbic acid on those parameters in mice fed the KD diet for a longer period.

In conclusion, results showed that the depressed Se-dependent GSHPx activity appeared to be the unique change in tissues among the parameters investigated in mice fed Se-deficient KD cereals as compared with the mice fed a non-KD diet.

ACKNOWLEDGMENTS

The authors are greatly indebted to Professor A. T. Diplock for his helpful gift of NADPH. The technical assistance of J.-J. Zhou, Y.-T. Yang, N. Yao, J.-C. Ma, and F.-M. Li is gratefully acknowledged.

REFERENCES

1. Xu, G.-L., Xue, W.-L., Zhang, P.-Y., Feng, C.-F., Hong, S.-Y., and Liang, W.-S. 1982. Selenium status and dietary selenium content of populations in the endemic and non-endemic areas of Keshan disease. Acta Nutr. Sin. *4*, 183.
2. Research Laboratory of Keshan Disease, Xi'an Medical College 1979. Observations on the effects of sodium selenite for preventing acute Keshan disease. Zhonghua Yixue Zazhi *59*, 457.
3. Rotruck, J. T., Pope, A. L., Ganther, H. E., Swanson, A. B., Hafeman, D. G., and Hoekstra, W. G. 1973. Selenium: Biochemical role as a component of glutathione peroxidase. Science *179*, 588.
4. Chow, C. K., and Tappel, A. L. 1974. Response of glutathione peroxidase to dietary selenium in rats. J. Nutr. *104*, 444.
5. Hafeman, D. G., Sunde, R. A., and Hoekstra, W. G. 1974. Effect of dietary selenium on erythrocyte and liver glutathione peroxidase in the rat. J. Nutr. *104*, 580.
6. McCord, J. M., and Fridovich, I. 1969. Superoxide dismutase, an enzymic function for erythrocuprein. J. Biol. Chem. *244*, 6049.
7. Aebi, H. 1974. Catalase. *In* Methods of Enzymatic Analysis. H. U. Bergmeyer and H. Ulrich (Editors), vcl. 2. p. 673. Verlag Chemie, Weinheim.
8. Prohaska, J. R., and Ganther, H. E. 1977. Glutathione peroxidase activity of glutathione *S*-transferases purified from rat liver. Biochem. Biophys. Res. Commun. *76*, 437.
9. Tappel, A. L. 1962. Vitamin E as the biological lipid antioxidant. Vitam. Horm. (N.Y.) *20*, 493.
10. Diplock, A. T., and Lucy, J. A. 1973. The biochemical modes of action of vitamin E and selenium: A hypothesis. FEBS Lett. *29*, 205.
11. Bonnett, R. 1981. Oxygen activation and tetrapyrroles. Essays Biochem. *17*, 1.
12. Keshan Disease Research Group, Xi'an an Medical Collge 1961. Investigations on clinical effects of large doses of ascorbic acid on acute Keshan disease patients and its mechanism. Chin. J. Intern. Med. *9*, 346.
13. Watkinson, J. H. 1966. Fluorometric determination of selenium in biological material with 2,3-diaminonaphthalene. Anal. Chem. *38*, 92.
14. Cohen, G., Dembiec, D., and Marcus, J. 1970. Measurement of catalase activity in tissue extracts. Anal. Biochem. *34*, 30.
15. Paglia, D. E., and Valentine, W. N. 1967. Studies on the quantitative and qualitative characteristics of erythrocyte glutathione peroxidase. J. Lab. Clin. Med. *70*, 158.
16. Habig, W. H., Pabst, M. J., and Jakoby, W. B. 1974. Glutathione *S*-transferases, the first enzymatic step in mercapturic acid formation. J. Biol. Chem. *249*, 7130.
17. Misra, H. P., and Fridovich, I. 1972. The role of superoxide anion in the autoox-

idation of epinephrine and a simple assay for superoxide dismutase. J. Biol. Chem. *247*, 3170.

18. Fried, R. 1966. Colorimetric determination of xanthine dehydrogenase by tetrazolium reduction. Anal. Biochem. *16*, 427.

19. Hansen, L. G., and Warwick, W. J. 1969. A fluorometric micromethod for serum vitamin A and E. Am. J. Clin. Pathol. *51*, 538.

20. Wilbur, K. M., Bernhein, F., and Shapiro, O. W. 1949. The thiobarbituric acid reagent as a test for the oxidation of unsaturated fatty acids by various agents. Arch. Biochem. Biophys. *24*, 305.

21. Lowry, O. H., Rosebrough, N. J., Farr, A. L., and Randall, R. J. 1951. Protein measurement with the Folin phenol reagent. J. Biol. Chem. *193*, 265.

22. Lawrence, R. A., and Burk, R. F. 1978. Species, tissue and subcellular distribution of nonselenium-dependent glutathione peroxidase activity. J. Nutr. *108*, 211.

23. Lawrence, R. A., Parkhill, L. K., and Burk, R. F. 1978. Hepatic cytosolic non-selenium-dependent glutathione peroxidase activity, its nature and the effect of selenium deficiency. J. Nutr. *108*, 981.

24. Xu, G.-L., and Diplock, A. T. 1983. Glutathione peroxidase, glutathione *S*-transferase, superoxide dismutase, and catalase activities in tissues of ducklings deprived of vitamin E and selenium. Br. J. Nutr. *50*, 437.

25. Locker, G. Y., Doroshow, J. H., Baldinger, J. C., and Myers, C. E. 1979. The relationship between dietary tocopherol and glutathione peroxidase activity in murine cardiac tissue. Nutr. Rep. Int. *19*, 671.

26. Lee, Y. H., Layman, D. K., and Bell, R. P. 1981. Glutathione peroxidase activity in iron-deficient rats. J. Nutr. *111*, 194.

27. Bonetti, E., Abbondanza, E., Corte, E. D., Novello, F., and Stirpe, F. 1975. Studies on the formation of lipid peroxide and on some enzymic activities in the liver of vitamin E-deficient rats. J. Nutr. *105*, 364.

77

The Protective Effect of Selenium against Viral Myocarditis in Mice

Ke-You Ge   *Shu-Qin Wang*
Jin Bai   *An-Na Xue*
Xue-Jun Deng *Cheng-Qin Su*
Shi-Quan Wu

Many selenium-responsive abnormalities including myocardial necrosis had been reported in a number of animal species (Frost and Lish 1975). Recent studies revealed that the prevalence of Keshan disease, an endemic cardiomyopathy in China, is related to the inadequate selenium intake of the local inhabitants. Oral administration of sodium selenite protected children from this disease (Chen 1980). Selenium deficiency alone, however, may not induce severe myocardial necrosis as observed in Keshan disease. Besides, the deficiency hypothesis seems not in conformity with some epidemiological characteristics, of which the notable annual and seasonal variations of the prevalence of this disease are illustrative examples (Sun 1982).

Clinical reports usually describe a mild fever before the appearance of heart symptoms (Cheng 1979). Virologists isolated a number of enterovirus strains from blood and tissue specimens obtained from Keshan disease victims (Su 1979). These findings suggest a possible viral infection along with this disease. Therefore, it is interesting to study the combined effect of selenium deficiency and viral infection.

EXPERIMENTS

There were three experiments in this series. The common procedures were as follows:

TABLE 1. The Composition of Low Selenium Basal Diets (%)

Sichuan diet		Semisynthetic diet	
Rice	62.7	Starch	61.0
Soybean	20.0	Petroleum yeast	29.7
Wheat	10.0	Methionine	0.3
Yeast	5.0	Vegetable oil	4.7
Cod liver oil	0.3	Cod liver oil	0.3
Vegetable oil	1.0	Mineral mixture	3.0
Mineral mixture	1.0	Vitamin mixture	1.0
		(vitamin E, 20 mg/kg)	

Weanling Kunming mice of both sexes were randomized into different dietary groups with or without selenium supplementation. Animals were caged five together, weighed weekly, and fed and given water *ad libitum*. The sucklings were inoculated intraperitoneally with either 0.07 ml CB-21 virus or virus-free culture on their seventh day of life, and sacrificed 1 week later by decapitation. Heart, lungs, liver, pancreas, spleen, kidney, and skeletal muscle were removed and fixed in Bouin's solution. Each heart was cut longitudinally into two pieces and embedded in the same block. Other tissues were processed routinely. Paraffin sections of all the specimens were prepared for histopathology. Selenium concentration was determined by a fluorometric method (Wang 1984). Blood samples were individually collected from the adults before feeding, mating, and at sacrificing, and those collected from sucklings were pooled by litter.

The stock diet was supplied by the Animal Center, Chinese Academy of Medical Sciences. Two low selenium diets were prepared in our laboratory. One, the Sichuan diet, mainly consisted of cereals produced in a Keshan disease area in Sichuan Province, southern China. Another was a semisynthetic diet based on starch and yeast (Table 1). Both diets contained about 14 g% protein and 40 mg/kg of vitamin E. The selenium concentration varied from 0.010 to 0.027 ppm. In all these experiments, sodium selenite was used for selenium supplementation.

CB-21 virus was isolated from the blood sample of a Keshan disease victim in Chuxong County, Yunnan Province. It was serologically identified as Coxsackie B_4 virus. Rhesus monkey kidney cell culture with a titer of $10^{7.8}$ to $10^{8.3}$ $TCID_{50}$ was used. It was thawed at room temperature and centrifuged before inoculation.

EXPERIMENT 1

Animals were assigned to three groups. The control group consumed the stock diet with 0.518 ppm Se. The Sichuan group consumed the

TABLE 2. Selenium Status and Virus-Induced Heart Lesions (Experiment 1)

Group	Dietary Se (ppm)	Blood Se (ppm)		Myocardial lesion	
		Adult	Suckling	Incidence	Percentage
Control	0.518	0.412	0.306	8/52	15.4
Sichuan	0.010	0.053	0.032	20/56	36.4
Sichuan + Se	0.140[a]	0.183	0.138	4/30	13.3

[a] Calculated average value.

natural cereal diet with 0.010 ppm Se. The third (Sichuan + Se group) consumed the same Sichuan diet, but was supplemented with 1.0 mg sodium selenium-kg body weight weekly by esophageal intubation.

Of the 46 females, 6 in the Sichuan group died of hepatic necrosis, but no deaths occurred in the control or Sichuan + Se group. The blood selenium concentration of both adults and sucklings was correlated with their selenium intakes. The intubation had been stopped since mating to avoid handling the pregnant mothers. Therefore the blood selenium level of the supplemented animals was much lower at sacrificing than at mating. However, it was still significantly higher than that of nonsupplemented animals. Similar results were also observed in sucklings.

Histopathological examination revealed myocardial lesions in virus-challenged sucklings in all dietary groups. As shown in Table 2, 20 of the 56 animals were attacked in the Sichuan group; 8 of 52 and 4 of 30 animals were attacked in the control and Sichuan + Se groups, respectively. The incidence of lesions in the Sichuan group was significantly higher than that of the control group ($\chi^2 = 5.80$, $P < .05$) and Sichuan + Se group ($\chi^2 = 4.86$, $P < .05$), but no difference was found between the control and Sichuan + Se groups ($\chi^2 = 0.06$, $P > .50$). Pathologically, focal myocardial necrosis played the principal part in the heart lesions. The necrotic foci were distributed throughout the myocardial tissue, but observed more in the left ventricular walls and the septa. Diffused myocardial degeneration was uncommon. The most extensive heart lesions were observed in selenium-deficient animals.

Pathological changes were also found in the liver, pancreas and skeletal muscle of some infected animals, but the incidences appeared not to vary with the selenium status. No myocardial changes were found in sucklings inoculated with virus-free cultures.

EXPERIMENT 2

Animals were assigned to five groups. The first three groups repeated exactly the design of experiment 1, except that the selenium con-

TABLE 3. Selenium Status and Virus-Induced Heart Lesion (Experiment 2)

Group	Dietary Se (ppm)	Blood Se (ppm)		Myocardial lesion	
		Adult	Suckling	Incidence	Percentage
Control	0.334	0.440	0.321	8/72	11.1
Sichuan	0.017	0.054	0.033	17/44	38.6
Sichuan + Se	0.147[a]	0.150	0.070	8/58	13.8
Semisynthetic	0.015	0.049	0.024	23/74	31.1
Semisynthetic + Se	0.145[a]	0.111	0.088	16/70	22.8

[a] Calculated average value.

tent of the Sichuan diet was 0.017 ppm here. The fourth group consumed a semisynthetic diet containing 0.015 ppm Se. The fifth group consumed this semisynthetic diet and received sodium selenite through esophageal intubation just as did the Sichuan + Se group.

Animals of all groups grew well. No death occurred due to hepatic necrosis in the low selenium intake groups. The blood selenium levels are given in Table 3 and, as in experiment 1, the blood Se of both adults and sucklings agreed well with their Se intakes. Myocardial necrosis was seen again in virus-inoculated sucklings of all dietary groups. Once more the selenium-deficient animals suffered a significantly higher incidence than did the selenium-adequate groups. The differences were analyzed with the Wilcoxon rank sum test. The difference between control and the two selenium-supplemented groups was not significant ($P>.05$). No difference was found between the two deficient groups ($P>.1$) or the two supplemented groups ($P>.1$). The difference between semisynthetic and the semisynthetic + Se group was not significant. The possible reason for this may be that the given amount of selenium was not enough to produce complete protection for the animals.

The morphological characteristics of the heart lesions and the changes of other organs were similar to those described in experiment 1.

EXPERIMENT 3

Animals were fed on Sichuan or semisynthetic diets, both containing 0.027 ppm Se. Four groups fed on either diet were supplemented with 0, 0.026, 0.10, and 0.33 ppm Se in drinking water. The calculated dietary selenium intakes of the four groups were 0.027, 0.062, 0.162, and 0.462 ppm, respectively. Blood selenium determination, virus and

TABLE 4. Selenium Status and Virus-Induced Heart Lesions (Experiment 3)

Group	Dietary Se (ppm)	Blood Se (ppm)		Myocardial lesion	
		Adult	Suckling	Incidence	Percentage
Sichuan diet					
A	0.027	0.075 ± 0.004	0.059 ± 0.003	73/182	40.1
B	0.062[a]	0.210 ± 0.009	0.146 ± 0.015	19/97	19.6
C	0.162[a]	0.477 ± 0.021	0.249 ± 0.019	24/115	20.9
D	0.462[a]	0.582 ± 0.021	0.316 ± 0.006	13/128	10.2
Semisynthetic diet					
E	0.027	0.103 ± 0.004	0.077 ± 0.007	22/68	32.4
F	0.062[a]	0.255 ± 0.023	0.166 ± 0.008	14/71	19.7
G	0.162[a]	0.517 ± 0.027	0.287 ± 0.009	17/85	20.0
H	0.462[a]	0.599 ± 0.020	0.328 ± 0.011	9/74	12.2

[a] Calculated average value.

virus-free culture inoculation, and morphological examination were conducted as in the forgoing experiments. The results are summarized in Table 4.

The blood selenium contents of animals on both Sichuan and semisynthetic diets were related to the selenium intakes. A linear relationship between blood selenium and the logarithmic values of selenium intake could be established for both adults and the offspring.

As in the previous experiments, myocardial necrosis was found only in virus-challenged sucklings and the incidence varied with the selenium status. The incidence of heart lesions in both dietary groups was reversely correlated with the logarithmic value of the blood selenium concentrations (Fig. 1). The correlation coefficients were $-.94$

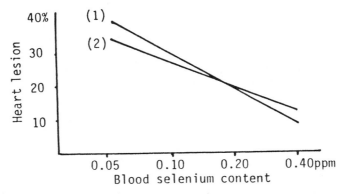

FIG. 1. Correlation between blood Se content and heart lesions in mice (experiment 3). (1) $\hat{Y} = -0.0623 - 0.03658x$ (Sichuan); (2) $\hat{Y} = 0.0187 - 0.2572x$ (semisynthetic).

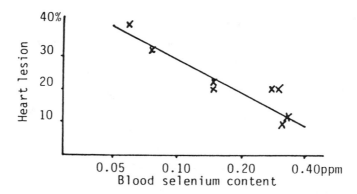

FIG. 2.   Correlation between blood Se content and heart lesions in mice (overall). $\acute{Y} = -0.0252 - 0.3177x;\ r = -.92.$

and $-.90$, respectively. The probabilities were both between .05 and .10. The regression lines were closely similar and no difference was found between the slopes and the intercepts of these two lines (.20 $< P < .50$). Therefore, all the groups should be regarded as samples from the same body and a joint regression line could be established to represent the correlation between the two variables of all the eight groups (Fig. 2). The correlation coefficient ($r = .92$) was statistically significant ($P < .01$).

DISCUSSION

Many pathological changes in the mice were related to selenium deficiency. As seen in experiment 1, lethal hepatonecrosis was observed in some animals on a diet containing 0.01 ppm Se, especially in females during pregnancy. If the diet was supplemented with sodium selenite, hepatic necrosis was no longer observed. In experiments 2 and 3, mice consumed a diet containing more than 0.015 ppm Se and no pathological changes were observed in major organs. However, 0.015 ppm Se is not considered adequate for animals. Whenever they were exposed to stress, functional defects became evident. As seen in this study, more myocardial necrosis occurred in selenium-inadequate animals.

It is well known that vitamin E is closely related to selenium in biological behavior. An adequate supply of vitamin E to animals in selenium deficiency studies is of much importance. The natural ingredient diet used by Knapka contained 36.7 mg tocopherol/kg Hurley's formula (Hurley and Knapka 1978) contained 32 mg/kg. These diets appeared to be adequate to both growth and reproduction of mice. The low selenium diets used in this study contained 40 mg vitamin E/kg, which should meet the minimal requirement of the experimental animals.

Coxsackie B viruses are cardiophyllic. Most strains in this family can induce heart lesions in baby mice (Grodums and Dempster 1962) and in human infants (Burch 1972). CB-21 virus was serologically identified as Coxsackie B_4 and induced spastic paralysis and myocardial necrosis in suckling mice (Su 1979), but appeared less virulent than did the prototype Coxsackie B_4 strain (K-Y Ge, unpublished).

The incidences of myocardial lesion of suckling mice inoculated with CB-21 virus ranged from 31 to 40% in selenium-deficient groups, which was significantly higher than that (10–21%) of the selenium-adequate groups. Selenium supplementation depressed the incidences of lesions, whether the animals were on the natural or semisynthetic low selenium diet. Selenium seems to strengthen the resistance of the mouse to viral infection. In other words, selenium-deficient animals are more vulnerable to viral injuries.

Many scientists in the field of Keshan disease research believe that pathogenic factors are present in foods in endemic areas, and various suspected factors have been tested (Yu and Su 1982; Wang 1982). Great amounts of analytic data indicate widespread selenium deficiency in endemic areas (Yang 1982). In connection with the results of these experiments, selenium deficiency could be assumed to be one pathogenic factor commonly present in Keshan disease areas, which may bring about an unfavorable environment for myocardial metabolism. On this basis, other factors such as viral infection then initiate or promote the occurrence of myocardial degeneration.

SUMMARY

Kunming mice were fed on natural or semisynthetic diets with or without selenium supplementation and bred after 6 weeks of feeding. Sucklings were inoculated with Coxsackie B_4 virus or virus-free culture.

The blood selenium concentration of both adults and sucklings was correlated with their selenium intake.

Myocardial lesions were seen only in virus-challenged sucklings. The incidence in low selenium intake groups was significantly higher than that in the selenium-adequate groups. Whenever selenium was given to the deficient groups, the incidence was reduced to a level comparable to that of the controls. A linear correlation was established between the blood selenium level and the incidence of heart lesions of virus-challenged suckling mice.

The results are discussed in connection with the etiology of Keshan disease.

REFERENCES

Burch, G. E. 1972. The role of viruses in the production of heart disease. Am. J. Cardiol. *29,* 231.

Chen, X. 1980. Studies on the relations of selenium and Keshan disease. Biol. Trace Elem. Res. *2,* 91.

Cheng, Y. 1979. Analysis of 1000 children's cases of Keshan disease in Sichuan Province. *In* Symposium of the Third Conference on the Etiology of Keshan Disease, p. 94.

Frost, D. V., and Lish, P. M. 1975. Selenium in biology. Annu. Rev. Pharmacol. *15,* 259.

Grodums, E. I., and Dempster, G. 1962. The pathogenesis of Coxsackie group B viruses in experimental infection. Can. J. Microbiol. *8,* 150.

Hurley, L. S., and Knapka, J. J. 1978. Nutrition Requirement of Laboratory Animals, 3rd Edition. National Academy of Sciences, Washington, DC.

Su, C. 1979. Preliminary results of viral etiology of Keshan disease. Chin. J. Med. *59,* 466.

Sun, J. 1982. A survey of the epidemic of Keshan disease in the north of China. Chin. J. Endemiol. *1,* 2.

Wang, F. 1982. Studies on the pathogenic factors of Keshan disease in the grain cultivated in endemic regions. Chin. J. Endemiol. *1,* 35.

Wang, G. 1984. Fluorometric determination of selenium in biological material, water and soil. Acta Nutr. Sin. (in press).

Yang, G. 1982. Relationship between selenium and the distribution of Keshan disease. Acta Nutr. Sin. *4,* 191.

Yu, W., and Su, Y. 1982. The relationship of water-soil-nutrition factors to the causation of Keshan disease. Chin. J. Endemiol. *1,* 71.

Selenium Distribution in Four Grassland Classes of China

Ji-Zho Ren
Zhi-Yu Zhou
Bim Pan
Wen Chen

Extensive research by nutritional scientists and hygienists suggests that selenium distribution always shows significant correlation with its ecological environment. It is also interesting for the grassland scientists to explore the relationships between selenium distribution and the grassland systematics in China. It is beneficial for pastoral animal production and to the grassland classification theory.

The grasslands in China have been classified by a comprehensive system involving three grades: the first grade is classes, the second is subclasses, and the third is types. The first grade is identified by bioclimatic conditions. This is the main grade in our grassland classification system.

We have studied the relationship between selenium distribution and grassland classes and their subclasses. Four classes of grassland (Table 1) have been selected for soil and forage plant sampling in order to study selenium distribution.

The sampling sites for the temperate arid grasslands are located in Alashan Banner, Inner Mongolia. It has an annual precipitation of 200 mm or less. The accumulative temperature above 0°C is about 3500°C, K value* 0.161. Soils are mainly light-brownish calci and gray-brownish desert soil.

*K value = annual precipitation/0.1 accumulative temperature above 0°C.

TABLE 1. Four Classes of Grassland Studied and Their Location in China

Class	Sites of sampling	Landscape
Temperate arid	Alashan Banner, Inner Mongolia	Desert
Subfrigid damp	Tianchou Alpine Grassland Experimental Station, Gansu Province	Alpine
Temperate humid	Loess Plateau Experimental Station, Qinyang County and Pingliang County, Gansu Province	Forest–steppe, loess plateau
Subtropic damp	South China Experimental Station, Hubie Province	Subtropic forest

The subfrigid damp grasslands are located in the northwest and southwest alpine area in China. It has an elevation of 3000 m (in the northwest) to 4000 m (in the southwest). There is no absolute frost-free period. The average annual temperature is no more than 1°C. The accumulative temperature above 0°C is about 1300°C. The annual precipitation is from 400 to 800 mm, and the K value is 3.18. Soils of this grassland class are alpine mat soil, mountain chestnut soil, and alluviation soil. The samples were taken at Tianchou Tibetan Autonomous County, Qilian Mountain, Gansu Province.

The temperate humid grasslands cover a transitional area between the steppe and foreregions. Generally, this is a forest–steppe landscape. Our sampling sites of this class were located on the Qingyang Loess Plateau at the Experimental Station of Gansu Grassland Ecological Research Institute, Gansu Province. Its annual precipitation is about 500 mm, and the accumulated temperature above 0°C is 3400°C, K value 1.7. Elevation is from 885 to 1080 m. The soils are yellowish silic loam and Black Lu loam, a kind of soil formed in a herbaceous vegetation environment. This has been one of the cultivated areas for thousands of years in China and severe water and soil erosion has occurred.

The subtropic damp grasslands are located in mid and south China. Our sampling sites were in Yizhang County and Enshi County, Hupei Province, where the annual precipitation is from 1200 to 1800 mm, accumulative temperature above 0°C is about 6170°–7344°C, and the K value is from 1.93 to 2.41. Soils are mainly brownish-yellow loam, pH 5.8–6.8.

SELENIUM DISTRIBUTION IN FOUR GRASSLAND CLASSES AND THEIR SUBCLASSES OF CHINA

The selenium distribution showed significant changes among different classes and their subclasses of grasslands. Although the selenium

TABLE 2. Selenium Content of Soil and Forage Plants in Four Grassland Classes (ppm)

Class	Soil (0–30 cm)		Forage plants	
	Range	Average	Range	Average
Temperate arid	0.100–0.210	0.153	0.011–0.517	0.121
Subfrigid damp	0.190–0.540	0.340	0.000–0.497	0.070
Temperate humid	0.100–0.180	0.129	0.000–0.018	0.011
Subtropic damp	0.120–3.390	0.827	0.000–10.80	0.646

contents range widely in soil and forage plants of every class, it is possible to distinguish the differences of selenium levels among the four classes according to the average levels of their selenium contents.

Selenium is plentiful in both soil and forage plants of the subtropic damp grasslands. Subfrigid damp grasslands are not as selenium rich as the subtropic damp grassland, but are not selenium deficient, on the average. The selenium content of temperate arid grasslands is above the deficiency level in forage plants, but it is low in the soil, suggesting high availability to plants. The temperate humid grasslands are selenium deficient both in soils and in forage plants.

The subclasses of subtropic damp grasslands are identified by soil and topographic characteristics. Table 2 shows that selenium content is not generally deficient in this class, but the selenium content of various subclasses is very different from each other. Table 3 reveals that the selenium deficiency occurs in the ridgeland subclass and slopes, but not in the valley subclass. Selenium content of the soil is increased as the leaching process weakens and the sedimental process begins to dominate. In the Tableland or Gulfy subclasses, severe selenium deficiency is evident (Table 4).

The subclasses of temperate arid grasslands are situated in the desert area. Soils from the three subclasses of this grassland class were analyzed, and none was above the level of selenium deficiency (Table 5). The rolling sandy land was the most severely deficient.

In the subclasses of subfrigid damp grasslands, the flood land subclass has 0.54 ppm selenium, the highest selenium content, followed

TABLE 3. Soil Se Content of Subtropic Grassland Subclasses (0–30 cm, ppm)

Subclass	Range	Average	Remarks
Ridge	0.160–0.210	0.185	Large and small ridges, leaching dominates
Slope	0.120–0.670	0.420	Slopes with leaching dominate and some retentive sedimentation
Valley	0.650–2.620	1.023	Sedimentation dominates

TABLE 4. Soil Selenium Content of Mild-Temperate Humid Subclasses (0–30 cm, ppm)

Subclass	Range	Average	Remarks
Tableland	0.100–0.140	0.122	Old cultivated field
Gulfy	0.110–0.180	0.145	

by 0.29 ppm in the secondary terrace, while the hill land has only 0.19 ppm, nearing selenium deficiency (Table 6). The selenium content evidently decreased as the leaching process increased.

The selenium content of forage plants in four grassland classes is influenced by many factors. Among them, the plants' ability to accumulate selenium may be the most important. The availability of selenium in the soil and other ecological circumstances also play significant roles.

Seventy forage plants growing in the four grassland classes were sampled and analyzed. Table 7 reveals that in the temperate arid class 65.7% of the forage plants have adequate selenium, 25.7% are below the selenium-deficiency level, and 8.6% are severely selenium deficient. For the subfrigid damp grasslands similar figures are 12.5, 50, and 37.5%, respectively, and for the tropic damp class they are 47.1, 48.0 and 4.8%. In the temperate humid grasslands, 100% of the forage plants are severely selenium deficient.

In order to achieve a general evaluation of the selenium contents, we have calculated a comprehensive criterion of selenium level (CCSL)

$$CCSL = (3X + 2Y + Z)/100$$

where X is the percentage of forage plants selenium content >0.050 ppm, Y is the percentage of forage plants selenium content 0.049–0.021 ppm, and Z is the percentage of forage plants selenium content <0.020 ppm.

The implications of the CCSL are as follows: CCSL >2 indicates no selenium deficiency, CCSL = 1.1–1.9 indicates selenium deficiency, and CCSL < 1 indicates severe selenium deficiency.

TABLE 5. Soil Selenium Content of Mild-Temperate Arid Grassland Subclasses (0–30 cm, ppm)

Subclass	Range	Average	Remarks
Rolling sandy land	0.100	0.100	Yellow desert soil
Basin	0.130–0.170	0.14	Water-collected area
Apron of the basin	0.190	0.19	

TABLE 6. Soil Se Content of Subfrigid Damp Grassland Subclasses (0–30 cm, ppm)

Subclass	Range	Average	Remarks
Flood land	—	0.54	Alluviation river bank
Second terrace	—	0.29	Alpine mat soil
Hill land	—	0.19	Mountain steppe soil

From the data in Table 7, comprehensive criteria of selenium levels (CCSL) have been derived, as shown in Table 8. The temperate arid grasslands and subtropic damp grasslands are not selenium deficient. The subfrigid damp grasslands are selenium deficient and the temperate humid grasslands are severely selenium deficient.

CONCLUSIONS

Four grassland classes and their eleven subclasses were studied to determine the relationship between the grassland classes and their selenium contents.

Soil selenium content within a certain subclass appears to be stable and not easily changeable. The subclass selenium levels are distinguishable. Subtropic damp grasslands and subfrigid damp grasslands are not selenium deficient, while temperate arid and temperate humid grasslands are selenium deficient.

Soil selenium contents of different subclasses usually differed due to leaching and soil erosion processes. The more leaching and erosion, the lower the selenium content.

Forage plant levels were quite different from the soil, however. At the same site forage selenium content may vary from 0 to more than 10.00 ppm selenium. Accordingly, one cannot predict selenium levels of forage plants from soil selenium levels of grassland classes and subclasses.

TABLE 7. Se Content of Forage Plants in Different Grassland Classes (%)

Class	>0.05 ppm (no Se deficiency)	0.05–0.02 ppm (Se deficiency)	<0.02 ppm (severe Se deficiency)
Temperate arid	65.7	25.7	8.6
Subfrigid damp	12.5	50.0	37.5
Temperate humid	—	—	100.0
Subtropic damp	47.1	48.9	4.8

TABLE 8. Comprehensive Criteria of Se Level (CCSL) of Grasses of Four
Grassland Classes

Class	CCSL	Evaluation
Temperate arid	2.57	No Se deficiency
Subfrigid damp	1.75	Se deficiency
Temperate humid	1.00	Severe Se deficiency
Subtropic damp	2.42	No Se deficiency

A comprehensive criterion of selenium levels (CCSL) for forage
plants was offered in this chapter. According to the CCSL system, the
temperate arid grasslands (CCSL = 2.57) and subtropic damp grasslands (CCSL = 2.42) are not selenium deficient, the subfrigid damp
grasslands (CCSL = 1.75) are selenium deficient, and the temperate
humid grasslands (CCSL = 1.00) are severely selenium deficient.

Selenium in the Soil–Plant System

G. Gissel-Nielsen

INTRODUCTION

Since Schwarz and Foltz (1957) showed that Se is an essential micro-nutrient in animal nutrition, the interest in this trace element has increased considerably, and the literature covering the biological aspects of Se has become very comprehensive. That is even the case for Se in soils and plants, although no one has ever succeeded in proving that Se is essential for higher plants. A beneficial effect of Se has been shown in some specific cases, but no vital function of Se in plant metabolism has ever been demonstrated. Consequently, the following is not an attempt to cover all aspects of Se in soils and plants, but to discuss some points of great immediacy. For a more comprehensive treatment of the subject, reviews like those of Sharma and Sing (1983) and Gissel-Nielsen *et al.* (1984) are recommended.

SELENIUM IN SOILS

For a number of years it has been known that the Se concentration of the soil in some areas is so high that plants growing there are toxic to animals. Much more common are areas where the crops contain so little Se that the local livestock suffer from Se deficiency. This was reported initially in about 1960 from the United States and New Zealand, and in China the deficiency is also well known. In Europe Se concentrations vary from deficient to toxic, as shown in Fig. 1. Crops of the Scandinavian countries are especially low in Se, as seen in Table 1. The reason for the low Se content in areas such as Scandinavia is not a

|⊞| INADEQUATE |▨| ADEQUATE

|⠿| SPOTWISE TOXIC |☐| NO INFORMATION

FIG. 1.   Se status of fodder crops in Europe.

low Se content of the soil, which is normally 0.1–0.4 ppm Se (Gissel-Nielsen *et al.* 1984), since in other places in the world, such as India, the same soil concentrations of Se result in 10 times higher concentrations in the plants than in Scandinavia (Patel and Mehta 1970). The difference is caused rather by a combination of the condition in the soil and the climate. In soils with, high pH, inorganic Se will mainly be present as selenate, which is hardly fixed at all in the soil. If the

TABLE 1. Se Content[a] of Crops in the Scandinavian Countries

Country	Grass	Cereals
Finland (1968–1969)	0.014	0.007
Sweden (1968)	0.027	0.011
Denmark (1972–1973)	0.040	0.018
Norway (1975–1978)	0.025	0.009

[a] ppm Se.

precipitation and thereby the leaching is low, most of the Se will be available to the plants. Contrary to that, low pH favors the selenite form, which is fixed strongly in the soil. If the precipitation and leaching is great, the remaining soil Se will have a very low availability to the plants. This last situation is descriptive of the condition in Scandinavia, resulting in the Se cycles shown in Fig. 2. As seen in the figure, the microbial activity in the soil might result in organic forms of Se that either can be fixed in the organic fraction, go into soluble forms that can be leached, or volatilized.

SELENIUM IN CROPS

In a number of countries the use of selenite-enriched premixed fodder has been legal for some years, and this has decreased the Se defi-

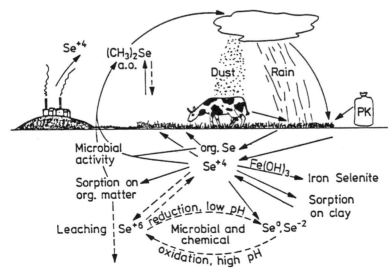

FIG. 2. Possible cycles for Se under field conditions.

TABLE 2. Results from Various Types of Se Supplementation in Barley (1983)

Treatment	Se concentration (ppb)
Control	13
10 g Se/ha as SeO_4^{2-} in PK fertilizer	69
20 g Se/ha as SeO_4^{2-} in PK fertilizer	137
120 g Se/ha as SeO_3^{2-} in PK fertilizer	88
5 g Se/ha as SeO_3^{2-} sprayed	69

ciency problem in livestock considerably. But not all problems have
been solved. Only some of the livestock are fed with premixed fodder,
and there are reasons to believe that in some cases the organic plant Se
of common crops has a higher biological value than selenite. Consequently, in many low Se countries the possibility of fertilizer or foliar
application of Se to crops has been studied. Many factors influence the
plant uptake and metabolism of added Se, such as clay and humus
content of the soil, soil pH, other nutrients, the chemical form of the
added Se, and climate, and these factors are discussed in a number of

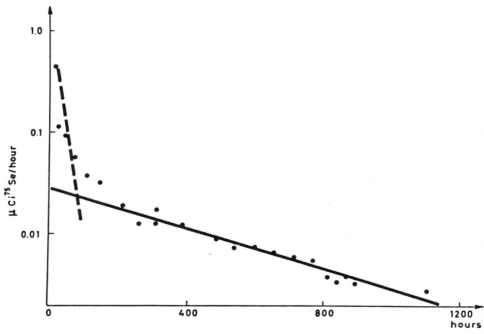

FIG. 3.   Elimination of ^{75}Se added with the food to a pheasant.

publications. The following is limited to problems concerned with the oxidation state of added Se and the influence of pH.

In most fertilizer experiments selenite has been used. Selenite has some advantages compared with selenate, but the strong soil fixation of selenite means that one needs to use much more Se per hectare than is necessary when selenate is used, and today the use of selenate-enriched fertilizers is legal in Finland and New Zealand. Table 2 shows some results from a field experiment with barley. It can be seen that the same Se concentration is obtained in the grains at foliar application of 5 g/ha as when 10 g Se as selenate or about 100 g Se as selenite was added along with the PK fertilizers. These results confirm earlier experimental findings, so which method one should prefer depends on, e.g., fertilizer and spraying practice; but all three methods are useful alternatives or supplements to Se addition to the fodder.

The effect on the environment of these measures has been studied as well, and the conclusion of these experiments is that when used correctly, they do not involve any threat to the fauna. The reason for this is a biological half-life for Se of only a few weeks (Fig. 3) (Gissel-Nielsen and Gissel Nielsen 1973). There is no bioaccumulation of Se in the food chain as had been observed for DDT and mercury, both of which have a very long biological half-life. Contrary to Se, they are not compounds with a positive biological function.

BIOLOGICAL VALUE OF SELENIUM

The biological value and function of Se is an area of great interest just now. The relative biological value of organic and inorganic compounds varies from one experiment to the other, and it is an open question which forms of Se are preferable in the fodder.

Another question is how much we can influence the chemical form of Se in plants. This has been studied in water-culture experiments with maize under varying nitrogen and sulfur levels, using selenite and selenate labeled with [75]Se (Gissel-Nielsen 1979). Using resin chromatography and precipitation techniques, the Se could be separated into selenoamino acid, selenite, selenate, and a residue that is mainly protein bound. One example of the results is given in Fig. 4. In this experiment, xylem sap was analyzed 10 min after addition of [75]Se. When selenite was used, more than 90% was translocated as amino acid-Se, while added selenate was mainly translocated as selenate. However, these results were influenced by the nitrogen and sulfur levels of greater interest than the results of such short-time experiments is to investigate whether there is any difference in the chemical

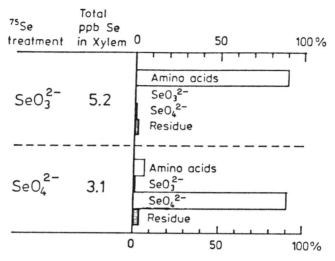

FIG. 4.   Se compounds in xylem sap of *Zea mays*.

form of Se in the grain and grass leaves, depending on the oxidation state of the added Se and the method of addition. These questions are subjects of present experiments, and preliminary results indicate that to a great extent selenate ends up in the form of selenoamino acids or protein.

ACID RAIN AND SELENIUM

The present discussion concerning the impact of acid rain on the environment has included a possible influence on plant availability of trace elements, including Se. The problem is complex because acid rain can exert an influence in different ways. Hurd-Karrer (1938) showed that sulfate can increase the solubility of Se in the soil, but Se is absorbed by the plants in competition with sulfate. Which one of these effects is dominant depends on a series of soil factors. In most cases the antagonistic influence seems to be the strongest, but most likely not of real importance.

Another way in which acid rain can influence the solubility of Se is through the soil pH influence on fixation of Se to clay minerals and iron oxides. This relation has been studied in laboratory experiments with solutions of ^{75}Se-labeled selenite (Hamdy and Gissel-Nielsen 1977). As seen in Fig. 5, acidification of the soil increases the fixation of selenite, but to only a small extent at the actual pH levels. All in all,

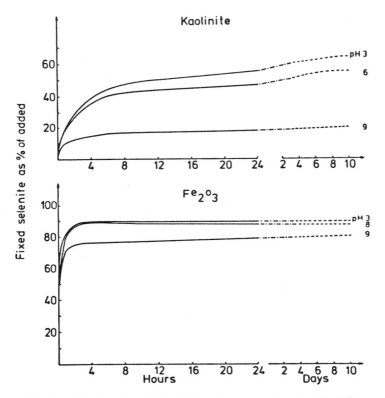

FIG. 5. Fixation of selenite. Laboratory experiments with [75]Se.

acid rain seems to have a negative, but small effect on the plant content of Se.

Another aspect of energy production from coal is the disposal of fly ash. When it takes place on or near farmland, leaching of mineral elements might influence the chemical composition of the crop plants.

TABLE 3. Effects of Fly Ash Supplementation of Soil on Barley Se Content (ppb)

Treatment	Barley (1982)		Winter barley (1983)	
	Grain	Straw	Grain	Straw
Control	26	25	28	28
25 t/ha	57	46	47	34
100 t/ha	147	80	121	55
Total Se in fly ash:	1400 ppb			
H_2O soluble:	32 ppb (2.3%)			

Experiments in different countries have all shown that up to 100 t fly ash/ha only had a positive effect on the Se content of the plants in areas of low-to-normal soil Se, as seen in Table 3, and fly ash has even been recommended as an Se fertilizer (Mbagwu 1983). However, other elements such as Cd might be the limiting factor for fly ash disposal on farm land.

CONCLUSION

The Se deficiency problem in many areas are caused by a decreased Se content of plants which is a result of several factors, among these a great increase in the plant production on the field. Consequently, it is logical to solve the problem where it is started, through fertilization or foliar application with Se, even if Se is not essential to the plants. Countries with low Se areas should therefore consider using these methods parallel to others in order to prevent Se deficiency in livestock and humans, as is now done in China, New Zealand, and Finland.

REFERENCES

Gissel-Nielsen, G. 1979. Uptake and translocation of selenium-75 in *Zea mays. In* Isotopes and Radiation in Research on Soil–Plant Relationships, pp. 427–436. IAEA, Vienna.

Gissel-Nielsen, G., and Gissel Nielsen, M. 1973. Ecological effects of selenium application to field crops. Ambio 2, 114–117.

Gissel-Nielsen, G., Gupta, U. C., Lamand, M., and Westermarck, T. 1984. Selenium in soils and plants and its importance in livestock and human nutrition. Adv. Agron. 37 397–460.

Hamdy, A. A., and Gissel-Nielsen, G. 1977. Fixation of selenium by clay minerals and iron oxides. Z. Pflanzenernaehr. Bodenkd. 140, 63–70.

Hurd-Karrer, A. M. 1938. Relation of sulfate to selenium absorption by plants. Am. J. Bot. 25, 666–675.

Mbagwu, J. S. C. 1983. Selenium concentrations in crops grown on low-selenium soils as affected by fly-ash amendment. Plant Soil 74, 75–81.

Patel, C. A., and Mehta, B. V. 1970. Selenium status of soils and common fodders of Gujarat. Indian J. Agric. Sci. 40, 389–399.

Schwarz, K., and Foltz, C. M. 1957. Selenium as an integral part of factor 3 against dietary necrotic liver degeneration. J. Am. Chem. Soc. 79, 3292–3293.

Sharma, S., and Singh, R. 1983. Selenium in soil, plant, and animal systems. CRC Crit. Rev. Environ. Control 13, 23–50.

80

Annual Topdressing of Pasture with Selenate Pellets to Prevent Selenium Deficiency in Grazing Stock: Research and Farming Practice in New Zealand

J. H. Watkinson

INTRODUCTION

Selenium was shown in 1958 to prevent certain stock disorders in New Zealand (Andrews *et al.* 1968), and within a year regulations had been promulgated allowing the Se dosing or injection of deficient stock using materials containing more than 3 mg/kg Se (Selenium Control Regulations 1959). It was not until April, 1982 that, in a notice issued in the New Zealand Gazette (1982), permission was given to topdress pasture with sodium selenate pellets (1% Se) at a rate not exceeding 1 kg/ha (10 g/ha Se) as an alternative treatment. This was more than 20 years after the first New Zealand field experiments were started. Although this is a long time, nevertheless New Zealand is evidently the first country where this pastoral treatment has been permitted. [In China the effects of Se sprayed into cereals on levels in the human population had been examined a little earlier (Cheng Bo-rong, personal communication).] The reason for the delay in approval is that Se is not only essential for animal life, but is also, as is well known, toxic in relatively small amounts, the ratio of dietary Se giving chronic toxicity and deficiency being about 100 times. The approved topdressing method has had to satisfy criteria for both effectiveness in preventing deficiency and safety in use. The latter includes handling of the product during mixing with fertilizer, transport, and distribution, as well as safety for the grazing animals and consumers of the animal products, not neglecting possible detrimental effects on the environment (Watkinson 1983a).

METHODOLOGY

Selenium topdressing is basically designed for the least annual application to the whole farm to prevent deficiency in all grazing stock for the same period. This ensures the use of a very small amount of Se from considerations of both toxicity and cost. The pasture absorbs about 15% of the Se applied as selenate, and grazing stock receive their Se continuously from the pasture. The concentration of pasture Se, however, varies quite markedly with time after application. It rises rapidly to a peak within a month, then decreases more slowly to almost the initial level (Watkinson 1983a) in as little as 9 months after topdressing of severely deficient soil (Fig. 1). In spite of the very large changes in pasture Se over quite short times, it has been well established that the method has a safety factor of about 20 times and that deficiency is prevented on the most severely deficient soil for 12 months (Watkinson 1983a). This may be contrasted with Se drenching

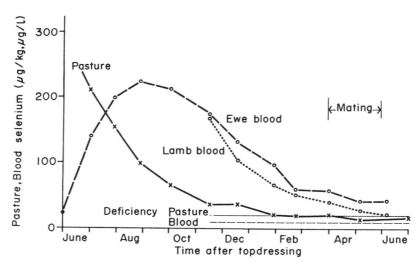

FIG. 1. Selenium in pasture, and blood of ewes and lambs after May selenate topdressing at 8.5 g/ha Se on a pumice soil.

where the safety factor is about 5 times, and deficiency is prevented
for only 2–4 months (Watkinson 1983a).

The main reason for the effectiveness of the method lies in the
buffering action of the grazing animal against the rapid changes in
pasture Se, as shown in the much smaller changes in blood Se values
(Figs. 1, 5). In terms of acute toxicity, even at the pasture Se peak the
animal consumes only a small amount of Se, being no more than 1 mg
Se/day for a 50-kg ewe, which can be compared with the drench
amount of 5 mg Se. Pasture Se increases linearly with application
rate, but peak blood Se (Fig. 1) increases only logarithmically with
peak pasture Se and so provides a wide safety margin against toxicity
through overapplication (Fig. 2) (Watkinson 1983a). At the other ex-
treme of the pasture Se minimum, the animal storage of Se and the
relatively slow excretion of absorbed Se ensures that the animal remains above deficiency for at least 12 months, even though the pasture Se consumed over the last 2 or 3 months could, under steady-state
conditions, be at deficiency level (Watkinson 1983a). This important
effect was first reported by Grant (1969). The experiment illustrated

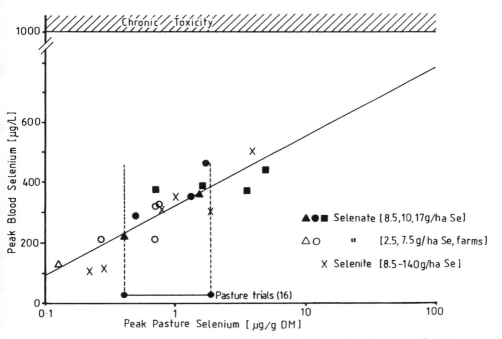

FIG. 2.   Peak Se in blood of sheep in relation to toxicity and peak pasture Se after
selenate and selenite topdressing as salt and pellets on several soil types. The
ranges are shown for peak pasture Se from 16 pasture uptake trials.

in Fig. 1 showed that 8.5 g/ha Se provided adequate Se in terms of animal production and blood Se over at least 12 months to both ewe and lamb. This was conducted on the most severely deficient kind of soil (pumice soil), and Se was applied at the least favorable topdressing time for preventing ewe infertility and lamb ill-thrift, i.e., just after mating. Other work has confirmed that the maximum approved rate of 10 g/ha Se is adequate for all soils and conditions and that Se can be applied at any time of the year (Watkinson 1983a).

In formulating an application rate that would be both effective and safe for stock, one is concerned with keeping well within the bounds of toxicity, on the one hand, and deficiency, on the other. The peak blood Se depends on the peak pasture Se concentration (Fig. 2), which in turn depends largely on the Se application rate and the amount of pasture per hectare (Watkinson 1983a).

The minimum blood Se, i.e., the value at the end of 12 months before the next topdressing, depends on the maximum amount of Se stored in the animal and its rate of excretion, which is modified somewhat by the low pasture Se intake at that stage (Fig. 3). The North Island experiments were on severely deficient soils, and the minimum blood Se depends very strongly on the peak blood Se because the pasture Se is at almost the initial level after 9 months (Fig. 1). Three experiments on less deficient South Island soils, in contrast, show higher blood Se on average after 12 months (Fig. 3). This is a result of the pasture Se dropping off much less rapidly with time. In fact, pasture Se and blood Se have been found adequate even after 2 years (Figs. 4 and 5). In this experiment the blood values of animals from the untreated area in-

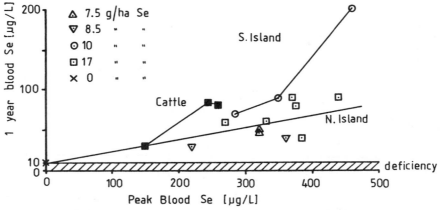

FIG. 3.   Sheep blood Se at peak and at 1 year after topdressing.

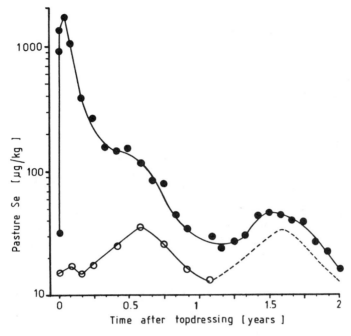

FIG. 4. Pasture Se at a South Island site after spring topdressing (●) at 10 g/ha Se (J. H. Watkinson and J. M. Munro,
unpublished). (○), Control.

creased appreciably when grazed on the Se-treated area in the second
year after topdressing (Fig. 5).

Both peak and minimum blood Se values as related to application
rate are given in Fig. 6. This and the data from Fig. 2 show that an
application rate of 10 g/ha Se gives adequate and safe blood Se values
for 1 year. If economically necessary, the rate could be reduced, particularly on the South Island soils shown.

TECHNOLOGY

The use of Se prills or pellets for increased safety over mixing Se
salts directly with fertilizer, as with cobalt topdressing, was first suggested in 1967 (Watkinson and Davies 1967). Some 10 years were to
elapse before the technique was examined and established as effective
and safe for use with sheep, cattle, and horses (Hupkens van der Elst
and Watkinson 1977). Of the experimental prills used, made from

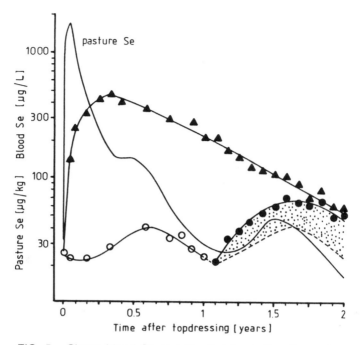

FIG. 5. Sheep blood Se at a South Island site after spring topdressing at 10 g/ha Se (J. H. Watkinson and J. M. Munro, unpublished). (▲), Topdressed area; (○), control area, then (●) topdressed area.

superphosphate, pumice, or clay granules, clay was the most satisfactory, being most resistant to abrasion and powder formation. The prills, containing 1% Se, may be regarded as a substance of relatively low toxicity to workers involved in mixing prills with fertilizer (Hupkens van der Elst and Watkinson 1977; Watkinson 1983a; Toxic Substances Regulations 1983). In one manufactured pellet (Selcote, registered under Patent Application No. 200421), the sodium selenate crystals are incorporated within the pellet, making the product even safer for handling.

Selenium is released on exposure to moisture, either by effusion from the granule impregnated with the selenate salt or, in one product, by dispersion of the outer protective layer. Selenate is weakly absorbed by all soil colloids and so is readily available for root absorption (Watkinson 1983a).

Another requirement is compatibility in mixing with fertilizer. The Se pellet at an application rate of 1 kg/ha provides an adequate amount of material for mixing compared with the rather small amount of sodium selenate it contains, approximately 20 g/ha. Furthermore, the

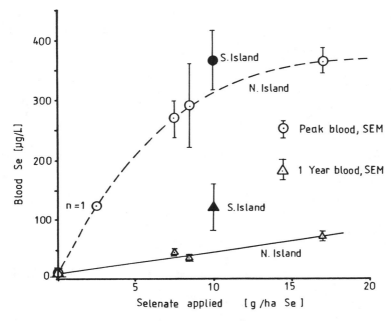

FIG. 6.   Sheep blood Se at peak and at 1 year after topdressing at different rates of Se.

similar particle size and density of pellets to the fertilizer ensures better mixing than the salt. However, because Se is needed by the grazing animal and not pasture, considerable unevenness in spreading within a paddock can be tolerated. Trials where Se prills were applied in strips at 20 times the uniform rate (17 g/ha Se) and covering 1/20 of each paddock gave blood Se levels in grazing stock not appreciably different from a uniform application (Hupkens van der Elst and Watkinson 1977). The Se pellets are relatively inert and have been safely mixed with all kinds of fertilizers. One fertilizer company offers Se pellets in 49 different fertilizer mixes (D. I. Glue, personal communication).

CONDITIONS

- The method of Se pellet application has been made as simple as possible, mainly in the interests of safety. As previously indicated, the application rate of no more than 1 kg/ha pellets is designed for annual application to the whole farm. After perhaps 30 years of accumulation of added Se to the topsoil, the rate could be lowered or pellets applied every second year (Watkinson 1983a).

- At low stocking rates, the topdressing of only a part of the farm can be useful management practice. For example, ewes grazing at mating on Se topdressed pasture for only 4 weeks have been given protection for about 8 months against ewe infertility, and white muscle disease and ill-thrift in lambs up to weaning (A. K. Metherell, personal communication).
- Restricting the rate to 1 kg/ha is easily ensured by mixing the pellets with the fertilizer in registered mixtures at dispatch. It is legally possible to apply pellets on their own at 1 kg/ha, but few if any farmers have been interested in this option. With a safety margin of about 20 times, toxicity from overapplication is very unlikely.
- No withholding period is necessary or desirable. Stock can graze paddocks being topdressed.
- Application can be at any time of the year, but spring gives higher animal levels at the critical periods than autumn.
- Application is effective on all classes of deficient soil and pasture, under low or high rainfall (or irrigation), and for sheep, cattle, and horses (Watkinson 1983a).
- Although there is competitive inhibition of selenate absorption by plants from sulfate, the effect is not great (Watkinson 1983a,b).
- Pellets can be safely mixed and applied with all kinds of fertilizer, including sulfate.
- Application can be safely made even if some animals have recently been dosed with Se, or even if Se is applied in error to soils with the highest natural content in New Zealand (Watkinson 1983a).
- New stock of unknown Se status brought in to a deficient area 3 months or more after topdressing should be dosed to ensure that they are adequate in Se.

FARMING PRACTICE

The pasture and blood of grazing stock have been monitored for a number of farms under normal farming conditions since 1982. All results have been within the ranges found under research conditions.

Topdressing has proved quite popular since its introduction, with Se pellets being produced by three New Zealand manufacturers.* The annual coverage in the first 2 years has been estimated at 0.13 and 0.37 million hectares, respectively, while the estimate for the current

---

*Agsel (Unitech Industries Ltd., PO Box 69050, Auckland), Selcote [Mintech (NZ) Ltd., PO Box 440, Nelson], Selsorb [Silicon Industries (NZ) Ltd., PO Box 10249, Hamilton].

year is over 0.5 million. This would represent over 10% of the deficient area in New Zealand, which would include a large area of sparsely stocked farms that would not be economic to topdress. Most farmers have had had pellets mixed in with their annual fertilizer topdressing.

The New Zealand Health Department monitors the conditions of mixing of the Se pellets in the fertilizer works and takes periodic urine samples from the workers involved. For purposes of transport and storage, the Se pellets at 1% Se are not regarded as a toxic material (Toxic Substances Regulations 1983). When mixed with fertilizer the Se concentration is only 20–40 mg/kg.

The cost of Se topdressing on a typical sheep farm on the deficient pumice soils is about 4% of fertilizer topdressing costs, or about six times the cost of materials for Se dosing (labor excluded). On less deficient soils in South Island, the application rate and cost could often be halved. Also, as indicated earlier, topdressing part of the farm could sometimes be practicable, reducing costs accordingly. The present costs are based on the use of a relatively pure grade of sodium selenate. A less pure and cheaper grade could be satisfactory, depending on the impurity. For example, a 10% selenite level would be acceptable.

Only two problems have been reported. In one, part of a farm was topdressed, in error, at 15 times the approved rate. The resulting pasture Se was in the predicted range, and no stock problems were expected or found with the stock rotations followed. In the second, a farmer left a spill of Se pellets on pasture grazed by sheep. Direct ingestion resulted in 10 deaths from a flock of 45 sheep.

Overall, the adoption rate of Se topdressing has been more rapid than expected, and it is likely to remain a firm alternative to other preventative methods for many New Zealand farmers on Se-deficient soil.

CONCLUSIONS

In New Zealand, research over 25 years and farming practice over 2 years have shown selenium topdressing as selenate pellets (containing 1% Se) at 1 kg/ha annually to be an effective, safe, and economic method for preventing deficiency in grazing stock.

ACKNOWLEDGMENTS

I am grateful to A. K. Metherell and J. M. Munro for permission to cite unpublished work, and to the Trimble Agricultural Research Trust, New Zealand Ministry of Agriculture and Fisheries, and Mintech (New Zealand) Limited for financial assistance.

REFERENCES

Andrews, E. D., Hartley, W. J., and Grant, A. B.   1968.   Selenium-responsive diseases of animals in New Zealand. N.Z. Vet. J. *16*, 3–17.

Grant, A. B.   1969.   Selenium topdressing—Results of some preliminary trials. Proc. Tech. Conf. N.Z. Fert. Manu. Assoc., 12th pp. 47–58.

Hupkens van der Elst, F. C. C., and Watkinson, J. H.   1977.   Effect of topdressing pasture with selenium prills on selenium concentration in blood of stock. N.Z. Exp. Agric. *5*, 79–83.

New Zealand Gazette   1982.   1235.

Selenium Control Regulations   1959.   N.Z. Statutory Regulation 1959/202.

Toxic Substances Regulations   1983.   N.Z. Statutory Regulation 1983/130.

Watkinson, J. H.   1983a.   Prevention of selenium deficiency in grazing animals by annual topdressing of pasture with sodium selenate. N.Z. Vet. J. *31*, 78–85.

Watkinson, J. H.   1983b.   Interactions between macro and micro elements—Sulphur and selenium. Seminar on Mineral Requirements for Ruminants, pp. 85–89. Univ. of Waikato.

Watkinson, J. H., and Davies, E. B.   1967.   Uptake of native and applied selenium by pasture species. IV. Relative uptake through foliage and roots by white clover and browntop. Distribution of selenium in white clover. N.Z. Agric. Res. *10*, 122–133.

Maternal Selenium Intake, Milk Content, and Neonatal Selenium Status in the Rat

Anne M. Smith
Mary Frances Picciano

The most extensive knowledge of selenium has come from animal studies, and its essentiality has been documented in many species including laboratory rats, chicks, pigs, lambs, and calves. Because of the limited knowledge base concerning human selenium metabolism, the human selenium requirement has not been determined with any degree of certainty. An "estimated safe and adequate daily dietary intake" for adults has been extrapolated from the dietary selenium level of 0.1 ppm required by most animal species (National Academy of Sciences/National Research Council 1980). The selenium requirement for laboratory rats of 0.1 ppm is based on the results of a study by Hafeman *et al.* (1974). These investigators demonstrated that 0.05 ppm selenium supported optimal growth, but a level of 0.1 ppm was necessary to meet the glutathione peroxidase (GSHPx) tissue requirement. This study, however, was performed on male rats only and comparable studies have not been conducted with pregnant or lactating rats, populations that may have altered selenium requirements.

Selenium deficiency has resulted in impaired reproductive performance in all animal species studied. Evidence also suggests that selenium metabolism is actually altered during gestation and possibly during lactation. Changes have been found in plasma selenium levels and GSHPx activity of pregnant and lactating rats (Behne *et al.* 1978a,b), and alterations in the selenium status of pregnant women have also been observed (Rudolph and Wong 1978; Behne and Wolters 1979; Perona *et al.* 1979; Butler *et al.* 1982; Swanson *et al.* 1983).

Information is necessary to estimate the selenium requirement for gestation or lactation in the rat such that optimum selenium nutrition can be maintained when using the laboratory rat for reproduction experiments. Just as the recommended selenium intake for adults has been extrapolated from animal data, the use of an animal model can also provide important information concerning selenium utilization during human reproduction.

This study was undertaken to determine the effect of various levels of selenium on maternal selenium status during pregnancy and lactation, the effect of maternal dietary selenium level on milk selenium content, and, finally, the effect of maternal selenium status and milk selenium content on the selenium status of the pups.

MATERIALS AND METHODS

Forty-eight nulliparous Sprague–Dawley female rats were housed in individual suspended stainless-steel mesh cages at a constant temperature (70°–76°F) and exposed to a 12-hr light-dark cycle. Animals were fed rat chow and tap water for no more than 1 week and then randomly assigned to an experimental group. At this point experimental diets and demineralized water were fed *ad libitum*. After a 14- to 20-day adaptation period and when their weight reached 220–240 g, 8 rats in each group were mated. Day 1 of pregnancy was determined by the presence of sperm in vaginal smears. Each group of animals remained on the appropriate experimental diet throughout gestation and lactation. Four nonpregnant rats were also fed each diet for 8 weeks to serve as controls. Pregnant rats were transferred to maternity cages on day 19 of gestation. Food dishes were positioned in the maternity cages such that only dams had access to food and maternal milk was the only nourishment for pups. Each litter was adjusted to 7 pups on day 2 of lactation. Body weights and food intakes were recorded weekly.

The composition of the basal diet used was as follows: casein 20%, sucrose 31.8%, cornstarch 31.5%, corn oil 10%, cellulose 2%, methionine 0.2%, vitamin mix (Teklad #40060) 1%, mineral mix (AIN-76, Se deleted) 3.5%. The mineral mix was supplemented to contain 250 ppm Fe and 10.5 ppm Cu. The basal diet contained 0.025 ppm selenium (marginally deficient). Three additional diets containing 0.05 (half requirement for adult rat), 0.10 (requirement), and 0.20 (twice requirement) were made by supplementing the basal diet with sodium selenite.

Tail blood samples were taken from dams on days 1, 8, and 15 of gestation and days 2, 8, and 18 of lactation and used to assess maternal selenium status by measurements of plasma selenium levels and plasma and erythrocyte GSHPx activities. On day 18 of lactation dams were sacrificed and tissue selenium and GSHPx activities were determined in liver, heart, and kidney. All selenium determinations were performed using gas chromatography with electron capture detection (McCarthy *et al.* 1981). Se-dependent GSHPx activity in all samples was determined by the coupled assay utilizing NADPH and *t*-butyl hydroperoxide (Paglia and Valentine 1967).

On day 18, dams were also milked prior to sacrifice and this milk was analyzed for selenium. The effect of maternal status and milk selenium content on the selenium status of the pups was assessed on day 2 and 18 of lactation. After adjusting litters to 7 pups on day 2, blood and tissues from the extra pups were used for analyses. Litters were sacrificed on day 18 of lactation prior to milking the dams. Plasma selenium concentration and plasma, erythrocyte, liver, heart, and kidney GSHPx activities were determined on the 2- and 18-day-old pups.

One-way analysis of variance was used to determine statistical significance of data from the four experimental groups (Steel and Torrie 1980). Pearson's correlation statistics were used to determine relationships between dam and pup selenium status (Nie *et al.* 1975). The probability level for significance was 5% or less.

RESULTS AND DISCUSSION

Pregnancy

During pregnancy, significant decreases in plasma selenium concentration and plasma and erythrocyte GSHPx activities were observed for all experimental groups. Despite these decreases during pregnancy, at term, no differences in these parameters were found among the groups receiving 0.050, 0.100, or 0.200 ppm selenium. These results suggest that the decreased selenium levels and GSHPx activities are primarily physiological effects of pregnancy and are not determined exclusively by dietary selenium intake. Animals receiving the diet marginally deficient in selenium (0.025 ppm), however, did have significantly lower plasma selenium levels and GSHPx activity at the end of pregnancy. Behne *et al.* (1978a,b) also found that serum selenium concentration and plasma GSHPx activity were reduced at term in pregnant rats receiving 0.3 ppm dietary selenium via laboratory chow. Conflicting reports have been made regarding changes in selenium status during human pregnancy. While the reports have all indicated that plasma selenium levels are depressed during pregnancy, some investigators have observed decreases in plasma or erythrocyte GSHPx activities (Rudolph and Wong 1978; Behne and Wolters 1979) and others have found either no change (Perona *et al.* 1979) or increases (Butler *et al.* 1982) in these enzyme activities. The reported fluctuations in these enzyme activities, however, cannot be compared with dietary selenium intake, since no estimation of intake was made in these investigations. In a more recent report of selenium metabolism during human pregnancy, an increase in retention of dietary selenium was observed, with a tendency for pregnant women to conserve selenium by decreasing urinary excretion (Swanson *et al.* 1983). This finding strongly suggests that selenium utilization is altered during pregnancy and that the selenium requirement may be elevated during this period.

Lactation

Plasma and milk selenium concentrations and plasma and erythrocyte GSHPx activities for dams on day 18 of lactation are listed in Table 1. Although there were significant differences in plasma selenium levels among dams receiving 0.025, 0.050, and 0.100 ppm selenium, no differences were found between dams receiving 0.100 ppm and those receiving 0.200 ppm selenium. Similarly, milk selenium levels rose with increasing amounts of selenium in the maternal diet, but the difference between the 0.100 ppm group and the 0.200 ppm group was not significant.

TABLE 1. Plasma and Milk Se (ng/ml) and Plasma and RBC GSHPx Activity (mU/mg Protein) for Dams on Day 18 of Lactation

	Se		GSHPx	
Group (ppm Se)	Plasma	Milk	Plasma	RBC
0.025	97(10.7)[a,b]	21(4.3)[b]	23(10.1)[b]	154(30.0)[b]
0.050	280(69.0)[c]	52(19.3)[c]	44(7.0)[c]	359(63.4)[c]
0.100	393(49.3)[d]	83(26.5)[d]	46(5.1)[c]	368(52.2)[c]
0.200	390(44.5)[d]	92(35.0)[d]	69(13.3)[d]	493(152.8)[d]

[a] Means (SD) for 8 dams in each column followed by an unlike superscript differ significantly at $P < .05$.

Despite similar plasma and milk selenium concentrations observed for the two groups receiving the higher levels of selenium, a level of 0.200 ppm selenium did result in significantly higher plasma and erythrocyte GSHPx activities by day 18 of lactation. Unlike animals fed the lower levels of selenium, dams fed 0.200 ppm selenium also had their plasma selenium concentrations and plasma and erythrocyte GSHPx activities return to their original (day 1 pregnancy) levels during lactation. Behne *et al.* (1978a,b) also observed increases in serum selenium concentration and plasma GSHPx activity after delivery in dams receiving 0.300 ppm selenium. Serum selenium concentration, but not plasma GSHPx activity, increased to its original level.

During lactation, dam organ selenium concentrations and GSHPx activities were also influenced by level of dietary selenium. On day 18 of lactation, mean (SD) liver selenium concentrations (ng/g wet tissue) for the groups fed 0.025, 0.050, 0.100 and 0.200 ppm selenium were 196 (62.1), 639 (171.6), 885 (156.2), and 1056 (217.6), respectively. Each increase in the level of dietary selenium resulted in a significant increase in liver selenium concentration. Dam organ GSHPx activities on day 18 of lactation are illustrated in Fig. 1. Liver GSHPx activity reflected liver selenium concentration, with significant increases for every increase in dietary selenium level. No differences in kidney GSHPx were found between dams fed 0.100 ppm and those fed 0.200 ppm selenium, and no differences in heart GSHPx activity were found among dams fed 0.050, 0.100, or 0.200 ppm selenium. These results suggest that during reproduction kidney and especially heart GSHPx activities may be under tighter homeostatic regulation than that of the liver.

A level of 0.200 ppm dietary selenium, therefore appears to have a beneficial effect on maternal selenium status during lactation, resulting in significantly greater liver selenium concentrations and plasma, erythrocyte, and liver GSHPx activities. More importantly, however, the appropriateness of maternal selenium intake must also be assessed in terms of the selenium states of the pups.

Pups

Plasma selenium concentrations and plasma and erythrocyte GSHPx activities for 2- and 18-day-old pups are listed in Table 2. On day 2, pups of dams fed 0.025 ppm selenium had significantly lower plasma selenium levels, but no differences were found among the other groups of pups. From 2 to 18, pups of dams fed 0.025 and 0.050 ppm selenium maintained low plasma selenium levels, whereas levels in the pups of dams fed 0.100 and 0.200 ppm selenium increased.

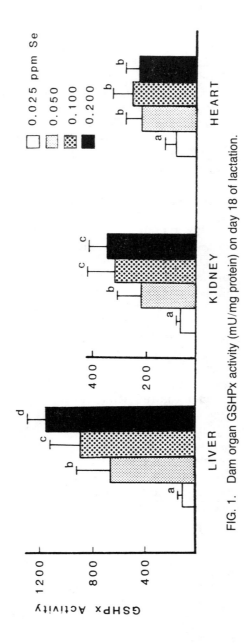

FIG. 1. Dam organ GSHPx activity (mU/mg protein) on day 18 of lactation.

TABLE 2. Plasma Se (ng/ml) and Plasma and RBC GSHPx Activities (mU/mg Protein) for 2- and 18-Day-Old Pups

| Group (ppm Se) | Plasma Se | | GSHPx | | | |
| | | | Plasma | | RBC | |
	Day 2	Day 18	Day 2	Day 18	Day 2	Day 18
0.025	77(13.1)[a,b]	61(5.4)[b]	30(15.5)[b]	37(12.4)[b]	374(76.5)[b]	150(32.8)[b]
0.050	106(21.3)[c]	138(27.9)[c]	47(15.2)[b,c]	62(12.2)[c]	470(125.6)[b,c]	236(59.4)[c]
0.100	116(32.2)[c]	192(24.6)[d]	86(26.0)[d]	85(23.5)[c]	506(139.2)[b,c]	228(14.3)[c]
0.200	120(16.2)[c]	194(14.2)[d]	69(14.3)[c]	83(26.6)[c]	539(177.0)[c]	287(51.0)[d]

[a] Means (SD) for 8 litters in each column followed by unlike superscripts differ significantly at $P < .05$.

Among groups, pup plasma selenium concentrations paralleled those of the dams on day 18 of lactation.

On day 2, plasma GSHPx activity was significantly higher in pups of dams fed 0.100 ppm selenium than in any other group. While this groups of pups maintained their level of plasma GSHPx activity from day 2 to day 18, the plasma enzyme activity increased in pups of dams fed 0.200 ppm selenium. On day 18, no differences were observed among pups of dams fed the three higher levels of selenium.

From day 2 to 18, erythrocyte GSHPx activity decreased for all groups of pups, suggesting that the decrease may be a physiological development not entirely dependent on dietary selenium. Pups of dams fed 0.200 ppm Se, however, did have significantly greater erythrocyte enzyme activity on day 18 than pups in all other groups. The maintenance of this higher level of erythrocyte GSHPx activity may be beneficial to the neonate in light of the known role of the enzyme in maintaining erythrocyte integrity (Rotruck et al. 1972).

Pup organ GSHPx activities are illustrated in Fig. 2. From day 2 to 18, pup GSHPx activity in all organs was significantly lower than activity in the organs of the dams. The level of selenium in the maternal diet did not consistently influence the organ enzyme activities on day 2; however, some effects were observed by day 18. On day 18, liver and kidney GSHPx activity in pups of dams fed 0.200 ppm selenium was significantly greater than the enzyme activity in pups in all other groups. Pup heart GSHPx activity on day 18 paralleled that of the dams, with no differences among the pups of dams fed the three higher levels of selenium.

These findings in 2- and 18-day-old pups indicate that the increase in maternal selenium intake from 0.100 to 0.200 ppm influenced pup selenium status primarily during lactation. A level of 0.200 ppm selenium in the maternal diet resulted in significantly greater erythrocyte, liver, and kidney GSHPx activities in 18-day-old pups. Significant correlations were also found between parameters of selenium status in dams and pups on day 18 of lactation (Table 3). A direct effect of maternal selenium status on that of her pups is suggested by a highly significant correlation between dam and pup plasma selenium concentrations. This effect seems to be mediated through milk selenium concentration and is supported by the presence of a significant correlation between pup plasma selenium and milk selenium and an even stronger correlation between dam plasma selenium and milk selenium concentration. Significant relationships were also found between milk selenium content and GSHPx activity in pup plasma, erythrocytes, liver, kidney, and heart.

The importance of maintaining optimal maternal selenium status

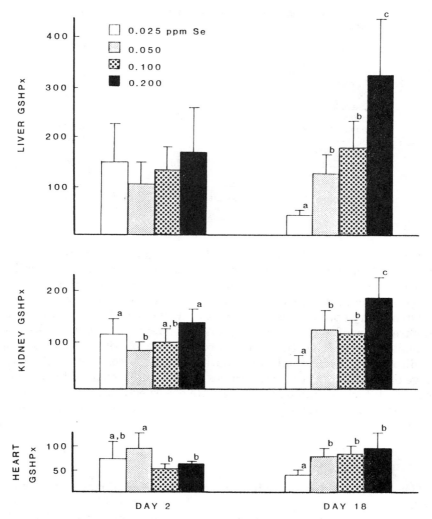

FIG. 2.   Pup organ GSHPx activity (mU/mg protein) on day 2 and 18.

during lactation should not be overlooked, especially when the neonate
is depending on maternal milk as its sole source of selenium. Very few
attempts, however, have been made at quantifying the selenium requirement for lactation in any species. For economic reasons, selenium
and milk production has been studied most extensively in dairy cows,
pigs, and sheep. In the sow, supplemental selenium during lactation is
crucial in preventing the onset of selenium deficiency in her progeny

TABLE 3. Dam–Pup–Milk Relationships—Results of Correlation Statistics[a]

Parameter	r	P value
Dam plasma Se with		
Pup plasma Se	0.9123	0.001
Milk Se	0.7427	0.001
Milk Se with pup		
Plasma Se	0.6639	0.001
Plasma GSHPx	0.5213	0.004
RBC GSHPx	0.4182	0.034
Liver GSHPx	0.6690	0.001
Kidney GSHPx	0.5886	0.001
Heart GSHPx	0.6239	0.001

[a] Correlation statistics (r values) are based on 32 observations per group.

(Mahan et al. 1975). Most investigations of selenium metabolism during lactation have focused on the transport of selenium into milk. Selenium in milk is largely protein bound and probably enters milk in some way connected with synthesis, transport, and secretion of protein by mammary epithelial cells (Allen and Miller 1981). In the ruminant the transport of selenium into milk is dependent on the level of dietary selenium as well as the form (Jacobsson et al. 1965; Jenkins and Hidiroglu 1971). The only information available concerning the passage of selenium into rat milk comes from a study reported by McConnell (1948) more than 35 years ago. He observed that injected labeled selenate was converted to organoselenium in the protein fraction of milk.

Summary

The results of this study indicate that the level of maternal dietary selenium affects maternal selenium status, especially during lactation. Effects on maternal plasma, erythrocyte, and liver GSHPx activity and liver selenium concentration suggest that a level of 0.20 ppm dietary selenium may be more appropriate than 0.10 ppm during gestation and lactation. The effect of dietary selenium on maternal selenium status is especially important due to the direct effect on pup selenium status. A selenium concentration of 0.200 ppm in the maternal diet provides more optimal selenium nutrition for the neonate than 0.100 ppm by maintaining adequate milk selenium levels as well as adequate maternal indices of selenium status during lactation. Further research is needed on the effect of higher levels of selenium in the maternal diet.

REFERENCES

Allen, J. C., and Miller, W. J. 1981. Mechanisms for selenium secretion into milk. Feedstuffs *53*, February 2, pp. 22–23.

Behne, D., and Wolters, W. 1979. Selenium content and glutathione peroxidase activity in the plasma and erythrocytes of nonpregnant and pregnant women. J. Clin. Chem. Clin. Biochem. *17*, 133–135.

Behne, D., von Berswordt-Wallrabe, R., Elger, W., Hube, G., and Wolters, W. 1978a. Effects of pregnancy and lactation on the serum selenium content of rats. Experientia *34*, 270–271.

Behne, D., von Berswordt-Wallrabe, R., Elger, W., and Wolters, W. 1978b. Glutathione peroxidase in erythrocytes and plasma of rats during pregnancy and lactation. Experientia *34*, 986–987.

Butler, J. A., Whanger, P. D., and Tripp, M. J. 1982. Blood selenium and glutathione peroxidase activity in pregnant women: Comparative assays in primates and other animals. Am. J. Clin. Nutr. *36*, 15–23.

Hafeman, D. G., Sunde, R. A., and Hoekstra, W. G. 1974. Effect of dietary selenium on erythrocyte and liver glutathione peroxidase in the rat. J. Nutr. *104*, 580–587.

Jacobsson, S. O., Oksanen, H. E., and Hansson, E. 1965. Excretion of selenium in the milk of sheep. Acta Vet. Scand. *6*, 299–312.

Jenkins, K. J., and Hidiroglu, M. 1971. Transmission of selenium as selenite and as selenomethionine from ewe to lamb via milk using selenium-75. Can. J. Anim. Sci. *51*, 389–403.

Mahan, D. C., Moxon, A. L., and Cline, J. H. 1975. Efficacy of supplemental selenium in reproductive diets on sow and progeny serum and tissue selenium values. J. Anim. Sci. *40*, 624–631.

McCarthy, T. P., Brodie, B., Milner, J. A., and Bevill, R. F. 1981. Improved method for selenium determination in biological samples by gas chromatography. J. Chromatogr. *225*, 9–16.

McConnell, K. P. 1948. Passage of selenium through the mammary glands of the white rat and the distribution of selenium in the milk proteins after subcutaneous injection of sodium selenate. J. Biol. Chem. *173*, 653–657.

National Academy of Sciences/National Research Council 1980. Recommended Dietary Allowances, 9th Edition. National Academy of Sciences, Washington, DC.

Nie, N. H., Hull, C. H., Jenkins, J. G., Steinbrenner, K., and Bent, D. H. 1975. SPSS: Statistical Package for the Social Sciences, 2nd Edition. McGraw-Hill, New York.

Paglia, D. E., and Valentine, W. N. 1967. Studies on the quantitative and qualitative characterization of erythrocyte glutathione peroxidase. J. Lab. Clin. Med. *70*, 158–169.

Perona, G., Guidi, G. C., Piga, A., Cellerino, R., Milano, G., Colautti, P., Moschini, G., and Stievano, B. M. 1979. Neonatal erythrocyte glutathione peroxidase deficiency as a consequence of selenium imbalance during pregnancy. Br. J. Haematol. *42*, 567–574.

Rotruck, J. T., Pope, A. L., Ganther, H. E., and Hoekstra, W. G. 1972. Prevention of oxidative damage to rat erythrocytes by dietary selenium. J. Nutr. *102*, 689–696.

Rudolph, N., and Wong, S. L.   1978.   Selenium and glutathione peroxidase in maternal and cord plasma and red cells. Pediatr. Res. *12*, 789–792.

Steel, R. D., and Torrie, J. H.   1980.   Principles and Procedures of Statistics. McGraw-Hill, New York.

Swanson, C. A., Reamer, D. C., Veillon, C., King. J. C., and Levander, O. A. 1983. Quantitative and qualitative aspects of selenium utilization in pregnant and nonpregnant women: An application of stable isotope methodology. Am. J. Clin. Nutr. *38*, 169–180.

Effect of Selenium and Vitamin C on the Myocardial Metabolism of Lipids in the Rat

Guang-Yuan Li
Jian-Guo Yang

Intravenous injection of large-dosage vitamin C in treating cardiogenic shock of acute Keshan disease (KD) has achieved an effective result since 1960 (*1*). In 1965, we experimentally prevented KD with sodium selenite on Huaishuzhuang Farm, Fu County, Shaanxi Province, and found its preliminary effectiveness after taking it successively for 2 years. Why could selenium prevent and vitamin C cure acute KD? It is well known that selenium deficiency may cause white muscle disease of animals and that arachidonic acid of skeletal muscle was elevated in white muscle disease lambs (*2*). We also found an increase of free fatty acids (FFA) in swine fed the selenium-deficient diet (*3*). We have known that vitamin C might increase glutathione peroxidase (GSHPx) activity, decrease the occurrence of exudative diathesis in chicks, and alleviate mortality in Pekin ducklings (*4*) fed a diet deficient in selenium and vitamin E. This chapter attempts to study the effect of selenium and vitamin C on myocardial metabolism of lipids and determine whether it can explain the interrelation between the prevention and cure of KD and the functions of selenium and vitamin C.

MATERIALS AND METHODS

Weanling rats (100) taken from the Animal Breeding Farm in our college, weight 40–60 g, were randomly divided into three groups and

TABLE 1. Percentage Diet Composition for Each Group

Group[a]	Maize powder	Soybean powder	Yeast[b]	Salt	Cod-liver oil	Sodium selenite (ppm)	Total Se content (ppm)
1	86.5	12.7	0.2	0.5	0.1		0.023
2	86.5	12.7	0.2	0.5	0.1	0.25	0.238
3	Animal Breeding Farm diet						0.184

[a] Group 1: Rats fed the basal diet from affected areas; group 2: rats fed basal diet + sodium selenite (0.25 ppm); group 3: rats fed Xian diet.
[b] Yeast: Petro-yeast.

TABLE 2. Contents of Total Lipid, FFA, and Selenium in Each Group

| Group | N | Myocardium | | Serum | | |
		Total lipid (mg/100 mg)	μm FFA/100 mg	N	FFA μEq/liter	Se (ppm)
1	23	3.81 ± 0.33[a]	281.16 ± 33.86	10	301.05 ± 36.15	0.083 ± 0.005
2	20	1.59 ± 0.15	182.88 ± 14.77	6	181.83 ± 26.84	0.422 ± 0.012
3	24	1.32 ± 0.11	200.68 ± 17.78	10	171.28 ± 23.78	0.497 ± 0.020

[a] Mean ± SD; N, number of rats.

received three kinds of diet formula, as noted in Table 1. All rats were allowed water and diet *ad libitum* for 2 months.

Free fatty acid, total lipid, and oxidation of butyrate and pyruvate in myocardium were measured in accordance with Nixon (5), Folch (6) and Umbreit's methods (7).

The blood GSHPx, serum thiobarbituric acid reaction value (TBA), and selenium were also determined according to the methods of the Research Group on Keshan Disease of the Chinese Academy of Medical Sciences (8,9).

RESULTS AND DISCUSSION

Total Lipid and FFA

The total lipid and FFA contents in rat myocardium and the serum FFA level are shown in Table 2. The total lipid and FFA contents in the affected diet group (group 1) were significantly higher than those of the other two groups ($P < .01$), but the selenium contents in group 1 were lower than the selenium-supplemented (group 2) and Xian group (group 3), as shown in Table 2.

We believe that the elevation of total lipid and FFA were induced by selenium deficiency, not by deficiency of vitamin E, because both groups were fed the same basal diet.

Oxidation of Sodium Butyrate and Pyruvate in Myocardium Homogenate

The oxidative rates of butyrate and pyruvate in the affected diet group declined, as shown in Table 3.

TABLE 3. Oxygen Uptake by Butyrate and Pyruvate in Rat Heart Homogenate

Group	N	Butyrate O₂ μl/100 mg/40 min	N	Pyruvate O₂ μl/100 mg/40 min
1	6	23.56 ± 1.50[a]	26	82.86 ± 4.03[a]
2	6	68.88 ± 10.45	26	124.54 ± 4.26
3	6	67.05 ± 8.35	26	107.26 ± 5.82

[a] Group 1 to 2, and group 1 to 3 were compared; there were significant differences at $P < .01$; *N*, number of rats.

TABLE 4. GSHPx, TBA, and Selenium Contents in Rat Blood, Serum, and Heart

| Group | N | GSHPx | | TBA | | Heart selenium (ppm) |
		Blood (units)	Myocardium (units)[a]	Serum (nM/ml)	Myocardium (MAD nM/mg Pr)	
1	12	11.80 ± 1.20	11.92 ± 2.02	11.97 ± 0.79	2.10 ± 0.15	0.083 ± 0.005
2	12	39.67 ± 1.93	37.72 ± 2.84	6.34 ± 0.59	1.09 ± 0.13	0.422 ± 0.012
3	12	34.33 ± 2.49	49.99 ± 2.88	7.81 ± 0.59	1.27 ± 0.20	0.497 ± 0.020

[a] One enzyme unit = decrease in μmol GSH/min/mg Pr.

TABLE 5. Results of Four Determinations in Children

	GSHPx (units)	Vitamin E (μg/ml)	TBA (nmol/ml)	FFA (mM/liter)
Affected area	71.08 ± 18.10 (31)[a]	8.56 ± 1.20 (20)	4.2 ± 1.1 (20)	0.71 ± 0.13 (18)
Nonaffected area[b]	90.16 ± 14.60 (31)	8.51 ± 1.90 (20)	2.4 ± 0.5 (20)	0.60 ± 0.12 (17)
t test	P < .001	P > .05	P < .001	P < .001

[a] Number of children in parentheses. Data are expressed as mean ± SD.
[b] Qian County, Shaanxi Province.

GSHPx Activity and TBA Value

Blood and myocardium GSHPx activities, serum, and myocardium TBA values of rats fed the three kinds of diet were measured, and their results are listed in Table 4.

The activities of blood GSHPx, the contents of serum FFA, and vitamin E and TBA values were also measured in children in Yangsou County, Shaanxi Province, which is a Kaschin–Beck disease area accompanying Keshan disease. The results are shown in Table 5.

In Tables 4 and 5, similar results were shown in the animals and children. When the Se status was poor, the activities of GSHPx were often lower, and the FFA and TBA values were higher than in the selenium-supplemented group or nonaffected diet group. These experiments indicated that the animals or children living in the selenium-poor areas could exhibit the disorder of lipid metabolism.

INTRAVENOUS INJECTION OF LARGE-DOSAGE VITAMIN C

An hour before sacrifice, the rats of groups 1 and 2 were injected with vitamin C, 2 g/kg, through their tail veins. The myocardial

TABLE 6. GSHPx Activities and TBA Values in Heart Homogenates after Injection of Vitamin C

Group	GSHPx (units)	TBA (nmol MAD/mg Pr)	P value
1	24.06 ± 2.42 (19)[a]	1.18 ± 0.17 (7)	<.01 (group 1 compared to 1 + C)
1 + vitamin C	51.21 ± 3.02 (11)	0.93 ± 0.13 (11)	
2	73.16 ± 5.08 (16)	1.09 ± 0.13 (12)	<.05 (group 1 compared to 2)
2 + vitamin C	—	0.81 ± 0.07 (10)	<.01 (group 1 compared to 2 + C)

[a] Number of rats in parentheses.

TABLE 7. Correlation Coefficients of Seven Kinds of Biochemical Determinations[a]

	$j=$	X_1 1	X_2 2	X_3 3	X_4 4	X_5 5	X_6 6	X_7 7
$i=1$	(X_1)	1.000	.625	−.400	−.485	−.468	−.585	.780
$i=2$	(X_2)	.625	1.000	−.357	−.497	−.555	−.698	.743
$i=3$	(X_3)	−.400	−.357	.999	.186	.259	.048	−.387
$i=4$	(X_4)	−.485	−.497	.186	1.000	.490	.285	−.484
$i=5$	(X_5)	−.468	−.555	.259	.490	1.000	.445	−.368
$i=6$	(X_6)	−.585	−.698	.048	.285	.445	.999	−.590
$i=7$	(Y)	.780	.734	−.387	−.484	−.368	−.590	.999

[a] X_1, Myocardium GSHPx; X_2, blood GSHPx; X_3, myocardium FFA; X_4, serum FFA; X_5, myocardium TBA; X_6, serum TBA; Y, myocardium selenium content. Progressive regression equation: $Y = -0.91 + 0.0061X_1 + 0.0051X_2 + 0.0037X_5$.

GSHPx activities and TBA value were determined, and results are shown in Table 6.

Obviously, the vitamin C decreased the TBA value and increased the GSHPx activity in the rat heart.

REGRESSION ANALYSIS

Progressive regression analyses of the seven biochemical determinations in the rats were calculated with a DJS-103 electronic computer (10). The correlation coefficients are shown in Table 7.

The results in Table 7 indicated that myocardium selenium content (Y) significantly correlated with myocardium GSHPx (X_1), blood GSHPx (X_2), and myocardium TBA values (X_5) ($F = 26.2745, P < .01$).

Myocardium selenium content (Y) and myocardium GSHPx (X_1) also showed a positive correlation ($r = .780, P < .05$).

According to these experiments and reports concerning selenium deficiency, especially A. T. Diplock's lectures in Beijing in 1981 (11), we suggested a hypothesis concerning the mechanism of the effect of selenium deficiency on myocardium, as shown in Fig. 1 (12). Figure 1 illustrates that if selenium is deficient in soil, it may result in selenium deficiency in the diet, and the activities of GSHPx in the organism are decreased. The antiperoxidation ability is consequently decreased. For this reason, the biomembranes are damaged, particularly the membranes of mitochondria, and the disorder of heart metabolism ensued. It seems possible that if appropriate supplementation with vitamin C and/or selenium is given, it could improve the lipid metabolism and eliminate some deleterious effects.

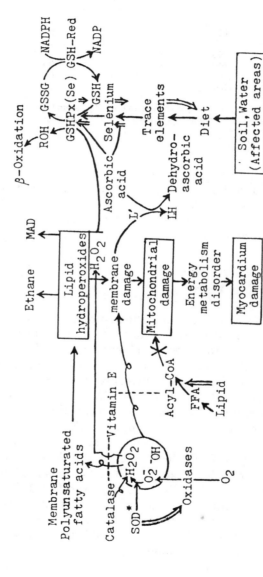

FIG. 1. Scheme of possible mechanisms in the effects of Se deficiency on the myocardium. SOD, Superoxide dismutase (12); L, lipid free radical; O₂, superoxide anion; (---), inhibition; (↓), decrease; (↑), increase; MAD, malondialdehyde, an intermediary production of lipid peroxidation.

REFERENCES

1. Wang, S.-C., and Cai, H.-J.   1961.   Chin. J. Intern. Med. *9,* 346–350.
2. Fischer, W. C.   1977.   J. Nutr. *107,* 1497–1501.
3. Li, G. Y. *et al.,* this volume, chapter 83.
4. Moran, E. T.   1975.   Poult. Sci. *54,* 266–269.
5. Nixon, M., and Chan, H. P. S.   1975.   Anal. Biochem. *97,* 403–409.
6. Folch, J.   1957.   J. Biochem. (Tokyo) *226,* 497–450.
7. Umbreit, W. W.   1972.   Manometric and Biochemical Techniques, 5th Edition, Chapter 10. Burgess Publishing Co., Minneapolis, MN.
8. Keshan Disease Research Group of the Chinese Academy of Medical Sciences 1977.   Compil. Res. Keshan Dis. *1,* 61–66.
9. Wang, G. Y.   1980.   Health Res. *1,* 73–78.
10. Lay, C. M., and Broyles, R. W.   1980.   Statistics in Health Administration, Vol. II, Chapter 8. Aspen Publication.
11. Diplock, A. T.   1981.   Lectures in Beijing: (1) Metabolic and functional defects in selenium deficiency. (2) The role of vitamin E and selenium in the prevention of oxygen-induced tissue damage.
12. Liu, C. B.   1983.   Acta Acad. Med. Xian *4,* 345–351.

83

Effect of a Selenium-Deficient Diet from a Keshan Disease Area on Myocardial Metabolism in the Pig

Guang-Yuan Li
Cong Han
Jian-Guo Yang

It is well known that Keshan disease (KD) is a chronic endemic cardiomyopathy and that the onset of acute-type KD can be prevented with sodium selenite. Why can selenium prevent KD? What role does selenium play in preventing KD? We used weanling piglets fed the low selenium grain from affected areas and a diet supplemented with selenium to study the effect of selenium on the pig's myocardial metabolism.

MATERIALS AND METHODS

In a KD severely affected area, the Shangzhenzi Farm, Huangling County, Shaanxi Province, 50 weanling piglets were fed a local staple grain mix low in selenium which consisted of 86% maize flour, 12% soybean flour, 1% selenium-poor petro-yeast and 1% salt as a basal ration (selenium, 0.021 ppm). In addition, 0.1% cod-liver oil and 10% maize stem powder were added to the ration. The drinking water was from a local well and it was mixed and boiled before use. This low selenium diet was supplemented with 0.25 ppm sodium selenite. The animals were randomly divided into two groups: one group (30 pigs) received the low selenium diet (Se-low group) and the other (20 pigs) received the selenium-supplemented diet (Se, 0.17 ppm).

After a breeding period of 2 months, the pigs were transferred to the city of Xian, slaughtered, and then myocardium homogenates were prepared according to Umbreit's method (1). The activities of succinate dehydrogenase (SDH), cytochrome-c oxidase (CO), and the oxidation of linoleic acid (LA) in the myocardial homogenate were measured by manometric and biochemical techniques (1). Free fatty acids (FFA) in the serum and myocardium were also measured using a colorimetric method (2). Oxygen content and its partial pressure in the artery and vein were determined by an ABL-2 acid and base instrument. The selenium content in the diet and the myocardium were analyzed using the 2, 3-diaminonaphthalene fluorescence method (3).

In the second experiment, in Xian, a batch of weanling pigs that were obtained from the same affected area were fed with the same ration, but drank Xian tap water.

RESULTS AND DISCUSSION

The growth, clinical features, and pathological changes in these experimental piglets have been reported by Dr. Zhu (4). The biochemical results concerning myocardial metabolism in both groups are discussed.

As shown in Table 1, the selenium contents of the myocardium were significantly lower in the Se-low group than in the Se-supplemented group, and the activities of SDH and CO were also low in the Se-low group.

In the second experiment, the selenium contents of the myocardium in the Se-low group were not as low as in the first experiment, but the selenium level was again lower than that in the Se-supplemented group. Serum and myocardium FFA contents in both groups showed a significant difference, as shown in Table 2.

A scattergram was drawn according to the myocardial selenium and

TABLE 1. Activities of Succinate Dehydrogenase and Cytochrome Oxidase in Myocardial Homogenate

Group	N^a	SDH [O_2 uptake (μl)]	CO [O_2 uptake (μl)]	Selenium (ppm)
Se-low	6	62.5 ± 21.0^b	206.4 ± 22.8^c	0.019 ± 0.009
Se-supplemented	16	1384.8 ± 263.4	1263.4 ± 243.8	0.184 ± 0.021

[a] Number of piglets.
[b] Mean $\pm$ SE, oxygen uptake μl/mg N/40 min.
[c] Mean $\pm$ SE, oxygen uptake μl/mg N/30 min.

TABLE 2. Serum and Myocardium FFA Contents of Both Groups

Group	N^a	Myocardium FFA (μM/100 mg)	Serum FFA (μM/liter)	Se (ppm)
Se-low	7	150.12 ± 6.29	304.15 ± 38.00	0.075 ± 0.030
Se-supplemented	8	93.38 ± 13.75[b]	147.70 ± 19.08[b]	0.455 ± 0.060[b]

[a] Number of pigs.
[b] Mean ± SE, significantly different from both groups, $P < .01$.

FFA contents. The correlation between selenium and FFA was negative ($r = .74$), as shown in Fig. 1.

Why were the FFA increased in the serum and myocardium? According to the studies on the pathohistological chemistry of the animals and the biopsy of the KD patient's myocardium in our laboratory, fatty infiltration in the myocardium and liver were often observed under either the electronic or light microscope. The membranes of the mitochondria were also deformed. These changes might be due to selenium deficiency, which caused a decline of the antiperoxidative ability. It has been shown that selenium is an integral part of glutathione peroxidase (GSHPx), which catalyzes the reduction of a large range of lipid hydroperoxides and hydrogen peroxide. If selenium is deficient in the body, GSHPx cannot catalyze the lipid peroxides and the biomembranes, especially those of mitochondria, could be damaged. Thus, FFA, the staple fuel of the myocardium, could not carry on mitochondrial oxidation and FFA was elevated in some tissues. This point of view is further supported by the results shown in Table 3.

The oxidative rates of linoleic acid in the Se-low group were lower

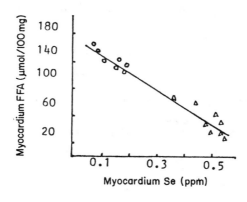

FIG. 1. Correlative scattergram of the heart Se vs FFA. $r -.74$; $P < .05$. (○), Se-low group; (△), Se-supplemented group.

TABLE 3. Oxidation of Linoleic Acid in the Myocardial Homogenate in Both Groups

Group	N^a	Oxygen consumption (μl/40 mg myocardium/min)			
		10 min	20 min	30 min	40 min
Se-low	9	37.96 ± 5.58	85.92 ± 11.09	123.92 ± 14.06	156.83 ± 18.70[b]
Se-supplemented	10	56.02 ± 3.52	120.06 ± 6.19	165.77 ± 7.80	213.88 ± 7.80[b]

[a] Number of pigs.
[b] $P < .01$.

TABLE 4. Comparison of the Arteriovenous Oxygen Difference in Both Groups

Group	N^a	Arteriovenous O_2 difference (O_2 ml%)	Decreased value in Se-low group (%)
Se-low	5	2.59 ± 0.54	49.2
Xian pigs	7	5.06 ± 0.87	

a Number of pigs.

than those in the Se-supplemented group. Thus, it seems possible that FFA were accumulated in the myocardium and serum. According to these experiments, we believe that the oxidative process of FFA is disordered because of selenium deficiency.

Our colleagues of the Department of Pharmacology used five of the experimental animals for the determination of oxygen content of the artery and coronary sinus blood with the ABL_2 acid–base instrument. The arteriovenous oxygen difference was calculated, and the results are shown in Table 4.

From Table 4, the utilization of oxygen in myocardium is seen to be decreased in the Se-low group. The partial pressures of the abdominal arterial and venous oxygen were also determined and similar results obtained. As shown in Table 5, there was no statistical difference between groups, but the arteriovenous PO_2 difference in the Se-low group tended to decrease.

The above experiments indicate that the selenium-deficient diet in KD areas could induce the disorder of myocardial metabolism in pigs evidenced by the decline of respiratory enzyme activity and FFA elevation. We consider that Keshan disease is substantially a kind of mitochondria disease due mainly to selenium deficiency and perhaps other factors.

TABLE 5. Comparison of the Partial Pressure of the Abdominal Arteriovenous Oxygen in Both Groups

Group	N^a	Artery (mm Hg)	Vein (mm Hg)	Arteriovenous PO_2 difference	Decrease
Se-low	7	65.5 ± 16.5	15.7 ± 9.0	49.7 ± 16.9	18%
Se-supplemented	7	76.3 ± 9.3	16.6 ± 9.5	60.5 ± 13.7	

a Number of pigs.

REFERENCES

1. Umbreit, W. W.   1972.   Manometric and Biochemical Techniques, 5th Edition. Burgess Publishing Co., Minneapolis, MN.
2. Nixon, M.   1979.   Anal. Biochem. *97,* 403.
3. Keshan Disease Research Group of the Chinese Academy of Medical Sciences 1977.   Compil. Res. Keshan Dis. *1,* 73–78.
4. Zhu, S.-Y.   1984.   Lesions of liver, skeletal and cardial muscle in piglings fed Se-low grains from Keshan disease areas. Unpublished.

84

Lesions of the Liver and Skeletal and Cardiac Muscles in Young Pigs Fed Low-Selenium Grains from Keshan Disease Areas

Shi-Ying Zhu

Studies on the relation between Se and Keshan disease and on the distributional relationship between Keshan disease and Se deficiency in animals in Shaanxi Province have been reported previously (*1–3*). Satisfactory results were obtained in trials to prevent Keshan disease by oral administration of sodium selenite (*4,5*). The present study presents the effect of low-Se grains from Keshan disease areas on young pigs.

MATERIALS AND METHODS

Two experiments were performed involving 140 weanling pigs (weight, 3–7 kg) obtained from a farm in a Keshan disease-affected area. They were housed separately in pens with brick or flagstone floors and were fed the designated diet and well water *ad libitum* for 10 weeks.

Fifty baby pigs in three groups were used in the first experiment. In group 1, 30 pigs were fed the low-Se grain (Se, 0.021 ppm) from the affected area as a basal diet, which consisted of 86% maize flour, 12% bean flour, 1% yeast, and 1% salt, as well as an adequate amount of maize straw flour. In group 2, 10 pigs were fed the basal diet supplemented with sodium selenite (Se, 0.17 ppm). In group 3, 10 pigs were given a semipurified diet (Se, 0.015 ppm) consisting of 24% petro-yeast,* 70% white potato starch,* 2% rapeseed oil, and mixed minerals and vitamins (5,6).

In the second experiment, 90 pigs were used: 53 pigs were fed the same basal diet as in experiment 1 (group 1), 18 pigs received the basal diet supplemented with sodium selenite (group 2), 10 the basal diet with 10% rapeseed oil for 4 weeks (group 4), and 4 the basal diet with silver sulfate (Ag, 2%) during the last 4 weeks (group 5). The remaining 5 pigs as a control were fed the diet from nonaffected areas with the same composition as the basal diet.

The animals were observed twice a day for clinical signs of the disease during the feeding period and were necropsied immediately after each died. At the end of the feeding period the remaining pigs were killed, and careful examination was made of the Se-deficient target organs, namely, the liver, heart, and skeletal muscle. Multiple blocks of each were collected in 10% neutral-buffered formalin solution, embedded in paraffin, sectioned, and routinely stained with hematoxylin and eosin for histopathological study. Selected sections were studied after special staining using Von Kossa's and Mallory's procedures.

RESULTS

Clinical Observations

The pigs fed the low-Se diet ate poorly and the growth rate fell. The manifestations seen in the diseased pigs several days before death included anorexia, preference to stand apart, reluctance to move, and eventually prostration, loss of coordination and sensitivity to touch, inability to rise, and occasionally diarrhea and foaming at the close proximity of the lips and nose. Deaths occurred as early as the twelfth day after feeding, and most animals (32 of 83 pigs) died from the fourth week to the sixth. The mean survival time was 38 days. In 7 of the 10 pigs fed the semipurified diet, death occurred from the fourth to the sixth week, with similar changes, mean survival being 31 days. As described previously 16 of the 38 pigs died of Se–vitamin E deficiency, mean survival time being 37 days. In 4 pigs fed the basal diet with silver, there was persistent diarrhea characterized by black watery feces. One pig died 25 days later.

*The petro-yeast and white potato starch were obtained with the help of the research group on Keshan disease of the Chinese Academy of Medical Sciences.

Gross Pathologic Changes

Death occurring in pigs of groups 1, 3, and 5 was related to gross pathologic characteristics of Se deficiency. In group 1, necropsy revealed that the surface and section of the liver showed petechial hemorrhage involving red congestion of scattered swollen, necrotic lobules adjacent to pale necrotic lobules and other reddish-brown normal lobules. Such changes were seen in 27 of 32 dead pigs. Accumulation of yellow watery fluid in the pericardial, thoracic, or abdominal cavities was seen in 16 pigs, and edema of the mesocolon sigmoideum in 13 pigs, gastroenteritis in 14 pigs, and marked "fish-flesh" changes of skeletal muscles in 12 pigs were seen. In group 3, hepatic petechial hemorrhages were found in 7 dead pigs, mesenteric edema in 3 pigs, excessive fluid in the pleural, pericardial, or peritoneal cavities in 2 pigs, and gastroenteritis in 4 pigs. In group 5, the same changes were found in 1 pig which died spontaneously. However, no gross pathologic changes were found in all the pigs that were killed. These gross pathologic changes are similar to those observed in previous studies on Se—vitamin E deficiency in pigs (8).

Microscopic Changes

Observations on the livers of all the dead pigs except for group 2 revealed acute hepatic centrolobular necrosis or associated interlobular infiltration by inflammatory cells. In the early stage of hepatocyte damage, the cords of centrovenous zones were separated, hepatocytes were swollen, basophilic granules of the cytoplasm decreased, and eosinophilic granules increased, with prominently stained cell membranes. The nuclei lost their vesicular appearance and became hyperchromatic. In advanced stages, the hepatocytes disintegrated in the centrolobular zones, accompanied by fresh hemorrhage in the central area of necrosis. In most instances, the degree of hemorrhage was proportional to necrotic damage. In the late stages, acute necrosis of the whole lobule occurred, accompanied by parenchymal lysis. The remaining stromal tissue was filled with blood, histocytes, and small numbers of neutrophils. In necrosis of the whole lobule, approximately 95% of the lobule was destroyed, resulting in complete disappearance of liver cells over wide areas. In some instances, infiltration by lymphocytes, mononuclear cells, and eosinophilic leukocytes was observed in the interlobular zones in the pigs that died as well as in the killed pigs. In addition, the phagocytic Kupffer cells contained hemosiderin pigment in the necrotic area and fascicular mineralization in the degenerative hepatocytes near the necrotic area. In addition, mineralization was sometimes seen in other lobules. In group 1, acute necrosis of

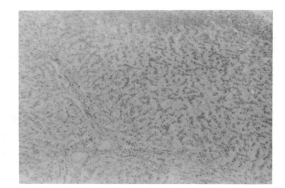

FIG. 1.   Liver from a pig that died after the basal diet was given for 49 days. There was acute extensive necrosis of whole hepatic lobules. From group 1, H&E stain (×400).

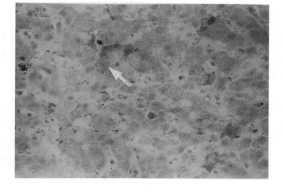

FIG. 2.   Extensive hepatocytic mineralization of a pig that died after receiving the basal diet for 22 days. Acute hepatic centrolobular necrosis was present to various degrees, with dense mineralized cytoplasmic granules (arrow) around necrotic zones. From group 1, Alizarin Red S stain (×400).

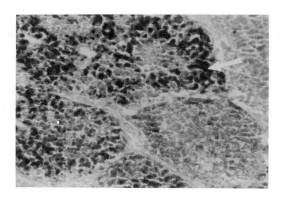

FIG. 3.   Liver of a pig receiving the basal diet with silver. The animal was killed after 23 days. Abundant hemosiderin and silver pigments were accumulated in the phagocytic Kupffer cells (arrow). From group 5, H&E stain (×1600).

hepatic centrolobules (Fig. 1) was detected in 32 of the dead pigs, which was associated with interlobular infiltration in 3 of 32 pigs, and extensive mineralized hepatocytes were filled with dense granules in 12 (Fig. 2), with interinfiltration seen in 5 of 20 killed pigs. In group 3, acute necrosis of centrolobules was found in 7 of the dead pigs and was associated with interlobular infiltration in 1 pig. In group 4, hepatic interlobular infiltration was observed in 4 of 10 killed pigs, accompanied by focal necrosis of hepatic centrolobules in 1 pig. In group 5, 1 dead pig had acute necrosis of hepatic centrolobules and hepatic interlobular infiltration, with both hemosiderin and silver pigment accumulated in the phagocytic Kupffer cells (Fig. 3).

Most cardial and skeletal muscular fibers were subjected to typical hyaline degeneration on early injury. They were swollen, their sarcoplasm was dark red, and cross-striations were obscure and eventually disappeared. The individual eosinophilic fibers were frequently in the normal muscle fibers, while fascicular hyalinized fibers were found in the myocardium. Occasionally, fascicular vacuolar degeneration occurred in the myocardium, myofibers were swollen with a larger vacuole in the sarcoplasm, but their nuclei were frequently intact (Fig. 4). In the advanced stages of the degenerative process, sarcoplasm was broken up into multiple disks and fragments, and eventually necrosis developed.

Necrobiotic changes found in many muscles were classified as a small or extensive focal necrosis: A small necrotic focus involved 4 or 5 fibers seen in a high-power visual field under the light microscope; an extensive focus was larger than the low-power visual field. Necrosis of myocardial and skeletal muscles was detected, by our standards, as follows: In group 1, small focal necrosis of the myocardium was detected in 9 of 32 dead pigs and extensive focus in 1 pig; small and extensive necrotic foci of skeletal muscles were in 4 of 32 pigs. In 20 killed pigs, small myocardial necrotic foci were found in 2 of 20 pigs, small skeletal muscular foci in 4 pigs, and extensive focus in 1 pig, with mineralization of skeletal muscular fibers in 1 pig. In group 3, small focal necrosis of the myocardium was found in 2 of 7 dead pigs, and extensive foci were associated with fascicular mineralization of myocardial fibers in another dead pig (Fig. 5). Small necrotic foci and three extensive foci of skeletal muscles (Fig. 6) were seen with mineralization in 1 pig (Fig. 7). In group 4, small necrotic foci in the myocardium were found in 9 of 10 pigs that were killed (Fig. 8), and two small foci were seen in the skeletal muscles. In group 5, extensive foci in the myocardium were detected in 1 dead pig (Fig. 9), and one small focus in the skeletal muscles, with mineralization in 2 of 3 killed pigs.

No changes were seen in the two control groups in which 28 pigs

FIG. 4. Vacuolar degeneration in the myocardium from a pig receiving the basal diet. The pig died after 51 days. Fascicular vacuolar degeneration occurred in the myocardium, and myofibers were swollen with a larger vacuole in the sarcoplasm. From group 1, H&E stain (×800).

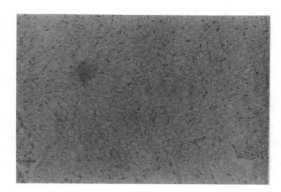

FIG. 5. Extensive myocardial necrosis of a pig fed the semipurified diet. The animal died after 25 days. Extensive myocardial necrosis was detected beneath the left ventricular epicardium, with mineralized myocardial fibers (arrow, ×1600). From group 3, H&E stain (×800).

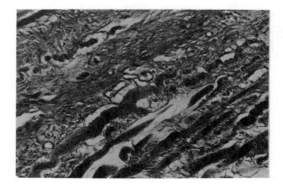

FIG. 6. Skeletal muscles from a pig fed the semipurified diet. The animal died 37 days later. There was extensive dense scar tissue in the skeletal muscles, and a few degenerated muscular fibers in the affected zone. From group 3, Mallory's stain (×400).

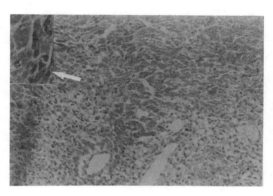

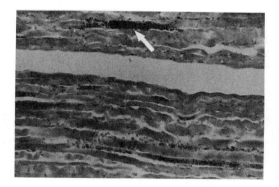

FIG. 7. Skeletal muscles of a pig that died 36 days after receiving the semipurified diet. Severely mineralized muscular fibers were located among the intact muscles. From group 3, Von Kossa's stain ($\times$800).

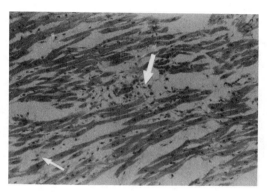

FIG. 8. Extensive myocardial damage in a pig that was killed 47 days after receiving the basal diet with rapeseed oil. There was a small myocardial necrotic focus in the left ventricular wall (arrow), and focal lysis of muscular fibers (arrow). From group 4, H&E stain ($\times$800).

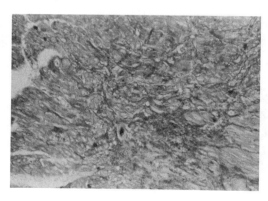

FIG. 9. Cardiac muscles of pig fed the basic diet with silver. The animal was killed 24 days later. Extensive myocardial necrotic fibers in the left ventricular wall were replaced by the scar tissue. From group 5, Mallory's stain ($\times$400).

were fed the basal diet supplemented with sodium selenite and 5 pigs were given grain from Xian, a nonaffected area. The results in pigs fed the Se-deficient KD diet are analogous to the results with the Se-supplemented semipurified diet in preventing the development of damage in pigs (9–11) and also were similar to lesions and mineralization in the myocardium in Keshan disease, which were previously studied (12,13). Our observations suggest that the main factor causing pathologic changes in pig cardiac muscles might be low-Se content of grains in Keshan disease areas.

REFERENCES

1. Xu, G.-L.   1982.   Acta Nutr. Sin. *4* (3), 183.
2. Xu, G.-L.   1981.   Chung-hua Yu Fang I Hsueh Tsa Chih *15* (4), 218.
3. Chen, J.-S.   1979.   Chung-kuo I Hsueh Ko Hsueh Yuan Hsueh Pao *1* (10), 135.
4. Song, H.-B.   1979.   Chung-hua I Hsueh Tsa Chih *57*, 457.
5. Qian, P.-C.   1981.   Chung-hua Yu Fang I Hsueh Tsa Chih *15* (4), 218.
6. Ruth, C.-K.   1974.   Am. J. Vet. Res. *35*, 237.
7. Whanger, P. D.   1977.   J. Nutr. *107* (1), 1288.
8. Van Vleet, J. F.   1976.   Am. J. Vet. Res. *37*, 911.
9. Van Vleet, J. F.   1977.   Am. J. Vet. Res. *38*, 1299.
10. Dvm, Y.-N.   1977.   Am. J. Vet. Res. *38*, 1479.
11. Forbes, K.-M.   1958.   J. Nutr. *65*, 585.
12. Zhu, S.-Y.   1965.   Xi'an I Hsueh Yuan Hsueh Pao, *14*, 35.
13. Li, X.-Z.   1981.   Chung-hua Hsin Hsueh Kuan Ping Tsa Chih *9* (2), 104.

Selenium Contents of Swine Liver in China

Xu-Jiu Zhai   *Yao-Hua Cheng*
Yu-Wen Wang   *Xiao-Fen Yan*
Yun Cai   *Zhi-Ming Qi*
Tai Gao

White muscle disease (WMD) in domestic animals occurs widely in the northeast and other regions of China and has brought serious losses to the animal-raising industry. The first clinical cases of WMD in pigs in Shanghai were reported by Xian-Fa Wu in 1955. Since then, considerable research has been undertaken and great successes were achieved in the prevention and cure of this Se-deficient disease using sodium selenite. The diagnosis of this disease was mainly based on clinical symptoms and pathological findings, but the content of Se in tissues of animals that suffered from the deficiency has never been analyzed until 1979.

As known, Se occurs in all cells and tissues of the animal body, and the liver and kidney usually carry the highest Se concentrations. The Se concentration in these organs may be an important criterion of Se status of the animal.

The purpose of the present work was to determine the Se status in swine in different provinces of China and to achieve an estimate of regional distribution of Se deficiency as well as the degree of deficiency.

MATERIALS AND METHOD

A total of 4547 liver samples obtained from clinically healthy adult pigs slaughtered at 129 abattoirs in 26 provinces of China were ana-

TABLE 1. Mean Concentrations of Se in Swine Liver (mg/kg dry matter)[a]

Abattoirs in provinces	N	$\bar{X}$	SD	L_{95}	L_{99}
Heilongjiang	154	0.15	0.069	0.28	0.32
Tangyuan	33	0.12	0.042	0.20	0.22
Hulan	31	0.21	0.071	0.34	0.39
Mudangiang	50	0.16	0.075	0.31	0.35
Beian	40	0.11	0.027	0.16	0.18
Jilin	94	0.35	0.16	0.67	0.77
Huaide	35	0.44	0.15	0.73	0.83
Shuangliao	31	0.31	0.15	0.61	0.71
Huinan	28	0.29	0.14	0.56	0.65
Liaoning	123	0.70	0.52	1.72	2.64
Bingyuan	23	1.10	0.63	2.33	2.72
Benai	35	0.17	0.078	0.32	0.37
Yingxai	29	1.19	0.20	1.59	1.72
Yixian	36	0.57	0.20	0.93	1.69
Neimeng	259	0.69	0.34	1.35	1.57
Keyouqianqi	35	0.41	0.26	0.93	1.09
Chifeng	45	1.03	0.27	1.57	1.73
Wengniuteqi	34	0.95	0.39	1.72	1.97
Siziwangqi	36	0.47	0.13	0.71	0.79
Guyang	36	0.57	0.28	1.11	1.29
Linhe	40	0.71	0.16	1.02	1.12
Hangjinhouqi	33	0.63	0.18	0.98	1.09
Hebei	145	0.50	0.31	1.11	1.30
Guyuan	36	0.38	0.14	0.66	0.75
Chengde	34	0.32	0.21	0.72	0.85
Zhangjiakou	40	0.39	0.22	0.82	0.96
Dingxian	35	0.94	0.18	1.29	1.40
Tianjin	28	0.89	0.24	1.37	1.52
Wuqing	28	0.89	0.24	1.37	1.52
Shandong	134	0.47	0.21	0.87	1.00
Pingyuan	36	0.37	0.13	0.62	0.70
Yantai	35	0.37	0.10	0.57	0.64
Qingdao	28	0.72	0.24	1.19	1.34
Taian	35	0.46	0.14	0.73	0.82
Shanxi	117	0.68	0.44	1.54	1.81
Datong	31	1.03	0.28	1.58	1.75
Yuci	31	0.69	0.37	1.42	1.65
Daning	36	0.19	0.14	0.46	0.55
Yuncheng	19	0.98	0.23	1.43	1.57
Jiangsu	136	0.57	0.26	1.68	1.24
Donghai	35	0.58	0.19	0.95	1.06
Yancheng	35	0.36	0.19	0.74	0.85
Gaoyou	36	0.56	0.25	1.06	1.21
Yixing	30	0.79	0.22	1.23	1.37
Anhui	102	0.53	0.27	1.09	1.22
Shangbu	31	0.36	0.11	0.58	0.65
Lainan	35	0.54	0.25	1.08	1.17
Shitai	36	0.67	0.32	1.30	1.49
Zhejiang	145	0.64	0.22	1.07	1.20
Chahgxing	35	0.58	0.24	1.05	1.19

(*continued*)

TABLE 1 (*Continued*)

Abattoirs in provinces	N	$\bar{X}$	SD	L_{95}	L_{99}
Yiaxing	35	0.61	0.13	0.87	0.95
Shaoxing	37	0.69	0.17	1.01	1.12
Jinhua	38	0.66	0.29	1.23	1.40
Jiangxi	191	0.60	0.24	1.08	1.21
Nanchang	40	0.62	0.21	1.03	1.16
Fengxin	42	0.62	0.15	0.92	1.01
Hengfeng	31	0.81	0.25	1.29	1.44
Jian	40	0.68	0.18	1.03	1.13
Guangchang	38	0.30	0.10	0.50	0.56
Fujian	102	0.66	0.24	1.12	1.27
Shunchang	27	0.57	0.25	1.06	1.21
Fuan	39	0.63	0.15	0.93	1.03
Zhangping	36	0.76	0.26	1.28	1.44
Henan	171	0.67	0.42	1.50	1.75
Linxian	31	1.43	0.22	1.86	2.00
Lingbao	35	0.32	0.22	0.75	0.88
Zhongmu	36	0.61	0.23	1.07	1.21
Shanggiu	35	0.58	0.19	0.96	1.08
Luoshan	34	0.47	0.12	0.72	0.80
Hubei	295	0.50	0.38	1.25	1.47
Zhushan	36	0.81	0.53	1.84	2.16
Fangxian	28	0.29	0.27	0.82	0.99
Badong	34	1.00	0.39	1.77	2.01
Lichuan	40	0.29	0.21	0.70	0.88
Jingmen	39	0.49	0.28	1.04	1.22
Xinzhou	41	0.53	0.15	0.83	0.92
Huanggang	39	0.27	0.12	0.50	0.57
Yangxin	38	0.33	0.19	0.71	0.83
Hunan	183	0.61	0.28	1.16	1.33
Yongshun	39	0.81	0.30	1.41	1.59
Liuyang	37	0.68	0.19	1.06	1.18
Shaoyang	35	0.43	0.13	0.69	0.78
Xintian	37	0.47	0.20	0.87	1.00
Ningyuan	35	0.64	0.32	1.26	1.46
Guangdong	355	0.94	0.37	1.67	1.89
Huaiji	37	0.89	0.24	1.36	1.50
Xingning	36	0.74	0.20	1.13	1.26
Zijin	34	0.81	0.32	1.44	1.64
Jiexi	35	0.49	0.26	1.01	1.17
Haifeng	35	0.97	0.52	1.98	2.30
Zhaoqing	36	1.12	0.24	1.59	1.74
Enping	41	1.08	0.22	1.52	1.65
Shunde	35	1.04	0.29	1.60	1.77
Wenchang	32	1.38	0.35	2.06	2.28
Changjiang	34	0.88	0.25	1.37	1.53
Guangxi	327	1.00	0.31	1.61	1.79
Nandan	34	0.98	0.43	1.82	2.09
Rongan	42	0.86	0.20	1.26	1.38
Gongcheng	36	0.89	0.28	1.44	1.61
Liujiang	36	1.31	0.21	1.73	1.86
Tianyang	36	0.90	0.36	1.61	1.84
Debao	36	1.02	0.23	1.48	1.62
Wuzhou	40	1.10	0.20	1.49	1.61

TABLE 1 (*Continued*)

Abattoirs in provinces	N	X̄	SD	L₉₅	L₉₉
Lingshan	31	1.13	0.25	1.62	1.77
Qinzhou	36	0.82	0.24	1.29	1.44
Sichuan	355	0.38	0.32	1.00	1.20
Guangyuan	35	0.35	0.34	1.02	1.23
Yunyang	35	0.12	0.10	0.32	0.38
Daxian	34	0.34	0.12	0.58	0.65
Guanxian	34	1.13	0.33	1.78	1.99
Luhuo	37	0.25	0.14	0.53	0.62
Yaan	39	0.43	0.11	0.63	0.70
Neijiang	31	0.31	0.14	0.59	0.68
Jiangjin	34	0.38	0.07	0.52	0.57
Xichang	33	0.24	0.14	0.50	0.59
Dukou	43	0.31	0.10	0.50	0.57
Guizhou	171	1.00	0.37	1.72	1.95
Tongzi	34	0.97	0.23	1.42	1.56
Tongren	34	0.75	0.28	1.30	1.47
Guiding	37	0.95	0.40	1.73	1.98
Longli	36	1.04	0.39	1.81	2.05
Qianxi	30	1.32	0.26	1.82	1.98
Yunnan	220	0.73	0.39	1.49	1.73
Zhaotong	45	0.75	0.36	1.46	1.69
Qujing	36	0.62	0.26	1.13	1.29
Guangnan	36	0.96	0.39	1.72	1.97
Gejiu	35	0.63	0.20	1.01	1.13
Yipinglang	34	0.41	0.27	0.93	1.10
Baoshan	34	1.03	0.41	1.83	2.09
Ningxia	82	1.30	0.26	1.81	1.97
Yinchuan	47	1.15	0.13	1.41	1.50
Pingluo	35	1.49	0.26	2.00	2.17
Gansu	240	0.64	0.47	1.56	1.84
Qingyang	44	0.14	0.047	0.23	0.26
Qingshui	35	0.24	0.13	0.49	0.57
Wudu	35	0.64	0.34	1.30	1.52
Yuzhohg	35	0.39	0.13	0.65	0.73
Wuwei	35	1.36	0.28	1.91	2.08
Zhangye	56	0.97	0.17	1.32	1.42
Qinghai	168	0.51	0.29	1.09	1.25
Huzhu	34	0.53	0.34	1.19	1.39
Xining	52	0.39	0.19	0.76	0.81
Huangzhong	37	0.38	0.20	0.78	0.91
Maqin	9	0.84	0.10	1.04	1.10
Geermi	36	0.71	0.31	1.32	1.52
Xinjiang	107	1.03	0.38	1.78	2.00
Weili	35	1.30	0.17	1.65	1.75
Hetian	35	0.57	0.24	1.04	1.18
Shuie	37	1.21	0.19	1.58	1.70
Shaanxi	143	0.81	1.37	3.50	4.32
Yanan	33	0.17	0.12	0.41	0.48
Huangling	38	0.18	0.051	0.28	0.31
Lantian	43	0.17	0.085	0.33	0.39
Ziyang	29	3.30	1.21	5.27	6.42
The nation:	4548	0.67	0.47	1.59	1.88

[a] *N*, Number of samples; L₉₅, upper limit of 95 in the distribution; L₉₉, upper limit of 99 in the distribution.

lyzed for Se concentration. The liver samples were dried at 105°C and were digested using $HClO_4$-HNO. A catalytic polarographic method was employed for the measurement of Se.

RESULTS

The mean levels of Se in swine liver at different abattoirs and in different provinces are presented in Table 1.

DISCUSSION

Lindberg (1968) reported that the mean Se level in the liver of normal pigs was estimated at 1.82 ± 0.16 ppm (dry matter) and that in pigs that had suffered from WMD 0.20 ± 0.05 ppm. In New Zealand, the Se content in swine liver was determined by S. R. B. Solly (1981) and co-authors. According to their data, the mean level of Se was 0.42 mg/kg (wet matter). Unfortunately, the Se level of swine liver in China has not been reported.

The results in the present work show that mean Se level in swine liver throughout China was 0.67 ± 0.47 mg/kg (dry matter), much lower than those mentioned above; i.e., Se-deficient regions are much more widely distributed in China.

Moreover, the close relationship between the degree of Se deficiency and the morbidity of WMD in animals was found in our observation, e.g., in Heilongjiang Province, where Se content was the lowest in Northeast China and the morbidity rate was estimated the highest in this region. Quite the reverse condition prevailed in Liaoning Province where the Se content was higher, but has not yet reached a normal level, and where the number of animals affected by WMD was relatively less than that in Heilongjiang Province. In regions other than the two provinces in question, this relationship was similar to that indicated above.

The Se content of liver in Ziyang County, Shaanxi Province, was significantly high, with a mean level of 3.30 = 1.21 ppm (dry matter), and serious Se intoxication in animals and human beings has occurred.

Effects on Chicks of Low-Selenium Feeds Produced in Mianning County of Sichuan Province

Yang-Gang Liu   *Giang-Zhu Zhang*
Kang-Ning Wang   *Feng Yang*
Ke-Ming Wu

The fact that Se is an essential trace element to animals has been generally recognized. The influences of Se deficiency on the health of animals and man have attracted worldwide attention. Se deficiency may result in Keshan disease in man (*1*), and white muscle disease in ruminant animals, liver necroses in pigs, and pancreatic dystrophy and exudative diathesis (ED) in chicks (*2*). Mianning County in western Sichuan Province is an area of high incidence of Keshan disease (*3*) where the grains and feeds have a very low Se content (about 0.02 ppm).

The Se requirements of chicks and laying hens were investigated by Combs *et al.* (*4,5*) and Latshaw *et al.* (*6*). The National Research Council (*7*) recommended that 0.1 ppm Se is adequate for chickens; however, the optimum Se level for local fowl is still not clear.

The following experiments were conducted to determine the responses of two breeds of chickens to low-Se feeds produced in Mianning and to different levels of supplementary Se. The response of the chickens to the low-Se feeds supplemented with vitamin E (VE) was also observed.

MATERIALS AND METHODS

Animals and Diets

White Rock (WR) and local Luning (LN) chickens, 326 and 36, respectively, were reared with low-Se diets alone or supplementary Se and/or VE from the age of 3 days. The WR were randomly divided into nine groups and the LN into two groups, each of which was housed in a wire cage for 8 weeks. The basal diet consisted of corn, soybean, wheat, and barley, supplemented with nutrients such as amino acids, minerals, and vitamins. Soybean, wheat, and barley were baked in order to destroy part of their natural VE and increase their palatability. The ingredients of the basal diet and its theoretical nutritional composition are listed in Table 1, while the treatment of every group is listed in Table 2. Selenium content of the basal diet, determined by fluorometric analysis, is 5.1 ppb. Both feeds and water were supplied *ad libitum*.

Indices and Analytical Methods

The birds were individually weighed at weekly intervals. Feed consumption was recorded. At 20, 30, and 42 days of the experiment period, blood glutathione peroxidase (GSHPx) activity was determined according to the method of Hafeman *et al.* (1974) (*7a*) as modified by

TABLE 1. Ingredients and Composition of the Basal Diet

Ingredients (%)	Theoretical composition (%)
Corn, 40.00	ME (kcal/kg), 3.08
Soybean, 32.00	Protein, N $\times$ 6.25, 17.60
Wheat, 14.27	Calcium, 1.10
Barley, 10.00	Inorganic phosphorus, 0.71
Calcium hydrophosphate, 2.68	
Calcium carbonate, 0.50	
DL-Methionine, 0.13	
Lysine, 0.05	
Salt, 0.40	
Trace minerals mixture, 0.04[a]	
Vitamin mixture, 0.01[b]	

[a] Supplied with the following chemicals per kilogram of diet: $CuSO_4 \cdot 5H_2O$, 10 mg; $FeSO_4 \cdot 7H_2O$, 337 mg; $MnSO_4 \cdot H_2O$, 150 mg; $ZnSO_4 \cdot 7H_2O$, 100 mg.

[b] Supplied with the following vitamins per kilogram of diet: vitamin A, 1500 IU; vitamin D_3, 200 IU; vitamin K_3, 1 mg; riboflavin, 3.6 mg; niacin, 27 mg; vitamin B_{12}, 10 μg; calcium pantothenate, 12 mg; biotin, 0.15 mg.

TABLE 2. Experimental Treatments (unit, mg/kg of Diet)

Group	WR chicks									LN chicks	
	1	2	3	4	5	6	7	8	9	10	11
Basal diet + Se (Na$_2$SeO$_3$)	0	0.05	0.10	0.15	0.20	0.30	0	0.05	0.10	0	0.10
Basal diet + vitamin E	0	0	0	0	0	0	20	20	20	0	20

TABLE 3. Influence of Se Level in Diets on Mortality Rates of Chickens

	Group										
	1	2	3	4	5	6	7	8	9	10	11
No. of birds[a]	39	36	41	35	37	31	35	39	33	18	18
No. of deaths	17	2	1	1	1	0	12	0	0	10	1
No. showing degenerated pancreas	17	0	0	0	0	0	12	0	0	10	0

[a] At the end of the 3 and 6 weeks, 6 birds from each WR group and 3 birds from each LN group were sacrificed.

Omaye (8), with H_2O_2 as the substrate. Results are expressed as units of enzyme activity. One enzyme unit (EU) represents a decrease in reduced glutathione concentration of 0.1 log units during a 9-min period at 37°C after subtraction of the blank rate.

After dying from disease, the chickens' body weights and pancreas weights were recorded. At the end of 3 and 6 weeks, 6 birds from each group of WR chick and 3 birds from the LN cage were sacrificed and their tissues, feathers, and blood samples were collected. At the end of the experiment, some chickens were examined.

Data were evaluated by Student's t test, analysis of variance, and the methods of least squares. When significant treatment effects were indicated, the differences between means were determined by Duncan's multiple range test.

RESULTS

Responses of the Chickens to Low-Selenium Diets

The development of the basal diet groups, whether WR or LN chicks, was severely affected by Se deficiency. Mortalities of the groups are shown in Table 3. Totals of 17, 12, and 10 birds died from Se deficiency in groups 1, 7, and 10, respectively, which were reared on low-Se diets. However, the mortality of those groups with supplementary Se was very low, and no significant indication of Se deficiency was observed. A significant number of the chickens in group 7, fed with the basal diet plus 20 ppm VE, died from Se deficiency. This shows that Se deficiency per se in the diets was a basal cause of the disease. The LN chicks suffered from Se deficiency earlier than WR (3 vs 5 weeks), suggesting that LN birds are more susceptible to Se deficiency.

The symptoms of Se deficiency observed are in agreement with Underwood's observations (2). The effects of Se deficiency on the pancreas and gallbladder are listed in Table 4. At the end of the experiment, all

TABLE 4. Effect of Se Deficiency on the Pancreas and Gallbladder

	Se supplement	Se deficient
Pancreatic weight (g)	1.80 ± 0.49	0.29 ± 0.09[a]
Pancreatic weight (mg/100 g body weight)	379 ± 76.3	165 ± 39.6[a]
Gallbladder + bile weight (g)	0.55 ± 0.24	1.42 ± 0.60[a]
Pancreatic fibrosis	No	Yes

[a] Significantly different ($P < .001$).

TABLE 5. Effect of Se Levels in Diet on GSHPx Activity[a]

Se levels (ppm)	WR (EU)	LN (EU)
0	3.86 ± 1.22	2.60
0.05	6.04 ± 0.42	
0.10	8.21 ± 2.67	9.12 ± 2.36[b]
0.15	10.30 ± 1.12	
0.20	12.39 ± 1.78	
0.30	13.23 ± 1.52	

[a] Measured at the end of 6 weeks.
[b] Basal diet + 0.10 ppm Se + 20 mg vitamin E/kg diet.

remaining birds from the basal diet groups had no signs of Se deficiency except that their pancreases showed severe fibrosis.

Activity of GSHPx

The activity of GSHPx in the blood of birds was markedly increased by Se supplementation (Table 5). There was a significant linear relationship between GSHPx activity and Se levels in the diets (r = .9680, P < .01). The results conform to the report of Omaye (8), showing that the enzyme activity is a sensitive index of Se nutritional status.

Growth and Feed Conversion

An Se deficiency has a severe effect on bird growth and development, but Se supplementation can correct it (Table 6). It should be noted that the addition of 0.20 ppm Se is adequate to obtain optimum growth when no vitamin E is supplemented. The growth rate at 8 weeks had the same tendency as at 6 weeks.

Table 7 presents the result of least-squares analysis of two factors. The results indicate that Se supplementation significantly promoted

TABLE 6. Influence of Se Levels on Chick Growth[a]

Se levels (ppm)	Group	WR (g/6 wks)	Group	LN (g/6 wks)
0	1	234 ± 79.9[d]	10	271 ± 67.6[b]
0.05	2	258 ± 70.5[c,d]		
0.10	3	266 ± 50.1[b,c,d]	11	353 ± 91.5[c,e]
0.15	4	296 ± 67.4[b,c]		
0.20	5	302 ± 77.4[b]		
0.30	6	290 ± 50.9[b,c]		

[a] Different letters in a column means significant difference (P < .05).
[e] Basal diet +0.10 ppm Se + 20 mg vitamin E/kg of diet.

TABLE 7. Effect of Se and Vitamin E Levels
on Bird Growth[a]

Se (ppm)	Least-square means [SE(g)]
0	217.7 ± 0.474[b]
0.05	279.1 ± 0.457[c]
0.10	278.9 ± 0.459[c]
Vitamin E	
(ppm)	
0	252.4 0.377 (NS)
20	264.7 0.381 (NS)

[a] Total gain in 6 weeks (g), the different letters show significant differences $(P < .01)$.

the growth of the birds $(P < .01)$, but VE supplementation did not. However, there was a significant interaction between Se and VE $(P < .05)$, and the result is shown in Table 8. The addition of the VE alone did not contribute to weight gain of the birds, but good results were obtained by the addition of both VE and Se.

Feed conversions of all pens of birds are shown in Table 9, which indicates that the feed conversions of both WR and LN are improved by the addition of Se.

DISCUSSION

In China, some areas in the northeast and northwest have been found to lack Se in the soil and grains produced thereon, while in Mianning County of Sichuan Province the grains and feeds have a very low Se content. This study shows that local low-Se feeds remarkably affected birds' health and production (Table 10). More than half of the chickens, including WR and LN, suffered from Se deficiency disease. Many chicks on a poultry farm in Mianning County died from edema under the ventral skin. After injecting Na_2SeO_3, 19 out of 24 birds rapidly recovered from the sickness. When Na_2SeO_3 was added

TABLE 8. Effect of Se and Vitamin E Combinations on Growth[a]

Vitamin E (ppm)	SE (ppm)		
	0	0.05	0.10
0	233.7[e]	257.5[c,d]	266.1[b,c]
20	201.0[e]	300.7[b]	291.6[b,c]

[a] This is the total gain in 6 weeks (g). Different letters mean significant differences $(P < .05)$.

TABLE 9. Effect of Se and Vitamin E Addition on Feed Conversion

Effect	Group										
	1	2	3	4	5	6	7	8	9	10	11
Total weight gain (kg/6 wks)	7.93	8.11	9.24	8.79	9.42	8.00	6.34	10.45	8.67	2.67	4.67
Total feed consumption (kg/6 wks)	243	21.6	21.4	20.5	22.6	21.2	18.9	27.7	23.9	8.6	12.9
Feed/gain	3.06	2.66	2.32	2.34	2.40	2.65	2.98	2.65	2.75	3.22	2.76

TABLE 10. Comparison of Responses in WR and LN Birds to Selenium Deficiency[a]

	WR			LN		
	Mortality (%)[b]	Weight gain (g/6 wks)	GSHPx (EU)	Mortality (%)[b]	Weight gain (g/6 wks)	GSHPx (EU)
Se supplement	0	292[c]	10.80[c]	8	353[c]	9.12[c]
Se deficiency	63	234[d]	3.86[d]	83	271[d]	2.60[d]

[a] Different letters in a column mean a significant difference ($P < .01$).
[b] Twelve killed birds have been subtracted from each WR group and 6 birds from each LN group.

841

to the drinking water of the chickens, no sick birds were found. Supplementing VE alone showed little effect on health and growth of the birds, and Se deficiency is considered the primary cause of the disease.

In the diet of young birds, supplementing 0.05 ppm of Se (as Na_2SeO_3) is enough to prevent the deficiency. But in order to sustain optimum growth rate, 0.20 ppm Se is probably essential when there is no supplementing VE in the diet. However, in the presence of adequate VE, the additional level of Se can be reduced to 0.05–0.10 ppm because of the interaction between Se and VE.

Luning birds produced in Mianning County have a greater hatching body weight and faster growth rate than White Rock ($P < .05$). The symptoms of Se deficiency in LN birds showed earlier than in the White Rock. It seems that the faster the growth of the birds, the earlier and more severe are the symptoms of Se deficiency disease.

REFERENCES

1. Xiang, R.-K. 1982. The Nutritional Physiological and Clinical Sense of Essential Trace Elements, pp. 296–317.
2. Underwood, E. J. 1977. Trace Elements in Human and Animal Nutrition, 4th Edition, pp. 324–326. Academic Press, New York.
3. Chen, X.-S. 1981. Relation of selenium deficiency to the occurrence of Keshan disease. *In* Selenium in Biology and Medicine, J. E. Spallholz, J. L. Martin, and H. E. Ganther (Editors), pp. 171–175. AVI Publishing Co., Westport, CT.
4. Combs, G. F., Jr. 1976. Selenium: Biochemical function and importance in poultry nutrition. Proc. Ga. Nutr. Conf. pp. 2–15.
5. Combs, G. F., Jr. *et al.* (1979). The selenium needs of laying and breeding hens. Poult. Sci. *58*, 871–884.
6. Latshaw, J. D. *et al.* (1977). The selenium requirements of the hen and effects of a deficiency. Poult. Sci. *56*, 1876–1881.
7. National Research Council 1977. Subcommittee of Poultry Nutrition: Nutrient Requirements of Poultry, 7th Edition. National Academy of Sciences, Washington, DC.
7a. Hafeman, D. G. *et al.* (1974). Effect of dietary selenium on erythrocyte and liver glutathione peroxidase in the rat. J. Nutr. *104*, 580–587.
8. Omaye, S. T. 1974. Effect of dietary selenium on glutathione peroxidase in the chick. J. Nutr. *104*, 747–753.

Selenium Deficiency in Horses (1981–1983)

Jiong Zhang   Yin-Jie Den
Zeng-Cheng Chen   Kang-Nan Zou
Su-Mei Liu   Xu-Jiu Zhai

An unknown endemic equine disease has occurred for many years in Qingshui County, Gansu Province. During the years 1981–1983, we examined the disease with regard to diagnosis and treatment. On the basis of evidence from etiological investigations, analysis of feed, drinking water, and blood and hair samples (both from affected and unaffected animals), and based on successful treatment and prevention, we have confirmed that the disease is related to Se deficiency.

EPIDEMIOLOGY

Qingshui County is situated in the southeastern part of Gansu Province, at 105° 45″–106° 30″ east longitude and 34° 34″–59″ north latitude, and 1500 meters above sea level. Topographically, it is a mountainous district of loess plateau, with serious soil erosion and poor vegetation. The annual average rainfall is 517–662 mm, 55% of which falls in July to September.

The disease occurs mainly in the western and southwestern parts of the country, where soil erosion is serious. It is localized and has distinct seasonal characteristics. The highest occurrence is from January to March. From 1978 to 1983, there were 412 cases, of which 152 were fatal. The average yearly incidence was 15-39%, with mortality at 40%. All species of animals as well as humans can be affected.

CLINICAL SIGNS

The disease appears in three types as follows:

Acute form: Affected animals may die suddenly without any premonitory signs, or with sudden onset of dullness, weakness, sweating, and trembling when moving or working. This is followed by severe respiratory distress, recumbency, and death. The temperature is usually normal, but the heart rate is increased up to 120/min and is often irregular, usually with systolic murmurs. The urine is thickened and brown in color. The animal usually dies within 24 hr. Mortality is above 95%.

Subacute form: This form is the most commonly observed, with interference in movement or brown-colored urine evident. There is dullness, weakness, and a staggering gait with frequent stumbling. The heart rate rises to 100–120/min with arrythmia and systolic murmurs. This is usually accompanied by reduplication and interruptions, rapid respiration, and swelling of the masseter and lingual muscles which leads to dysphagia. After 3–4 days, the animal becomes recumbent, lacks strength to chew, and salivates. Food boluses may be thrown up, and drinking water is expelled from the nostrils due to dysphagia. If correct medical attention and proper care are given, the animal may recover gradually. Mortality is between 30 and 45%.

Chronic form: There is depression, anorexia, emaciation, weakness, tachycardia (up to 80–100/min), and diarrhea. The vesicular murmur may be increased.

Laboratory findings included a total red blood cell count of 4–5.4 million/mm^3). The white blood cell count is usually in the normal range, but the granulocytes often shift to the left, and the monocytes are markedly decreased. The erythrocyte sedimentation rate is 60–90 mm in 15 min. The SGOT and SGPT levels are increased in affected animals (Table 1). Urine samples from 30 animals showed positive protein, hemoglobin, and myohemoglobin.

TABLE 1. Effects of Selenium Status on Equine Blood Enzymes

	Enzyme unit levels	
Animal condition	SGOT	SGPT
Affected horses (30)[a]	419	174
Unaffected horses (11)		
With Se supplement	332	28
Without Se supplement	297	21

[a] Number of animals.

PATHOLOGY

Gross changes in necropsy are bilaterally symmetrical lesions, especially in the musculus masseter, triceps brachii, and quadriceps femoris. There are localized streaked or radially striated gray or grayish yellow areas of degeneration. The liver and kidneys are enlarged by degeneration and hemorrhages, and cut surfaces have a red and yellowish brown color. The lungs are edematous and hyperanemic. The heart is in hypertrophy with a softening of the left ventricle and is grayish red in color. There is a gray or grayish yellow degenerated area on the endocardium of the left ventricle and multiple hemorrhages of various sizes in the epicardium. The pericardial sac is filled with light yellowish fluids. The bladder is distended with urine, which is brownish in color.

Histologically, the characteristic lesions are manifested by degeneration, hemorrhage, necrosis, and edema of the liver and kidney, and by degeneration, necrosis, and histocytosis of heart and skeletal muscles.

ETIOLOGICAL STUDIES

Differential diagnoses were made between equine infectious anemia, botulism, leptospirosis babesiases, azoturia, and toxic plant poisoning. None of the evidence from epidemiological investigations, causative organisms, and mycological examinations indicated that the disease was contagious.

A total of 582 samples of feed, water, blood, and tissues of equines from affected and nonaffected areas were analyzed for trace elements, including Cu, Mn, Fe, Zn, Mo, Se, Co, Ni, and for macroelements K, Mg, Ca, and Na. In affected areas the Se content of the samples was found to be corn, vetch (*Vicia sativa*), and winter wheat straw, all below 0.01 ppm; alfalfa hay and wheat bran were below 0.03 ppm; drinking water was below 0.001 ppm; serum and hair of diseased animals was 0.0137 ± 0.01 ppm and 0.085 ± 0.026 ppm, respectively (Tables 2 and 3). Se contents of six species of cereals were below 0.015 ppm. The difference of Se contents in the hair of groups with and without Se supplements was highly significant ($P < .001$; Table 2). The amounts of all other elements were in the normal range.

TREATMENT AND PREVENTION

A total of 109 subacute cases have been treated symptomatically since 1981 and this method was ineffective; 14 cases were treated

TABLE 2. Selenium Levels (ppm) in Equine Tissues[a]

Samples	Group	Mule	Donkey	Horse
Hair	Affected animals	0.082 ± 0.027 (6)	0.072 (2)	0.08 ± 0.03 (3)
	Animals in affected area (without Se supplements)	0.075 ± 0.03 (27)	0.08 ± 0.028 (21)	0.08 ± 0.027 (17)
	Animals in affected area (with Se supplements)	0.148 ± 0.049 (34)	0.123 ± 0.04 (27)	0.178 ± 0.06 (11)
	Healthy animals (unaffected area)	0.322 ± 0.11 (11)	0.24 ± 0.09 (11)	0.246 ± 0.15 (4)
Serum	Affected animals	0.018 ± 0.008 (4)	0.0044 (1)	0.006 (1)
	Animals in affected area (without Se supplements)	0.014 ± 0.005 (10)	0.0158 ± 0.0127 (14)	0.0149 ± 0.004 (12)
	Animals in affected area (with Se supplements)	$(21) < 0.001$ 0.048 ± 0.015 (30)	$(13) < 0.001$ 0.025 ± 0.01 (38)	$(6) < 0.001$ 0.041 ± 0.02 (14)
	Healthy animals (unaffected area)	0.064 ± 0.02 (7)	0.062 ± 0.03 (15)	0.052 ± 0.019 (6)

[a] $\bar{X}$ ± SD; numbers in parentheses are number of specimens analyzed.

TABLE 3. Selenium Content (ppm) in Feeds, Cereals, and Water[a]

Samples	Affected area	Unaffected area
Corn	<0.001 (17)	0.086 (1)
Wheat	<0.015 (11)	0.09 ± 0.0078 (4)
Wheat bran	<0.014 (1)	0.031 (1)
Wheat straw	<0.006 (11)	0.066 ± 0.029 (4)
Vetch	<0.001 (14)	
Peas	<0.001 (5)	0.046 ± 0.031 (4)
Alfalfa hay	<0.03 (9)	0.144 ± 0.13 (4)
Millet	<0.013 (5)	0.041 ± 0.13 (4)
Vetch hay	<0.001 (2)	
Oats	<0.001 (1)	
Water	<0.001 (5)	

[a] $\bar{X}$ ± SD; numbers in parentheses are number of specimens analyzed.

causatively with sodium selenite (Na_2SeO_3) at a dose of 0.2 mg/kg body weight intramuscularly (0.25 mg/kg orally). All treated animals recovered in a week after one or two treatments. The color of urine changed within 3 days and the swelling of the muscles subsided, stiff movements were alleviated, and appetite and vigor were improved. Safe dosage intramuscularly and orally was 0.12 mg/km body weight. If undesirable side effects occurred, chloropromazine hydrochloride was used intramuscularly.

Sodium selenite can also be used for prevention. In an affected area, 97 horses were injected with sodium selenite at a dosage of 0.12–0.15 mg/kg/animal. This procedure was continued once a year for 3 years. In 1983, the number of animals that were injected was increased to 319. All these animals remained healthy, but in a control (untreated) group, 25 out of 437 animals (5.7%) suffered from the disease.

CONCLUSION

This research confirmed that the unknown endemic equine disease in Qingshui County was related to Se deficiency. The epidemiology, clinical signs, and pathological characteristics of the disease were observed and summarized, and effective treatment and prevention methods have been adopted.

The main pathological changes are degeneration, hemorrhage, necrosis, and edema of the liver and kidney as well as degeneration, necrosis, and histocytosis of the heart and skeletal muscles.

The Se contents in the liver, heart, hair, and serum of affected equines were all lower than normal. The differences in the amount of

Se in affected and unaffected animals and in supplemented and control groups were statistically significant. These differences were important in predicting and diagnosing the disease.

The Se deficiency in animals in affected areas was associated with the lower level of Se in feed and water. Other elements investigated were all at normal levels.

Based on the Se level in serum, we can identify subclinical cases. This contributes greatly to the diagnosis of the disease.

Selenium in the Retina and Choroid of Some Animal Species

Dubravka Matešić
Branko Gavrilović
Vesna Kornet

In an attempt to explain the mechanism of the transformation of photon energy into a nerve impulse in the eye, i.e., the way in which the retina is excited, Siren (1964) postulated a hypothesis that, besides rhodopsin, a transducer is involved in the process. Since it was known that the shape of the curve of spectrum sensitivity of a selenium photocell resembles the shape of a curve characteristic of the human eye within the same wavelength range (Ferencz and Urbanek 1935), selenium was chosen as a presumptive transducer. In order to verify the hypothesis, Siren analyzed the amount of selenium in the retina of some animals species and obtained results indicating that animals with well-developed sight have 630–810 ppm selenium (tern, roe deer), while the retina of the guinea pig, which has poor visual acuity, contains a smaller (100 times) amount of selenium. However, there still remains the question whether we are dealing with a correlation between visual sensitivity and the amount of selenium in the retina or one between a ratio of the number of rods to the number of cones in the retina and the amount of selenium.

Although the form in which selenium exists in the eye tissues is unknown, activity of Na_2SeO_3 has been indicated by some experiments. It is possible to induce cataract by the subcutaneous injection of Na_2SeO_3, as Bhuyan *et al.* (1981) showed in their study. Although the biochemical mechanism of the cataractogenesis induced by selenium has not been explained, the authors suppose that some structural and functional changes in protein take place in the lens, resulting from the formation of disulfide and selenotrisulfide. This is obviously a question of too large a quantity of added selenium because cataract has been determined in rats with an advanced lack of selenium (Frost and Lish 1975). However, some positive effects have been obtained by adding Na_2SeO_3; selenium inhibits development of retinal dystrophy induced by monoiodoacetic acid (Suleimanov et al. 1981). The authors have tried to explain this effect by its inhibition of the transformation of monoiodoacetic acid in the vascular membrane.

The described relationship between selenium and some tissues in the eye prompted us to examine the level of this element in the eye tissues of some animal species differentiated by visual acuity and by age.

MATERIALS AND METHODS

Eyes of dogs and cats were obtained from the Department of Pathology, Veterinary Faculty, University of Zagreb, and eyes of cattle and pigs from the slaughterhouse of the Agriculture and Food Processing Organization in Kutjevo. Eyeballs were immediately cut up and lenses, retinas, and choroids separated.

Selenium was determined by the hydride process of atomic absorption spectroscopy (AAS). The apparatus consisted of AAS (Perkin–Elmer), electrodeless discharge lamp for selenium, deuterium background corrector, and a hydride generation sampling system.

RESULTS AND DISCUSSION

Although the levels of selenium that we have determined have been considerably lower than that obtained by Siren (1964), they still support the author's hypothesis. Analyzing the eye tissues of animals with different visual acuity, we have found that the level of contained selenium varies. The highest selenium level has been found in the cat's retina (11.38 ppm), then in dog's choroid (8.96 ppm), while in cattle and pigs the level of selenium in identical tissues has not exceeded 1.58 ppm (Table 1).

The amount of selenium in the lens is fairly characteristic within a species: In cats, the range is from 0.44 to 0.47 ppm, in pigs 0.28 to 0.32 ppm, in cattle 0.18 to 0.30, and in dogs (at the age of 10 years) 0.21 to 0.33 ppm.

Increased selenium content has been noted, both in the lens and in the retina of dogs over 10 years of age. Selenium in the lens closely

TABLE 1. Quantity of Selenium in the Eye Lens, Retina, and Choroid of Some Animal Species (μg Se/g Dry Weight)[a]

Animal	Age	Lens	Retina	Choroid
Cats	3 months	0.47	—	6.31
	1 year	0.44	11.38[b]	6.20
	2 years	0.44	7.08	8.87
Dogs	7 months	0.25	2.34	2.80
	9 months	0.21	0.87	1.12
	1 year	0.22	2.87	—
	3 years	0.21	2.29	3.87
	4 years	0.23	1.80	2.54
	8 years	0.22	2.78	2.52
	9 years	0.33	1.70	8.96
	10 years	0.30	2.32	7.98
	11 years	0.55	4.44	4.06
	15 years	0.75	4.62	4.38
Cattle	3 months	0.23	0.75	0.43
	4 months	0.29	0.88	1.23
	10 months	0.18	0.64	0.55
	>10 years	0.18	—	0.59
	>10 years	0.21	1.31	0.61
	>10 years	0.30	1.08	0.59
Pigs	7 months	—	0.44	0.48
	2 years	—	1.04	1.02
	2 years	0.28	0.80	1.58
	3 years	0.32	0.92	1.06
	3 years	0.32	1.08	0.65

[a] Each sample contained identical tissue of both eyes of the animals.
[b] Represents highest value.

approximated a level of 0.75 ppm, which is 2–3 times the level determined in tissues of younger animals of the same species. One may speculate from these data whether there is a connection between this element and the occurrence of cataract at advanced age. An amount of 4.62 ppm selenium has been found in the retina of a 15-year-old dog, a quantity twice the average value characteristic of this animal species.

Before tissues were analyzed, the sex of each animal had been determined and there was no significant difference regarding this characteristic.

Since we have no data about the amount of selenium in feed consumed by the subject animals, we can only suppose that variations of selenium level in identical tissues could be due to this factor.

REFERENCES

Bhuyan, K. C., Bhuyan, D. K., and Podos, S. M. 1981. Cataract induced by selenium in rat. I. Effect on the lenticular protein and thiols. IRCS Med. Sci.: Libr. Compend. 9, 194.

Ferencz, E., and Urbanek, J. 1935. J. Rev. Opt. *14,* 317 (cited by Siren 1964).

Frost, D. V., and Lish, P. M. 1975. Selenium in biology. Annu. Rev. Pharmacol. *15,* 159.

Siren, M. J. 1964. Is selenium involved in the excitation mechanism of photoreceptors? Sci. Tools *11,* 37–43.

Suleimanov, N. S., Gasanova, S. A., and Gadjiyeva, N. A. 1981. Depressive selenium effect on the development of the experimental retina dystrophy evoked by mono-iodacetic acid injection. Selen. Biol., Mater. Nauchn. Konf., 3rd, 1977.

Experimental Induction of Cardiomyopathy in Young Bovine

S. Kennedy
D. A. Rice
C. H. McMurray

A number of vitamin E and selenium deficiency diseases occur in Northern Ireland; among these are nutritional degenerative myopathy (NDM) in cattle and dietetic microangiopathy in pigs. The feedstuffs available for feeding ruminants are deficient in selenium, ranging from <0.01 to 0.05 (mean 0.023) mg/kg in hay and grass silage. In barley and oats, the range is <0.01–0.04 (mean 0.02) mg/kg. This places Northern Ireland among the low-selenium regions of the world. Because of the low selenium levels, the erythrocyte glutathione peroxidase (GSHPx) levels of cattle fed these deficient diets are also low (mean 21.5 IU/g Hb in a survey of 177 herds) (1).

NDM does not appear to be a problem in these herds, unless the intake of vitamin E (α-tocopherol) is reduced. NDM occurs indoors at the end of winter feeding and more frequently at turnout to grass in the spring. A variety of clinical signs are present, depending on the muscle systems affected. Among these is sudden death resulting from cardiomyopathy.

In order to examine in detail the factors involved, we have developed a diet which is naturally low in selenium and made deficient in α-tocopherol by treatment of barley with sodium hydroxide (2). Polyunsaturated fatty acids can be added to this diet in the form of formaldehyde and casein-coated linseed oil. This, like grass, allows the unsaturated fatty acids, in this case linolenic acid, to bypass the hydrogenation mechanisms in the rumen.

Animals maintained solely on the deficient diet will develop NDM spontaneously. Adding PUFA to the diet will induce or trigger the disease in animals before they develop the disease spontaneously. The time scale, clinical signs, and gross and microscopic pathology are similar to that observed in the natural disease.

CARDIOMYOPATHY

The gross appearance of the cardiomyopathy is shown in Fig. 1. Macroscopically the cardiac lesions appeared as white striae or diffuse pale areas in the left ventricular and interventricular septal myocardium. Microscopic lesions were only rarely observed in the right ventricular and atrial myocardium.

Histopathological examination revealed that changes in the Purkinje fibers (Fig. 2) preceded the appearance of the myocardial lesions. The first evidence of cardiomyopathy was the development of intracytoplasmic lipofuscin droplets in the Purkinje fibers of both bundle branches and all levels of the intramyocardial ramifications of the conduction system. Lipofuscinosis was followed by hyaline degeneration and eventually coagulative necrosis of these fibers. There was extensive proliferation of the peri-Purkinje connective tissue in the region of the necrotic fibers.

Myocardial lesions (Fig. 3) were characterized by diffuse hyaline degeneration and multifocal myocardial necrosis. The repair response was characterized by fibroblastic proliferation resulting in infarctoid foci of fibrosis. The inflammatory reaction was sparse and consisted of focal accumulations of lymphocytes and diffuse myocardial infiltration of lymphocytes, plasmacytes, and macrophages. Fibrinoid necrosis and thrombosis of myocardial blood vessels were frequently observed.

Polyunsaturated fatty acids potentiated the effect of vitamin E and selenium deficiency by accelerating the development and increasing the severity of the cardiomyopathy.

DISCUSSION

This description of the experimental cardiomyopathy in young cattle is similar to what has been observed in naturally occurring cases in cattle. The cardiomyopathy would also appear to have many similarities to Keshan disease—a fatal cardiomyopathy of children and women of child-bearing age in China (*3,4*). The bovine cardiomyopathy

FIG. 1.   Left ventricular cardiomyopathy.

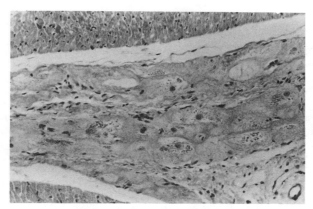

FIG. 2.   Purkinje fiber lipofuscinosis.

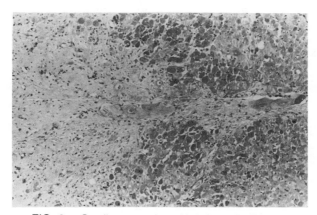

FIG. 3.   Cardiomyopathy with infarctoid fibrosis.

may in fact be a suitable experimental model for the study of Keshan disease.

We know that selenium deficiency by itself will not explain the occurrence of the disease under our conditions in cattle and requires at least a simultaneous deficiency of vitamin E (α-tocopherol). Polyunsaturated fatty acids will make the dual deficiency worse by inducing tissue damage—as shown by elevation in plasma creatine phosphokinase (CK)—before the dual deficiency has progressed far enough to cause damage (5). We conclude therefore that the nutritional intake of at least these three nutrients must be considered simultaneously when examining their etiological or predisposing roles in this disease.

NDM in cattle can be prevented by administration of both vitamin E and selenium alone or in combination before turnout to spring pasture (6).

REFERENCES

1. McMurray, C. H., and Rice, D. A. 1980. Vitamin E and selenium deficiency diseases. Ir. Vet. J. *36*, 57–67.
2. McMurray, C. H., Rice, D. A., and Kennedy, S. 1983. Experimental models for nutritional myopathy. *In* Biology of Vitamin E, Ciba Found. Symp. 101, pp. 201–223. Pitman Books, London.
3. Gu, B.-Q. 1983. Pathology of Keshan disease. Chin. Med. J. *96*, 251–261.
4. Chen, X., Chen, X., Yang, G. Q., Wen, Z., Chen, J., and Ge, K. 1981. Relation of selenium deficiency. *In* Selenium in Biology and Medicine. J. E. Spallholz, J. L. Martin, and H. E. Ganther (Editors), pp. 171–175. AVI Publishing Co., Westport, CT.
5. McMurray, C. H., Rice, D. A., and Kennedy, S. 1983. Nutritional myopathy in cattle from a clinical problem to experimental models for studying selenium, vitamin E, and polyunsaturated interactions. *In* Trace Elements in Animal Production and Veterinary Practice. N. F. Suttle *et al.* (Editors), Occas. Publ. No. 7, pp. 61–73. British Society of Animal Production, Edinburgh, UK.
6. McMurray, C. H., and McEldowney, P. K. 1977. A possible prophylaxis and model for degenerative myopathy in young cattle. Br. Vet. J. *133*, 535–542.

Section VIII

Selenium in Human Disease

Selenium Ecological Chemicogeography and Endemic Keshan Disease and Kaschin—Beck Disease in China

Tan Jian-An   *Li Ri-Bang*
Zheng Da-Xian   *Zhu Zhen-Yuan*
Hou Shao-Fan   *Wang Wu-Yi*
Zhu Wen-Yu

Ecological chemicogeography deals with the distribution and motion of chemical elements, especially life-related elements, in the geographic environment, and their effects on organisms, including health effects on humans.

Living beings and their environment interact with each other, thereby determining the process of evolution. If the ecological balance of energy and material exchange between living beings and the environment is upset, the living beings fail to adapt themselves to the special characteristics and changes of the external environment. Organisms degenerate and even species disappear. Of course, man is not an exception to this case. Physical, chemical, or biological factors in different environments all have influence on humans or other organisms. However, research on the effects of chemical factors in the environment on humans so far is less than that regarding other factors. Advances in analytical techniques and biological sciences in the past two decades have made it possible to push studies on the relationship between environmental chemical factors and human health to new frontiers. We have been studying the relation of some endemics to the chemical geographic environment for more than 10 years. Therefore, a new research field, ecological chemicogeography, has been developed for medical purposes.

At the end of the 1960s, in response to some departments, we started to engage in environmental pathological studies for both endemic diseases, Keshan disease, and Kaschin–Beck disease. The fact that the distribution of both endemic diseases is related to some special characteristic of the geographic environment, especially soil characteristics, was identified. Hence, it was realized that both endemic diseases were mostly concerned with abnormal chemical factors in the geographic environment (1). Multielement analysis, including selenium, in water, soils, and grains in some affected and nonaffected regions was conducted. In 1973, after a sequence of investigations in Shaanxi, Sichuan, Yunnan, Jilin, and Gansu Provinces, lasting some years, it was confirmed that in large areas selenium, zinc, and some other elements in affected regions were lower in the environment than in nonaffected areas. Among all the elements studied, the selenium content showed the strongest differences (2). Similar results were obtained from a successive investigation. At the same time, the molybdenum content in food grains was also found to be low in some affected regions (3). In 1977, our research and sample collection sites expanded to every province of China, except Taiwan Province (Fig. 1), covering the

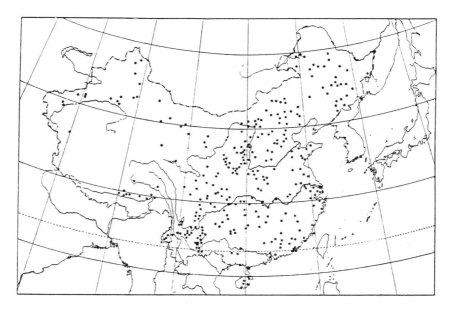

FIG. 1.   Sample collection sites, 1977.

main types of environments in China. The results fully proved that extensive areas in which Keshan disease and Kaschin–Beck disease occur coincide with the low environmental selenium regions (4–6). Some typical affected and nonaffected regions were investigated and analyzed and at the same time, tests for prevention, and treatment of Kaschin–Beck disease were carried out using sodium selenite (7). These research results proved the idea that the diseases were related to selenium deficiency in the ecological environment.

RESEARCH METHODS

Research was carried out by analyzing life-related elements, by studying their similarities and differences in the geographic distribution to explore the connection of elements in the environment with human health, seeking clues for prevention and treatment of disease.
Methods included the following:

1. Geographic comparison and cartography of the disease incidence, environmental types of chemical elements, and their content in the environment and in people;
2. Correlative analysis of various factors in the environment between man and disease;
3. Typical environmental comparisons; and
4. Determining selenium by fluorescence spectrophotometry.

RESULTS

The geographic distribution of Keshan disease, Kaschin–Beck disease, and white muscle disease in China is shown in Figs. 2 and 3. The pathology of both Keshan disease and Kaschin–Beck disease has not yet been fully understood. White muscle disease in animals has been known to be an enzootic disease related to environmental selenium deficiency. Figures 2 and 3 show that the diseases distribute themselves mainly in a distinct wide belt, running from northeast to southwest China. They are located just in the middle transition belt from the southeast coast to the northwest inland areas. The belt is mainly characterized by temperate forests and forest–steppe soils which belong to the brown drab soil system. The distribution and tendencies of the above three diseases are identical. This belt is called the disease belt, while the belts adjacent to it are called the southeast nondisease belt and the northwest nondisease belt, respectively. Though in most

Keshan Disease
Kaschin-Beck Disease
Both Diseases

FIG. 2.   Distribution of Keshan disease and Kaschin–Beck disease.

cases the three diseases occur together in the same region within the disease belt, they also exist independently in some regions.

The geographic distribution of soil selenium content, which is the basis of the selenium ecological cycle, is given in Fig. 4 and Tables 1 and 2. Figure 4 indicates that the distribution of total selenium content is closely related to that of Keshan disease and Kaschin–Beck disease. The soil selenium content in different areas within the non-disease belts is above 0.125 ppm, while that in various areas within the disease belt is mostly below 0.125 ppm. The water-soluble selenium content in soil is similar, being in affected areas mostly below 3 ppb Se, while in nonaffected areas it is likely to be above 3 ppb Se.

The geographic distribution of selenium in food grains is the key link in the ecological selenium food chain. Food grains must be a key link in the ecological cycle. The main distribution of food grain selenium content is given in Fig. 5 and Tables 3 and 4. The low food grain selenium belt, like soil, appears in the middle of the country running from the northeast to the southwest and basically coincides

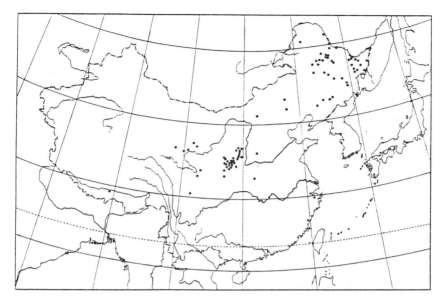

FIG. 3.   Distribution of white muscle disease.

with the distribution of the three endemic diseases. The selenium content of food grains increases regularly toward both the southeast and the northwest. Thus, we have divided China into seven selenium nutritional background belts according to the staple food grain selenium content (8). The grain selenium content within the low selenium belt is mostly less than 0.025 ppm (see Table 3). Crops within the low selenium belt almost always contain lower selenium than those in the other two belts.

TABLE 1.   The Proportional Distribution of Soil Samples for Different Total Se Levels

Se (ppm)	Nonaffected belt (%)[a]	Affected belt (%)[b]
<0.125	19	80
0.125–0.175	22	16
0.175–0.40	4	4
0.40 −1.50	12	0
>1.50	2	0
Average Se ± SD:	0.332 ± 0.357	0.087 ± 0.054

[a] Number of samples: 233.
[b] Number of samples: 68.

TABLE 2. The Proportional Distribution of Soil Samples for Water-Soluble Se Content (%)

Se (ppm)	SE belt, nonaffected (n = 128)	Middle belt, affected (n = 84)	NW belt, nonaffected (n = 127)
<0.003	36	84	7
0.003–0.006	36	10	39
0.006–0.008	14	5	24
0.008–0.011	8	1	13
>0.011	6	0	17
Average Se ± SD:	0.005 ± 0.003	0.003 ± 0.002	0.007 ± 0.003

The geographic distribution of selenium by the content of selenium in the hair of 5- to 15-year-old children has been studied. Hair selenium content reflects not only the selenium metabolism level in the human body, but also the ecological selenium cycle between humans and the environment. The geographic distribution of selenium in hair in China has the same correlation as that in topsoil and food grains (see Fig. 6 and Table 5).

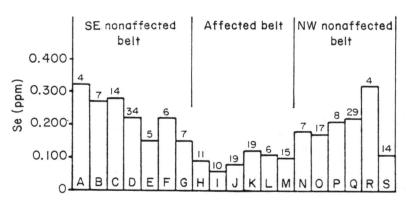

A, Sand desert and steppe soils
B, Solonchakes and saline meadow
 soils
C, Cultivated desert soils
D, Desert soils
E, Chestnut earths and desert-steppe
 soils
F, Chernozems
G, Subalpine meadow soils
H, Black soils
I, Dark brown earths

J, Drab earths and hu lu-tu
K, Purplish soils
L, Red-brown earths and red-drab
 earths
M, Cultivated soil on North China plain
N, Paddy soils on Yangtze plain
O, Paddy soils on southeast coastal
 plain
P, Paddy soils on red and yellow earths
Q, Yellow earths
R, Red earths
S, Lateritic soils

FIG. 4.   Distribution of selenium content in topsoils of main soil types.

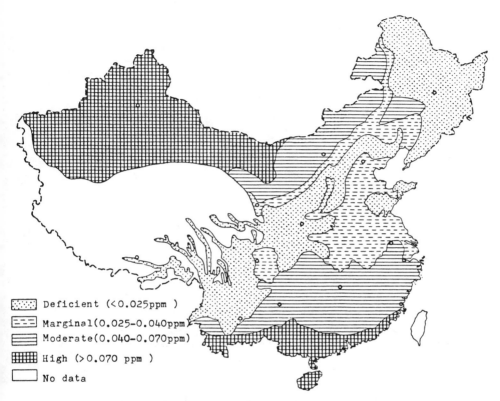

Deficient (<0.025ppm)

Marginal(0.025-0.040ppm)

Moderate(0.040-0.070ppm)

High (>0.070 ppm)

No data

FIG. 5.   Selenium nutritional background in China (grades). Nutritional background value means the weighted means of selenium for three staple food grains from different regions.

The average selenium content in hair for each site investigated within the low selenium belt is mostly below 0.200 ppm. Thus, hair selenium becomes an important indicator for studying the ecological connection of environmental selenium with Keshan disease, Kaschin–Beck disease, and other diseases.

DISCUSSION

Geographic Conjugation, Ecological Connection, and Health Effects of Environmental Selenium

Apparent geographic distribution of disease incidence, soil selenium, grain selenium and hair selenium indicate the following: (1)

TABLE 3. The Proportional Distribution of Food Grain Samples for Se Content (%)

Grain	Se (ppm)	SE belt (nonaffected)	Middle belt (affected)	NW belt (nonaffected)
		$n = 71^a$	$n = 259$	$n = 157$
Wheat	<0.025	18	82	17
	0.025 ± 0.040	36	15	17
	>0.040	47	4	67
	Average Se ± SD	0.052 ± 0.026	0.018 ± 0.010	0.106 ± 0.091
		$n = 16$	$n = 253$	$n = 69$
Maize	<0.025	31	86	20
	0.025–0.040	13	11	26
	>0.040	56	3	54
	Average Se ± SD	0.053 ± 0.028	0.016 ± 0.009	0.049 ± 0.031
		$n = 256$	$n = 101$	$n = 25$
Rice	<0.025	3	77	5
	0.025–0.040	20	16	4
	>0.040	78	7	92
	Average Se ± SD	0.064 ± 0.031	0.020 ± 0.015	0.087 ± 0.046

[a] Number of samples.

There is a selenium ecological connection between environmental elements and the environment and the humans. The selenium metabolic levels in living beings are strongly affected by the environment in which they live. (2) There is a geographic connection between the three endemic diseases (Keshan disease, Kaschin–Beck disease, and white

TABLE 4. Se Content of Grains and Beans in the Affected and Nonaffected Belts

Item	Nonaffected belt		Affected belt	
	Se mean (ppm)	Number of samples	Se mean (ppm)	Number of samples
Wheat	(1) 0.052[a]	71	0.018	259
	(2) 0.106	157		
Maize	(1) 0.053	16	0.016	253
	(2) 0.049	69		
Rice	(1) 0.063	256	0.020	101
	(2) 0.087	25		
Buckwheat	0.039	1	0.017	4
Millet	0.097	11	0.018	51
Broom corn millet	0.061	13	0.014	5
Potato	0.032	4	0.017	26
Sweet potato	0.035	15	0.016	40
Soybean	0.059	31	0.019	37
Pea	0.055	18	0.013	27
Broad bean	0.051	15	0.021	6

[a] (1), SE belt; (2), NW belt.

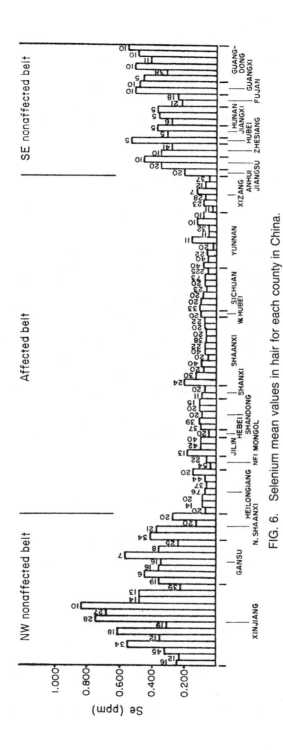

FIG. 6.   Selenium mean values in hair for each county in China.

TABLE 5. The Proportional Distribution of Hair Samples for Se Content (%)

Se (ppm)	SE belt, nonaffected (n = 177)[a]	Middle belt, affected (n = 547)	NW belt, nonaffected (n = 451)
<0.200	5	81	14
0.200–0.250	29	12	30
0.250–0.500	48	7	36
>0.500	19	1	20
Average Se ± SD	0.352 ± 0.124	0.138 ± 0.087	0.329 ± 0.145

[a] Number of sampling sites.

muscle disease of animals) and selenium in the environment and hair. (3) Selenium is an essential element. The geographic connection of selenium with the endemic diseases implies that selenium might present an important health effect. The correlation calculations for selenium concentration in soils, grains, and hair indicate this connection (Table 6). The correlation calculations and regression analysis using reliable data on Kaschin–Beck disease and hair selenium in some typical regions, e.g., Yongshou County, Shanxi Province, show a significant negative correlation (correlation coefficient = −0.478, $\alpha_{.01}$ = 0.319), indicating the correlation of environmental selenium with the diseases (9).

In addition, the nutritional levels of selenium at different developmental and physiological ages have been studied (10). These studies

TABLE 6. Correlation Matrix for Se in Soil Grains and Hair

	Total soil	Water soluble in soil	Rice	Wheat	Maize	Hair
Total soil		0.341[a] (328)[c]	0.208[b] (85)	0.503[a] (113)	0.191 (99)	
Water-soluble in soil			0.276[b] (84)	0.538[a] (118)	0.202[b] (102)	
Rice						0.698[a] (66)
Wheat						0.658[a] (93)
Maize						0.654[a] (81)

[a] Significant at P < .01.
[b] Significant at P < .05.
[c] Numbers in parentheses are sample numbers for the correlation calculation.

show that the ages in which Kaschin–Beck disease is likely to occur are right at the time when hair, blood, or some other tissues have lower selenium levels. In regard to Keshan disease, the correlation of environmental selenium and morbidity has not been conducted owing to lack of available data.

The Influence of the Geographic Environmental Elements on the Formation of Low Selenium

Occurrence of low selenium environments and their regular distribution is not accidental. It is strongly related to the formation, development, and features of the physical landscape. For this reason, the principle of regional differentiation and the theory of geographic zones can explain the causes above. It is clear that the low selenium ecological environments occur mainly in the area around temperate forest and forest–steppe landscapes. The medium and high selenium ecological environments usually exist in the areas of arid and semiarid desert and steppe landscape, as well as hot humid forest landscapes. Selenium excess is generally caused by nonzonal factors, such as land formations and lithologic characteristics which could cause selenium deficiency. For example, the low selenium ecological environment in eastern Sichuan Province results from the purple rock formation which has a low selenium content. It is evident that the formation and distribution of different selenium ecological environments are the consequences of action of both zonal and nonzonal factors. Zonal factors may be the dominant ones in one place, while the nonzonal factors may dominate somewhere else. Generally speaking, the distribution of low selenium ecological environments is mostly controlled by zonal factors. Zonal factors are water, heat, (climate), organism, and soil systems, and their distribution is controlled by planetary factors such as the shape of the earth and the distribution of sea and land. Zonal factors, especially water and heat, play a key role in selenium being leached, its transport, and its enrichment of landscapes, since these factors determine the physical, chemical, and biological properties of ' soil in landscapes which affect the motion and cycling of selenium in the environment. Nonzonal factors, i.e., landform and rocks, still play an important role in the selenium regional distribution in young landscapes. Therefore, on the basis of natural landscapes, selenium ecological landscapes may be divided according to the selenium ecological transport flux or selenium content in organisms. Man and animals are the main indicators which reflect the health situation of local inhabitants or animals in various areas with different environmental selenium levels.

Selenium Threshold Values and Ecological Landscapes

It is very important to determine selenium threshold values, not only for medicine, and agricultural and pastoral production, but also for dividing selenium ecological landscapes. The selenium threshold values for dividing ecological landscapes are determined by geographic comparison and correlation analysis between diseases, selenium concentration in environmental materials, and the human body. Five selenium ecological types include deficient, marginal, medium, high, and excessive landscapes. The selenium threshold values of each type for different environmental materials (soil, grains) and human hair are listed in Table 7. These selenium threshold values are the basis of dividing selenium ecological landscapes. The selenium ecological landscapes in China are divided by the principle mentioned above and are shown in Fig. 7. This kind of map is beneficial to exploring the relation of the low selenium environment with the diseases concerned, researching causes of the formation of low selenium environments, and predicting the problems of health in relation to environmental selenium.

Low Selenium Landscape Zones of the World

Considering the physical geographic analysis principle, it is not difficult to comprehend that there are two low selenium zones in the world. One is located in the Northern Hemisphere, around the areas of the temperate forest and forest-steppe in North America and Eurasia and has an annual precipitation of around 450–1100 mm. Another is located in the Southern Hemisphere, because the continents in this hemisphere, Africa, Australia, and South America, are not only smaller than those in the Northern Hemisphere, but also their lands are situated at low latitudes with humid-hot climates. Low selenium zones distribute themselves mainly in the southern ends of these three continents. In western and southern Australia, low selenium environments and white muscle disease appear in coastal zones with more than 500 mm annual precipitation (*11,12*). Previous reports concerned with white muscle disease and the selenium content in the environment and in living beings also provide an inference regarding the two low selenium zones mentioned above. For example, first, the countries in which white muscle disease occurs are almost always situated in the two low selenium zones, such as the United States, Canada, Italy, Britain, France, Germany, Norway, Finland, Switzerland, Greece, Turkey, Hungary, USSR, Japan, and China in the Northern Hemisphere, and New Zealand, Argentina, Australia, and South Africa in

TABLE 7. Threshold Values (ppm) for Se Dividing Ecological Landscape

Se	Total in topsoil	Water-soluble in topsoil	Food grains	Hair (children)	Effect
Deficient	<0.125	<0.003	<0.025	<0.200	Se-responsive diseases
Marginal	0.125–0.175	0.003–0.006	0.025–0.04	0.200–0.250	Potential Se deficiency
Moderate	0.175–0.40	0.006–0.008	0.040–0.070	0.250–0.500	
High	0.40+	0.008+	0.070+	0.500+	
Excessive	≥3.0	≥0.02	≥1.0	≥3.0	Se poisoning

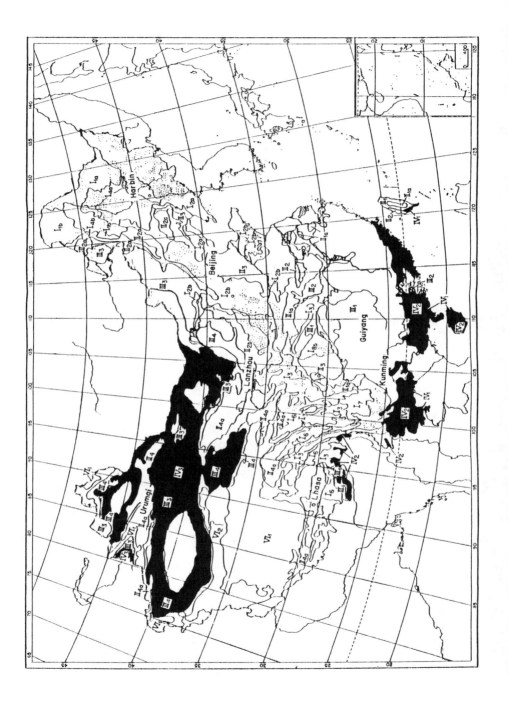

the Southern Hemisphere. Second, Gissel-Nielsen *et al.* reported that the selenium content in maize and wheat in some countries is approximately the same as the selenium levels in the corresponding grains within the low selenium zones in China. Third, Masironi (*13*) and Schrauzer (*14*) reported that the selenium content in blood and some tissues has the same trend as that mentioned above. However, why are there only animal selenium responsive diseases and no Keshan disease or Kaschin–Beck disease in America, Europe, and Australia? There are three reasons: (1) These areas are highly urbanized and people

---

FIG. 7.   The types of selenium ecological landscapes.

I. Se-deficient ecological landscapes
  1. Temperate forest soil, Se-deficient landscapes
    a. Dark brown earths
    b. Brown taiga soils
  2. Warm temperate forest soil, Se-deficient landscapes
    a. Brown earths
    b. Drab soils
  3. Vertical zone temperate forest soil, Se-deficient landscapes
  4. Temperate forest-steppe soil, Se-deficient landscapes
    a. Black soils
    b. Gray-black soils
    c. Meadow chernozems
  5. Warm temperate forest-steppe soil, Se-deficient landscapes (Clayization dark loess soils)
  6. Subalpine meadow steppe soil, Se-deficient landscapes
  7. West section of north subtropic, forest soil, Se-deficient landscapes (drab-red soil)
  8. Purplish soil, Se-deficient landscapes
II. Marginal level Se ecological landscapes
  1. East section of north subtropic, forest soil, Marginal level Se landscapes
    a. Yellow-brown earths
    b. Yellow-drab soils
  2. Temperate (warm) steppe soil, marginal level Se landscapes
    a. Chernozems
    b. Dark loess soils
    c. Chestnut earths
  3. Temperate (warm) alluvial plain soil, marginal level Se landscapes
  4. Alpine meadow soil, marginal level se landscapes
III. Moderate level Se ecological landscapes
  1. Middle subtropic, moderate level Se landscapes
  2. Subtropic alluvial plain, moderate level Se landscapes
  3. Temperate (warm) dry-steppe soil, moderate level Se landscapes
  4. Temperate (warm) desert-steppe soil, moderate level Se landscapes
IV. High level Se ecological landscapes
  1. Tropic forest soil, high level Se landscapes
  2. South tropic forest, high level Se landscapes
  3. Temperate (warm) desert soil, high level Se landscapes
V. Se—excessive ecological landscapes
VI. Se—unknown landscapes
  1. Alpine steppe soil landscapes
  2. Alpine cold desert landscapes

there have more chance to move from place to place; (2) the populations rely upon foods derived from a variety of regions. In contrast,
Chinese rural residents depend heavily upon grains that are restricted
to local products; (3) the composition of these populations' diets is
different from that of Chinese rural peoples' diets. All these examples
enable the people who live in the low selenium environments in America, Europe, Australia, and so on to decrease the effect of local low
selenium conditions. In spite of all these facts, the selenium concentration in human blood in these areas is still lower than that in tropic and
subtropic areas, and the morbidity from cardiovascular diseases in
these areas is obviously higher than that in tropic and subtropic areas
(*13*). Epidemiological investigation of cardiovascular diseases has also
proved that the morbidity from cardiovascular disease in some temperate urban areas of north China is higher than that in some tropic and
subtropic urban areas of southern China (*15*). Hence, it is worthwhile
to further study whether the present nutritional selenium levels in
some areas in America and Europe are adequate.

Causes of Keshan Disease and Kaschin–Beck Disease and Low Selenium Ecological Environments

Kaschin–Beck disease and Keshan disease were first recognized in
1844 and 1935, respectively. Since then, 150 and near 50 years have
passed. Many different hypotheses about the diseases have been suggested, and there is one common point in all of them: Scientists, in
spite of their different viewpoints, mostly realize that both of these
diseases are endemic, which is related to the local natural environment. Any kind of hypothesis about causes of endemic diseases should
be confirmed by external environmental investigations. Our studies
were aimed first at providing the ecological chemicogeographic background to search for causes of these two endemic diseases, and second,
to suggest countermeasures of regional prevention and treatment for
controlling and eliminating them. So far, great progress has been
made on understanding causes. The extensive geographic and ecological links between environmental selenium, human selenium, Keshan
disease, Kaschin–Beck disease, and white muscle disease have been
confirmed to a great extent in the whole country. As a result, we
consider that selenium deficiency in the environment is an important
factor for the occurrence of these two endemic diseases. We have also
noticed that although Keshan disease and Kaschin–Beck disease generally appear only in low selenium environments, in some low selenium areas either Kaschin–Beck disease or Keshan disease occurs
alone. In some other areas, neither occurs. As a matter of fact, the Se

responsive symptoms in animals in low selenium environments are similar to those above, i.e., effects of Se deficiency on different animals may be different, so different animal species have different selenium responsive symptoms. It must be considered that the low selenium ecological environment is the basic factor causing these diseases. But there are some other environmental factors which must be taken into account, especially in the case of Kaschin–Beck disease. Physiological and biochemical functions of an element are always affected by some other elements, either antagonistically or synergistically. Besides selenium in the environment, there might be some other factors acting together in the occurrence of these endemic diseases. Several combinations of factors with Se are shown in Fig. 8. This will provide an explanation of why in some low selenium areas both endemic diseases occur, while in some other areas only one disease or even none occurs. There is still much internal and external environmental work to do in order to achieve the final remedy for both endemic diseases. First, it is necessary to know the biological function of selenium and its related factors at the level of molecular biology. Second, combination models of selenium and related factors in different ecological environments need to be understood for their health effects. Therefore, a multidisciplinary approach is required for the final answer to the problems. To date, encouraging advances in the exploration of the relation of selenium to these endemic diseases would not have been made without a multidisciplinary approach to research.

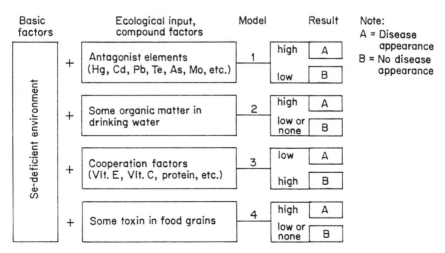

FIG. 8.   Combination models of selenium deficiency with compound factors and Keshan disease or Kaschin–Beck disease.

REFERENCES

1. The Group of Endemic Diseases and Environment, Institute of Geography, Academia Sinica 1973. Research reports. Unpublished. 1974. 1976.
2. The Group of Endemic Diseases and Environment, Institute of Geography, Academia Sinica 1974. Research reports. Unpublished.
3. The Group of Endemic Diseases and Environment, Institute of Geography, Academia Sinica 1976. Research reports. Unpublished.
4. The Group of Endemic Diseases and Environment, Institute of Geography, Academia Sinica 1979. The Keshan disease in China: A study of the geographic epidemiology. Acta Geogr. Sin. *34* (2).
5. The Group of Endemic Diseases and Environment, Institute of Geography, Academia Sinica 1981. Relationship between the distribution of Keshan disease and selenium content of food grains as a factor of chemical geographic environment. Acta Geogr. Sin. *36* (4).
6. The Group of Endemic Diseases and Environment, Institute of Geography, Academia Sinica 1982. Geographic distribution of selenium content in human hair in Keshan disease and nondisease zones in China. Acta Geogr. Sin. *37* (2).
7. The Group of Endemic Diseases and Environment, Institute of Geography, Academia Sinica, and Henan Province Antiepidemic Station 1982. Research report. Unpublished.
8. The Group of Endemic Disease and Environment, Institute of Geography, Academia Sinica 1982. The relation of Keshan disease to the natural environment and the background of selenium nutrition. Acta Nutr. *4* (3).
9. The Reports of Scientific Investigation 1984. Kaschin–Beck Disease in Yongshou County, p. 55. People's Medical Publishing House.
10. Hou, S.-F. 1984. The relationship between the selenium dynamics in the course of human body growth and the Kaschin–Beck disease epidemiology. Acta Geogr. Sin. *39* (1).
11. Steele, P. 1980. Aust. Vet. J. *56*, 529.
12. Reuter, D. J. 1973. Selenium in soil and plants: A review in relation to selenium deficiency in South Australia. Agric. Rec. *2*, July.
13. Masironi, R. 1976. Selenium and Cardiovascular Diseases, Preliminary Results of the WHO/IAEA Joint Research Programme, Proceedings of a Symposium on Selenium-Tellurium in the environment. Selenium-Tellurium Development Association Inc., Darien, CT.
14. Schrauzer, G. N. 1976. Anticarcinogenic Action of an Essential Trace Element, Proceedings of a Symposium on Selenium-Tellurium in the Environment. Selenium-Tellurium Development Association Inc., Darien, CT.
15. Wu, Y. 1979. Unpublished.

Selenium and Keshan Disease in Sichuan Province, China

Cheng Yun-Yu

Sichuan Province is one of the most highly endemic provinces with Keshan disease in China. Currently, the incidence rates of counties such as Mianning and Xichang are still the highest in the country. Since 1974, selenium intervention by oral administration and fortified salt as well as investigations on the possible relationships between epidemiological characteristics of Keshan disease and the selenium status of local inhabitants and the environment have been carried out in Sichuan Province. The results are summarized here.

THE RELATIONSHIP BETWEEN THE EPIDEMIOLOGICAL CHARACTERISTICS OF KESHAN DISEASE, THE SELENIUM STATUS OF THE LOCAL INHABITANTS, AND THE SELENIUM CONTENT OF GRAINS

Selenium Content of Grains in Endemic and Nonendemic Areas

The inhabitants of endemic areas live mainly on grains. Therefore, five varieties of grains were collected from 10 highly endemic, 10 low endemic, and 5 nonendemic counties and analyzed for their selenium content by the fluorometric method (*1*). The results are given in Table 1.

The selenium content of rice, wheat, and corn, the three most important grains in the local diet, was significantly lower in endemic coun-

TABLE 1. Se Content of Grains in Endemic and Nonendemic Areas ($\bar{X} \pm$ SE; ppm)

Areas	Rice	Wheat	Corn	Potato	Soybean
Highly endemic	0.0104 ± 0.005 (27)[a]	0.0119 ± 0.006 (21)	0.0112 ± 0.005 (21)	0.0045 ± 0.0045 (4)	0.0254 ± 0.011 (19)
Low endemic	0.0183 ± 0.009 (61)	0.0160 ± 0.006 (40)	0.0155 ± 0.004 (55)	0.0132 ± 0.004 (15)	0.0301 ± 0.004 (39)
Nonendemic	0.0450 ± 0.008 (5)	0.0448 ± 0.01 (6)	0.0451 ± 0.02 (5)	—	0.0449 ± 0.009 (5)

[a] Numbers in parentheses are number of samples.

878

ties than those of the nonendemic counties, the latter being more than 0.040 ppm Se, while the former were less than 0.02 ppm Se. Among the endemic counties, the grain selenium content of the highly endemic counties was lower than that of the low endemic counties. In endemic areas, the selenium content of soybeans was higher than that of the rice, wheat, corn, or potato. However, people in endemic areas live mainly on rice, wheat, and corn, and soybeans make up only 0–3% of the diet of the local inhabitants. In Liangshan, a highly endemic area, the potato consumption of the Yi minorities forms about half of their annual food consumption. However, the average selenium content of 19 potato samples collected was less than 0.014 ppm Se. It is believed that constantly relying on the above low selenium foods in the endemic areas will inevitably produce a great impact on the selenium level of the local inhabitants.

Hair Selenium Content of Inhabitants in Endemic and Nonendemic Areas

Hair selenium content was used as an indicator for human selenium status. Hair samples were collected from 3 to 6-year-old male preschool children in endemic and nonendemic areas. According to the morbidity, the endemic areas under investigation were divided into two categories, i.e., highly endemic areas with incidence rates above 0.1%, which are also referred to as active endemic areas, and low endemic areas with incidence rates below 0.05%, which are referred to as nonactive endemic areas. The nonendemic areas were also further divided into areas far from and areas close to endemic areas.

The results showed that the average hair selenium content in highly endemic sites was less than 0.12 ppm, those in the nonendemic sites far from endemic areas exceeded 0.2 ppm, and those in the low endemic and nonendemic sites close to endemic areas fell into the 0.12–0.18 ppm Se range.

Hair Selenium Content of Susceptible and Nonsusceptible Children

Children from farmers' families are most prone to this disease. They make up 99% of the total Keshan disease cases. Hair samples from children from farmers' families and nonfarmers' families (commune staffs, factory workers, doctors, teachers, etc.) in the same endemic area were collected, and their hair selenium content is summarized in Table 2.

The results in Table 2 show that the hair selenium levels of pre-

TABLE 2. Se Content of Children's Hair from Farmers' and Nonfarmers' Families

Site	No. of samples	Children	Hair Se content ($\bar{X} \pm$ SE,ppm)	P (farmers' vs nonfarmers')
Active (1)	78	Nonfarmer	0.01455 ± 0.004	<0.01
	59	Farmer	0.0647 ± 0.003	
Active (2)	10	Nonfarmer	0.219 ± 0.05	<0.01
	10	Farmer	0.064 ± 0.01	
Active (3)	21	Nonfarmer	0.127 ± 0.006	<0.01
	25	Farmer	0.072 ± 0.012	
Nonactive	20	Nonfarmer	0.1904 ± 0.008	<0.01
	20	Farmer	0.1362 ± 0.008	

school children from farmers' families were significantly lower than those from nonfarmers' families. The hair selenium content of primary school pupils was the same. This is probably due to the difference in the composition of their family diets and the source of food. Children from farmers' families live on rather monotonous diets and the foods were mainly produced and prepared locally. Therefore, the selenium content of their foods is greatly affected by the selenium content of local soil and water. On the other hand, the diets of children from nonfarmers' families are more diversified and their foods were produced in different areas. As a result, their hair selenium status is less affected by the local soil and water selenium content.

In the same endemic area, hair selenium of adult farmers was significantly higher than that of their children, when compared to those of the preschool children (Table 3). Although food sources are the same, the difference in hair selenium might be due to the monotonous grain consumption and the possibly lower selenium availability to children. This difference is consistent with the nonsusceptibility of adults in endemic areas.

TABLE 3. Hair Selenium Content of Children and Adults in Farmers' Families

	Mianning County		Xide County	
Group	No. of samples	Hair Se content ($\bar{X} \pm$ SE; ppm)	No. of samples	Hair Se content ($\bar{X} \pm$ SE; ppm)
Preschool children	15	0.0565 ± 0.005	32	0.0495 ± 0.003
Primary school children	11	0.0774 ± 0.007	32	0.0669 ± 0.004
	11	0.0955 ± 0.015	32	0.0680 ± 0.004
Adult	$F = 4.24$	$P < .05$	$F = 8.06$	$P < .01$

TABLE 4. Hair Se Content of Patients and Healthy Children in Xide County

Group	No. of samples	Hair Se content ($X \pm$ SE; ppm)	
Healthy children	20	0.0510 ± 0.004	$\left.\right\}$ $P > .05$
Patients	20	0.0497 ± 0.003	

In order to compare the selenium status of Keshan disease patients and healthy children, hair samples from patients and healthy children of the same age were collected and analyzed for selenium content. The results in Table 4 show that there were no significant differences in their hair selenium content.

Seasonal Prevalence and Variation of Hair Selenium

Prevalent seasonal variation is one of the unique epidemiological characteristics of Keshan disease. Although it can occur at any time of the year in Sichuan Province, the peak season in most endemic areas is summer. It was reported that the incidence between April and September makes up 85–98% of the total incidence in a year, while that of June, July, and August accounts for 45–73% of the total. Three counties in the eastern and southwestern part of Sichuan Province were selected for the collection of hair samples from the same male primary school children in different seasons of the year.

Results in Table 5 show that there were no significant differences in hair selenium content between different seasons.

EFFECT OF SODIUM SELENITE IN THE PREVENTION OF KESHAN DISEASE

Field studies on the effects of sodium selenite in the prevention of Keshan disease in highly endemic areas of Sichuan Province were started in 1974. Since then, approximately 620,000 people in 24 counties have been administered sodium selenite. Remarkable effects on the lowering of incidence rates have been observed. The compound has been proved to be effective and practical in the control of the disease and in reducing fatalities.

Effects of Sodium Selenite Tablets

The effects of sodium selenite tablets administered orally in four of the most highly endemic communes from 1974 to 1976 have been

TABLE 5. Seasonal Prevalence of Keshan Disease and Variations of Hair Se Content

	Shizhu County		Ningnan County		Dokou County	
	No. of samples	Hair Se content ($\bar{X} \pm$ SE; ppm)	No. of samples	Hair Se content ($\bar{X} \pm$ SE; ppm)	No. of samples	Hair Se content ($\bar{X} \pm$ SE; ppm)
Prevalent season	29	0.1750 ± 0.0067	20	0.1649 ± 0.01	24	0.1413 ± 0.0077
	28	0.1481 ± 0.005	20	0.1358 ± 0.0074	24	0.1571 ± 0.0077
Nonprevalent season	28	0.1559 ± 0.0057	20	0.1631 ± 0.01	25	0.1420 ± 0.008
	26	0.1584 ± 0.0071	20	0.1419 ± 0.0085	20	0.1509 ± 0.0066

reported (2). Since then, selenium administration has been continued until 1983. When comparisons were made, the annual incidence rate of selenium-treated children with that of the control children of the same age (≤10 years old) in the nontreated neighboring communes (Table 6) was significantly lower in each succeeding year. However, before the selenium intervention (1972–1973), the incidence rates of the four treated communes were not significantly different from those of their nontreated neighbors.

This observation is also true for the whole of Mianning County (Table 7). In 1979 and again in 1982, the number of Se-treated children has been reduced because some of the communes failed to meet requirements of the working protocol.

At the same time, selenium intervention was carried out in counties within the Liangshan Prefecture other than in Mianning County. Table 8 shows the results observed from four highly endemic counties. The children (≤10 years old) in the most highly endemic communes were given sodium selenite tablets and their low endemic communes remained untreated.

It is obvious that in each county, the incidence rate of the treated population was significantly lower than that of the untreated population.

Effects of Selenium-Fortified Salt

In order to study the effects and feasibility of using sodium selenite-fortified table salt in the prevention of Keshan disease, Mianshan and Shengou communes of Xide County were supplied with sodium selenite-fortified salt (15 ppm) and their neighboring communes, Lake and Guangming, served as controls with ordinary table salt. Supplementation began in 1977.

It is shown in Table 9 that before the use of selenized salt, the average annual incidence rate of Keshan disease total within the population from 1974 to 1976 was 3.19 and 1.11 per 1000 in the treated and untreated populations, respectively. After using selenized salt, between 1977 and 1983, 11 subjects in the treated population suffered from Keshan disease. The average annual incidence was 0.195 per 1000. In the untreated population, there were 76 cases, with the average annual incidence rate being 0.86 per 1000 ($P < .01$). As for the 11 cases found in the treated population, most of them occurred in the first 3 months after the supplementation program was begun. These people might not have taken enough selenized salt. Among them, the hair selenium content of four of the cases has been analyzed, and the content in three was of the same level as the untreated controls; the

TABLE 6. Incidence Rate of Se-Treated and Nontreated Children in Mianning County

Year	Treated			Nontreated			Treated vs nontreated
	Subjects	Cases	per 1000	Subjects	Cases	per 1000	
Before intervention							
1972–1973	24,225	172	7.10	3580	29	8.10	$P > .05$
After intervention							
1976	12,578	4	0.32	1833	33	18.00	$< .01$
1977	12,747	1	0.08	1855	10	5.39	$< .01$
1978	12,465	1	0.08	1878	31	16.51	$< .01$
1979	11,146	3	0.27	1901	22	11.57	$< .01$
1980	10,624	1	0.09	1915	18	9.40	$< .01$
1981	10,282	0	0	2013	16	7.95	$< .01$
1982	9,801	0	0	2063	12	5.81	$< .01$
1983	8,730	1	0.11	2093	4	1.91	$< .01$
Total:	88,373	11	0.12	15,553	146	9.39	$< .01$

TABLE 7. Incidence Rate of Se-Treated and Nontreated Children in Mianning County

Year	Treated			Nontreated			Treated vs nontreated
	Subjects	Cases	per 1000	Subjects	Cases	per 1000	
1977	32,194	1	0.03	22,525	83	3.6	$P < .01$
1978	29,332	8	0.27	22,315	104	4.28	$< .01$
1979	22,582	22	0.97	24,616	126	5.12	$< .01$
1980	25,364	17	0.67	25,051	99	3.25	$< .01$
1981	24,991	6	0.24	25,616	77	3.01	$< .01$
1982	17,027	5	0.29	25,954	55	2.12	$< .01$
Total:	151,490	59	0.39	146,071	544	3.72	$< .01$

TABLE 8. Effects of Se in Prevention of Keshan Disease in Four Counties of Liangshen Prefecture

County and years	Treated			Nontreated			Treated vs nontreated
	Subjects	Cases	per 1000	Subjects	Cases	per 1000	
Mianning, 1977–1982	151,490	59	0.39	146,071	544	3.72	$P < .001$
Linnan, 1976–1981	16,132	15	0.93	13,119	72	5.49	$< .001$
Xichang, 1974–1980	118,933	16	0.13	30,332	76	2.51	$< .001$
Leibo, 1979–1982	7,691	0	0	18,240	63	3.45	$< .001$

TABLE 9. Effect of Selenized Salt in Prevention of Keshan Disease

Group	Before supplementation (1974–1976)			After supplementation (1977–1983)			Before vs after supplementation (P)
	Total subjects	Total cases	Annual incidence per 1000	Total subjects	Total cases	Annual incidence per 1000	
Treated	19,122	61	3.19	56,440	11	0.195	<.01
Nontreated	29,850	33	1.11	87,888	76	0.86	>.05

fourth case was 0.15 ppm. Before starting supplementation with selenized salt, 30 hair samples were collected from preschool children,
primary school students, and adults, respectively, in the treated communes and analyzed for their selenium content. The average content
of selenium was 0.0495, 0.0669, and 0.0680 ppm, respectively. Four
months after giving the supplement, the corresponding hair selenium
contents were increased to 0.1127, 0.1266, and 0.1670 ppm, respectively, and after 1 year, they were 0.2365, 0.2581, and 0.2979 ppm Se,
respectively.

The above results demonstrated that selenized salt is as effective as
sodium selenite tablets in the prevention of Keshan disease.

DISCUSSION

Selenium Deficiency—An Important Factor in the Etiology of Keshan Disease

From the above investigations, we have found that (1) the selenium
content of grains in the endemic areas was significantly lower than
that in the nonendemic areas; (2) the hair selenium content of inhabitants in the endemic areas was significantly lower than that in the
nonendemic areas; (3) in the same endemic area, the hair selenium
content of children from farmers' families was significantly lower
than that of nonfarmers' families; (4) in farmers' families, the hair
selenium content of preschool children was significantly lower than
that of their parents; (5) sodium selenite intervention, whether in the
form of selenium tablets or selenized salt, produced remarkable effects
in the prevention of Keshan disease. All these facts indicate that the
poor selenium status of local inhabitants resulting from the consumption of low selenium grains plays an important role in the etiology of
Keshan disease.

However, we have also found that the following phenomena are not
consistent with the selenium theory. In a few nonendemic areas which
were located close to endemic areas, the hair selenium content of local
inhabitants was found to be quite low. For example, in Nanbu County,
the hair selenium content was as low as that of the endemic area.
Further, the selenium status of Keshan disease patients was not significantly different from that of the healthy subjects in the same endemic area, and no relationship has been found between the seasonal
prevalence of Keshan disease and the variation of hair selenium content. Finally, sodium selenite has been shown to have no effect in

treating Keshan disease patients. All this indicates that selenium deficiency is not the only cause of this disease.

The epidemiological characteristics of Keshan disease in Sichuan Province suggests that a microorganism, such as a virus, might play a role in the occurrence of this disease, since its prevalence has pronounced seasonal and yearly variations and the majority of patients are children. Although certain strains of virus (such as Coxsackie and Echo) have been isolated from the blood and hearts of deceased patients, they have not yet been shown to be related to the cause of Keshan disease. However, Bai and Ge (3) found that baby mice deficient in selenium suffered more severely from myocardial necrosis than those of the controls after an injection of Coxsackie B_4 virus. This indicates that in the presence of selenium deficiency, the organism might be more susceptible to a viral infection.

Owing to the multicausal etiology of Keshan disease, we suggest that comprehensive measures in the control of this disease be carried out. In addition to selenium intervention, emphasis should be placed on improving the dietary pattern (i.e., eating more foods rich in selenium and having a diversified diet), the sanitary quality of drinking water, and personal hygiene.

Criteria for Applying Sodium Selenite to Prevent Keshan Disease

The 10-year observation on the effects of selenium supplementation in the prevention of Keshan disease in Sichuan Province covered 620,000 people in 24 counties. The results indicate that selenium supplementation by using selenite is safe and effective. However, since the prevalence of Keshan disease has significant yearly variation and regional differences, the necessity of selenium supplementation in endemic areas should be determined by recent prevalence. For the active endemic areas, it is necessary to supplement the whole population, or at least the susceptible children. For the nonactive endemic areas, since it is neither feasible nor economical to carry out massive selenium supplementation, comprehensive prevention is recommended instead of selenium supplementation.

A number of epidemiological investigations indicated that areas with an average hair selenium content in children of more than 0.2 ppm Se were free from Keshan disease, and some cases occurred when the selenium content was between 0.12 and 0.20 ppm Se. When the selenium content fell below 0.1 ppm, Keshan disease may occur. Therefore, we suggest using hair selenium content of local children as a criterion to monitor and predict the prevalence of Keshan disease. Whenever and wherever the hair selenium level drops below 0.1 ppm, prompt supplementation of selenium should be adopted immediately.

Advantages and Disadvantages of Selenium Supplementation by Selenium Tablets and Selenized Salts

So far, the principal methods of selenium supplementation adopted in this province are oral administration of selenium tablets and selenized salt. The dosage of selenium tablets is 0.5 mg every 7–10 days for children of 1 to 5 years of age and 1 mg for 6- to 10-year-olds. In 3 months, the hair selenium content can be increased to 0.2 ppm. Selenium supplementation must begin 3 months before the epidemic season. Usually the administration of selenium tablets should last approximately 8 months, i.e., children taking the tablets 32 times. In cases where the beginning of selenium supplementation is delayed, a quick supplementation method has been developed. By this method, children take their weekly dose of selenium every day for 10 days, then once a month with a double dose. The advantages of oral administration of selenium tablets are (1) the dosage can be controlled accurately; (2) protection of the susceptible children is effective; and (3) enhancement of the human selenium level occurs in a short time. However, this program has at least two disadvantages: First, doctors and medical workers have to deliver the selenium tablets to every family every week, and second, they have to watch the children take the tablets. This makes it difficult to carry out massive selenium supplementation programs continuously for many years.

Selenized salt contains 15 ppm of sodium selenite in the table salt. According to our investigation, the amount of salt consumed daily per capita in the endemic areas is about 10 g, and this contains 68 μg of selenium. Selenium supplementation by means of selenized salt has several advantages. It is convenient, safe, and produces no side effects. Long-term and massive supplementation of selenium can be easily carried out, especially in thinly inhabited districts. By this method, hair selenium can be increased to 0.12 ppm and 0.2 ppm Se after 4 months and 12 months of Se supplementation, respectively. In order to quickly prevent Keshan disease and carry out prevention over a long period, it is recommended that people be given selenium tablets every day for a short time before they begin to consume selenized salt. The amount of selenium supplemented can be adjusted according to the background selenium level of the local inhabitants.

REFERENCES

1. Wang, G.-Y., and Yang, G.-Q. 1983. Determination of microamount of selenium in biological, water, and soil samples by the fluorometric method. Health Res. (Weisheng Yanjiu) *12* (3), 1.
2. Keshan Disease Research Group 1979. Observations on the effect of sodium selenite in prevention of Keshan disease. Chin. Med. J. *92* (7), 471.
3. Bai, J., and Ge, K.-Y. 1982. Myocardial necrosis induced by viral infection in mice. Acta Nutr. Sin. *4* (3), 241.

Histochemical Observations Related to Keshan Disease

Chen Xiao-Shu   *Xie Yu-Hong*
Meng Guang-Shan   *Chen Xue-Cun*
Li Wen-Xian

The main pathological feature of Keshan disease (KD), an endemic cardiomyopathy of unknown cause, is the multifocal necrosis of heart muscle. The earliest ultrastructural change of heart muscle cells was found in the mitochondria. Oral administration of sodium selenite was highly effective in prevention of KD. This study explored the relationship between the function and morphological changes of the heart tissue of patients who died of KD and the myocardial histochemical differences between KD patients and young pigs reared on low-selenium cereals grown in KD endemic areas.

MATERIALS AND METHODS

The investigation was conducted on 11 subacute patients, 3 chronic-type patients who died of KD, and 6 controls who died of other diseases without gross or microscopic changes in heart muscle. The changes in activity and localization of acid phosphatase (Holt's method) and succinic dehydrogenase (Nachlas' method) were observed. Blocks of heart tissue were dissected within 6 hr after death, frozen immediately with solid carbon dioxide, and sectioned at $-20°C$.

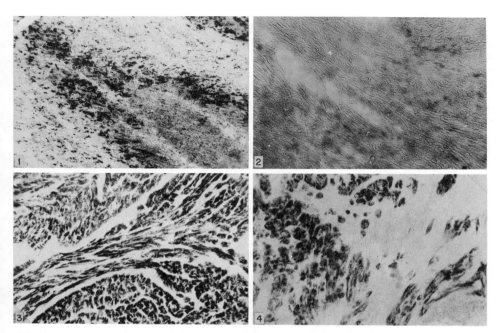

FIG. 1.   Activity of ACPase increased in subacute KD patient's heart muscle (3.2 × 10).

FIG. 2.   No ACPase activity in control patient's heart muscle (3.2 × 10).

FIG. 3.   Pattern of SDHase in control patient's heart muscle (3.2 × 10).

FIG. 4.   Activities of SDHase decreased in damaged heart of subacute KD patient (3.2 × 10).

RESULTS

Acid phosphatase (ACPase) activity, shown as dark brown particles in myoplasm, increased sharply in myocardiocytes around the necrotic foci in the heart of all subacute KD cases (Fig. 1). However, the particles could not be found in the heart of chronic-type KD patients or controls (Fig. 2).

Succinic dehydrogenase (SHDase) activity was shown as fine violet granules arranged in rows along muscle fibrils in normal heart muscle (Fig. 3). The intensity of color was greatly reduced or completely disappeared in damaged myocardiocytes of all 11 subacute KD cases (Fig. 4). On the other hand, the large spherical blue formazan granules observed in the myocardium of KD patients seem to be located in

TABLE 1. Histochemical Changes of SDHase, ACPase in 18 Patients

Number	Sex	Age	Clinical diagnoses	Myocardial changes	SDHase[a]	ACPase activity around necrotic focus	
						Fresh tissue	Fixed tissue
3	F	6 yr	Subacute KD	Conform to subacute KD	+++	↑	
4	M	2½ yr	Subacute KD	Conform to subacute KD	+	↑	↑
5	F	3 yr	Subacute KD	Conform to subacute KD	+++	↑	↑
6	M	3 yr	Subacute KD	Conform to subacute KD	++	↑	↑
7	M	5 yr	Subacute KD	Conform to subacute KD	+	↑	↑
10	M	½ yr	Cardiomyopathy	Conform to subacute KD	+++	↑	↑
12	F	4 yr	Subacute KD	Conform to subacute KD	+	↑	↑
14	M	2½ yr	Subacute KD	Conform to subacute KD	+	↑	↑
17	M	3 yr	Subacute KD	Conform to subacute KD	±	↑	↑
18	F	4 yr	Subacute KD	Conform to subacute KD	±	↑	↑
13	M	8 yr	Chronic KD	Conform to chronic KD	++	↑	↑
15	M	6 yr	Chronic KD	No obvious changes	+		
11	F	8 day	Pneumonia	No obvious changes	++		
1	F	48 yr	Cor pulmonale	No obvious changes	±		
2	M	2½ yr	Toxic dysentery	No obvious changes	+		
8	F	26 yr	Intracranial tumor	No obvious changes	+		
9	M	¾ yr	Measles and pneumonia	Interstitial myocarditis	++++		
16	M	4 yr	Malnutrition III°	Interstitial myocarditis	++++		

[a] +, ++, +++, and ++++ represent the amount of large coarse blue granules.

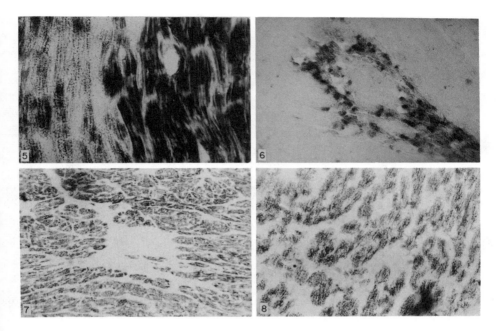

FIG. 5. The activity of SDHase showed as large blue granules in heart of sub-acute KD patient (3.2 × 40).
FIG. 6. Activity of ACPase increased in the heart of young pig fed low-selenium, low vitamin E diet (3.2 × 40).
FIG. 7. The pattern of SDHase in the heart of young pig fed low-selenium and low vitamin E diet, showing large blue granules (3.2 × 40).
FIG. 8. Same as Fig. 7, but showing decreased activity, (3.2 × 10).

mitochondria (Fig. 5). However, the large and spherical granules were fewer than those of acute feverish disease such as bacterial dysentery. The results are summarized in Table 1. ACPase activities increased significantly around the necrotic foci of myocardium in young pigs fed the low-selenium and low vitamin E semisynthetic diets or low-selenium diets composed of cereals grown in the KD endemic area (Fig. 6). The change of ACPase activity in the heart of these young pigs is similar to that of KD patients. The pattern of SDHase in both damaged and normal heart muscle of pigs is similar to that of human heart (Figs. 7 and 8).

Based on the histochemical changes, it is suggested that the necrotic changes in the heart muscle of KD patients might be related to selenium deficiency.

93

Pathogenic Factors of Keshan Disease in the Grains Cultivated in Endemic Areas

Wang Fan   An Ru-Guo
Li Guang-Sheng   Yang Tong-Shu
Li Cai   Zhu Ping

Based on the morphological features of myocardial lesions and the epidemiological behaviors of Keshan disease (KD), the authors considered that the disease might be caused by some geochemical agents acting upon the human body through grains, vegetables, and drinking water (1–3). A series of experiments designed in accordance with the biogeochemical hypothesis on the etiology of KD were performed. It was shown that the pathogenic factors of KD existed in the grains cultivated in endemic areas (endemic grains). These factors had an ability to elicit some metabolic derangements in the myocardium and induce some latent damages in myocardial cells. The main points of these results are summarized in brief here.

PATHOGENIC FACTORS OF KD IN ENDEMIC GRAINS (4)

It was found by a series of experiments that multifocal myocardial necrosis could be produced in rats by gavage of $NaNO_2$ (75 mg/kg body weight twice daily for 10 consecutive days). When weanling rats were kept on rations of endemic grains for 3 months, their myocardium became more vulnerable to the damages induced by $NaNO_2$. With the same dosage and administration of $NaNO_2$, the incidence of myocar-

TABLE 1. Comparison of Myocardial Necrosis of Various Groups of Rats

Groups	No. of animals	No. of animals with myocardial necrosis	Mean areas of myocardial necrosis[a]
1. Daxing (nonendemic) grains	29	2	0.014
2. Jinxian (nonendemic) grains	29	2	0.02
3. Huinan (endemic) grains	30	7	0.22
4. Shangzhi (endemic) grains	29	12	0.42
5. Huangling (endemic) grains	29	10	0.24

[a] Measured by means of a square micrometer under the microscope; treated statistically by Wilcoxon's two-sample rank test. P value: Group 1:Group 4 and Group 2:Group 4, $P < .01$; Group 1:Group 5 and Group 2:Group 5, $P < .05$; Group 1:Group 3 and Group 2:Group 3, $P > .05$.

dial necrosis was distinctly higher, and the necrotic area was much larger in the animals kept on endemic grains than those on grains from nonendemic areas (nonendemic grains), as shown in Table 1.

No significant difference of myocardial lesions was found microscopically between the groups kept on endemic grains and those on nonendemic grains when $NaNO_2$ was not used or isoproterenol and cold baths were applied instead. These results suggested that there existed in endemic grains some sort of factors potentially harmful to the myocardium and that there were some specific relations between these factors and the nitrite. With the synergism of nitrite and these factors, more serious myocardial necrosis could be produced.

INFLUENCE ON TRACE ELEMENT METABOLISM (5)

Some element contents in the myocardium of rats kept on endemic and nonendemic grains for 3 months were determined by atomic absorption spectrophotometry. Using $NaNO_2$ twice for only 1 day (75 mg/kg body weight by gavage each, at an interval of 6 hr) as a provoking factor, the calcium content in the myocardium was strikingly higher in the animals kept on endemic (Shangzhi County, Heilongjiang) grains than those on nonendemic (Daxing County, Beijing) grains. The results are shown in Table 2.

These results suggested that some latent metabolic derangement of calcium can be elicited by feeding endemic grains. With the synergism of some factors able to cause myocardial ischemia and anoxia (such as $NaNO_2$), myocardial calcium overload may occur, resulting in myocar-

TABLE 2. Calcium Contents in the Myocardium of Rats[a]

Groups	No. of animals	Mean ± SD (μg/g dry wt)	Median (μg/g dry wt)
1. Endemic grains	10	249 ± 31	246
2. Endemic grains + ISP[b]	10	684 ± 609	429
3. Endemic grains + NaNO$_2$	10	302 ± 34	310
4. Nonendemic grains	10	303 ± 86	265
5. Nonendemic grains + ISP	10	442 ± 87	392
6. Nonendemic grains + NaNO$_2$	10	263 ± 23	264

[a] P value: Group 1:Group 4, $P > .05$; Group 2:Group 5, $P < .05$; Group 3:Group 6, $P < .05$.
[b] ISP, Isoproterenol.

dial necrosis. These results are consistent with those of our morphological studies above.

INFLUENCE ON SARCOLEMMAL MEMBRANE PERMEABILITY OF MYOCARDIAL CELLS

Using horseradish peroxidase (HRP) as a tracer, the influence of endemic grains upon sarcolemmal membrane permeability of myocardial cells was studied in rats. Weanling male rats were divided into an endemic (Dunhua County, Jilin) grain group and a nonendemic (Daxing County, Beijing) grain group. After 2 or 3 months, HRP was injected intravenously and localized in the ventricular myocytes by light microscopy. The number of myocytes containing HRP-reactive product was much more in the endemic grain group than that in the control (Table 3), indicating that sarcolemmal membrane permeability of myocardial cells was markedly increased by feeding endemic grains.

The number of cardiac myocytes containing HRP-reactive product could be obviously decreased by adding Se (0.1 mg Na$_2$ SeO$_3$/kg diet) or

TABLE 3. Number of Cardiac Myocytes Containing HRP-Reactive Product

Groups	Feeding for 2 months		Feeding for 3 months	
	No. of animals	No. of myocytes containing HRP-reactive product[a]	No. of animals	No. of myocytes containing HRP-reactive product[a]
1. Nonendemic grains	7	5(a)	16	63.5(c)
2. Endemic grains	7	52(b)	16	180.5(d)

[a] Median. P value: a:b, $P < .05$; c:d, $P < .01$, treated by Wilcoxon's two-sample rank test.

TABLE 4. Number of Cardiac Myocytes Containing HRP-Reactive Product[a]

Groups	No. of animals	No. of myocytes containing HRP-reactive product		
		Cumulative number	Mean	Median
1. Endemic grains	14	1721	122.9	94
2. Endemic grains + Se	14	755	53.9	42.5
3. Endemic grains + Mo	14	552	39.4	32

[a] P value: Group 1:Group 2 and Group 1:Group 3, $P < .01$; Group 2:Group 3, $P > .05$.

Mo [100 mg $(NH_4)_6Mo_7O_{24} \cdot 4H_2O$/kg diet] to the endemic grains (Table 4), suggesting that the membrane damage elicited by endemic grains can be prevented by supplementation of Mo as well as Se.

INFLUENCE ON SOME MYOCARDIAL ENZYMES AND MYOSIN

The activities of cardiac succinic dehydrogenase (SDH) and cytochrome oxidase (Cyt Ox) were determined by oxygen electrode in rats fed endemic (Shangzhi County, Heilongjiang) and nonendemic (Daxing County, Beijing) grains for 3 months (6). As shown in Table 5, there was an activity decrease of both enzymes in the endemic grain group, especially the activity of Cyt Ox (a statistically significant difference).

The activity of myocardial Ca-activated myosin (B) ATPase was determined in rats fed on similar rations mentioned above for 3 months. The specific activity of the enzyme was markedly lower in the endemic grain group than that in the control (Table 6).

The myosin ATPase isoenzymes of ventricular myocardium were determined by polyacrylamide gel electrophoresis in rats. As seen in the electrophoretograms, a high V_1 peak, a wide V_2 peak, and a narrow

TABLE 5. Activities of Cardiac SDH and Cytochrome Oxidase[a]

Groups	No. of animals	SDH (ΔO_2 mm Hg/6 min/2 mg wet wt, 37°C)	Cyt Ox (ΔO_2 mm Hg/3 min/2 mg wet wt, 37°C)
1. Nonendemic grains	7	53.9 ± 19.2	34.8 ± 7.9
2. Endemic grains	7	50.3 ± 15.5	24.0 ± 13.9

[a] P value: SDH Group 1:Group 2, $P > .05$; Cyt Ox Group 1:Group 2, $P < .05$.

TABLE 6. Specific Activity of Myocardial Ca-Activated Myosin (B) ATPase

Groups	No. of animals	Specific activity (μM P_i/mg/min)	t	P
1. Nonendemic grains	10	0.75 ± 0.15	2.036	$>.05$
2. Endemic grains	12	0.58 ± 0.18		

and low V_3 peak made their appearance in rats kept on nonendemic grains, while for those kept on endemic grains, the V_3 became higher associated with an obviously lowered V_1, suggesting a shift of the isoenzymes from V_1 toward V_3 under the influence of feeding endemic grains.

In addition, a quantitative analysis was made on myocardial structural proteins of rats by sodium dodecyl sulfate–polyacrylamide gel electrophoresis. It was found that the light chain 2 (Lc_2) was markedly increased in rats kept on endemic grains ($Lc_2/Lc_1 = 0.73 \pm 0.10$) as compared with those kept on nonendemic grains ($Lc_2/Lc_1 = 0.50 \pm 0.02$). A significant difference existed statistically between the former and the latter.

LIGHTENING THE PATHOGENIC ACTION OF ENDEMIC GRAINS BY AMMONIUM MOLYBDATE (7)

Some known agents were added to endemic grains to study their protective action on rat myocardium. It was found that the addition of ammonium molybdate [$(NH_4)_6Mo_9O_{24} \cdot 4H_2O$] produced some prophylactic effect in reducing the myocardial injury caused by the synergism of nitrite and feeding on endemic grains (Table 7).

In view of the fact that ammonium molybdate had been proved to be an effective trace element fertilizer, we had tried fertilizing cereals

TABLE 7. Prophylactic Effect of Ammonium Molybdate[a]

Groups	Incidence of myocardial necrosis	Area of myocardial necrosis
1. Endemic grains + $NaNO_2$	$\frac{16}{28}$(a)	0.90(c)
2. Endemic grains + $NaNO_2$ + $(NH_4)_6Mo_7O_{24} \cdot 4H_2O$	$\frac{6}{30}$(b)	0.17(d)

[a] P value: a:b, $P < .01$; c:d, $P < .01$.

and vegetables with it to prevent KD in an endemic area of Yunnan Province. In the years from 1976 to 1979, a comparative observation was carried out between two production brigades: one as a fertilized area and the other as a control. During the 3 years only one of the 2149 inhabitants in the fertilized area suffered from KD, with a morbidity of 0.47%, while 11 of the 1746 inhabitants in the control area contracted the disease, the morbidity being 6.30%. The difference was highly significant statistically ($P < .01$).

In 1979, the original control area was changed into a fertilized area and a neighboring production brigade was designed as a new control area. From 1979 to 1982, the morbidity of KD in the two fertilized areas was found to be 0.90% and 1.13%, respectively, while that in the new control area was 2.73%. There still existed a highly significant difference ($P < .01$) between the fertilized and the control areas. The results of 6 years of continuous observation indicated that the fertilization with ammonium molybdate could lower the morbidity of KD.

The essence of the pathogenic factors in endemic grains has not been clarified. Although Se content was lower in endemic grains, the pathogenic action was not eliminated by adding fish meal to supplement the Se deficiency in our earlier experimental pathological studies. Mo content was also lower in the rations with endemic grains, but we did not think the deficiency of Mo to be a fundamental factor in the occurrence of KD according to the present data. It is likely that some underlying factor or factors besides Se and Mo can exist in endemic grains, which has yet to be identified.

REFERENCES

1. Wang, F., and Li, G.-S. 1962. A discussion on the etiology of Keshan disease from morphologic aspects. Chin. Med. J. *48* (1), 17.
2. Wang, F., and Li, G.-S. 1963. The etiology concept of Keshan disease from a biogeochemical viewpoint. Chin. Med. J. *49* (2), 112.
3. Wang, F., and Li, G.-S. 1973. The biogeochemical hypothesis on the etiology of Keshan disease. *In* Symposium of the Third Conference on the Etiology of Keshan Disease, pp. 27–33.
4. Wang, F. 1979. Experimental studies on the etiology of Keshan disease. Chin. Med. J. *59* (8), 471.
5. An, R.-G. 1982. Influence of the grains cultivated in endemic areas of Keshan disease upon the metabolism of certain elements in the myocardium. Chin. J. Endemiol. *1* (1), 19.
6. Yang, T.-S. (1982). Effect on respiratory chain system by feeding grains from Keshan disease areas. Chin. J. Endemiol. *1* (2), 108.
7. Wang, F. 1982. Studies on the etiologic relationship of molybdenum deficiency to Keshan disease. Acta Nutr. Sin. *4* (3), 271.

94

Effect of Long-Term Selenium Supplementation on the Incidence of Keshan Disease

Wen Zhi-Mei   Qian Peng-Chu
Chen Xiao-Shu   Liu Ren-Wei
Fu Ping   Huang Jia-Hong

Broad observations suggested that Se supplementation is remarkably effective in reducing the incidence of the acute and subacute type of Keshan disease (KD) which have been reported by the Keshan Disease Research Group, Chinese Academy of Medical Sciences, since 1972 (Chen 1981). The intervention study carried out from 1974 to 1975 showed that the incidence rate of the treated group was 2.2 and 1.0 per 1000, while that of the nontreated group was 13.5 and 9.5 per 1000 in 1974 and 1975, respectively. The intervention measure of using Se supplements has occurred for 7 years among 1 million people living in KD endemic areas. The question of whether Se can completely protect the heart from the slightest damage or only can decrease the occurrence of acute and subacute attacks in KD endemic residents is a matter of interest to many people.

A survey was conducted in October, 1983 and January, 1984 in three groups of the population that resided in a KD severe endemic area. Group 1 has been supplemented with Se tablets for 10 years, Group 2 with Se in table salt for 7 years, and Group 3 lived in a closed region and never received any supplementation. All children from 3 to 10 years old were placed in the survey, which included physical examination, electrocardiography (ECG), and roentgenography. The occurrence of abnormal changes upon ECG or X-ray examinations, or both, but with no clinical symptoms was designated as diagnostic criteria for

TABLE 1. Hair Selenium Content and Changes of ECG and Heart Roentgenography in Selenium-Treated and Nontreated Children

Group	Hair Se (ppm ± SE)	Number of subjects	Heart roentgenograph and ECG changes					
			H:C = 0.50[a]		H:C = 0.51–0.55		H:C = 0.56–0.60	
			Nor ECG	Abn ECG[b]	Nor ECG	Abn ECG	Nor ECG	Abn ECG
Se tablets treated (1)	0.107 ± 0.008 (27)[c]	194	190	1	3	0	0	0
Se salt treated (2)	0.155 ± 0.007 (16)	211	206	2	3	0	0	0
Nontreated (3)	0.059 ± 0.008 (20)	218	204	5	4	3	1	1
Before treatment (4)	0.072 ± 0.002 (20)	183	175	1	2	3	2	0

[a] H:C, Heart:chest ratio.
[b] Nor, Normal; Abn, abnormal.
[c] Numbers in parentheses are number of samples.

a latent type of KD, according to the Diagnosis Working Standard Criteria set by the Committee for the Diagnosis of Keshan Disease in 1982. Because some of the changes on ECG, such as right bundle branch block and heart enlargement, are permanent, surveys were conducted only in children born after Se intervention in the treated group and children of the same age in the nontreated group.

Hair Se contents, number of patients, and their main manifestations are listed in Table 1 and compared with the data examined in 1974 before the supplementation of Se tablets (Group 4).

The results listed in Table 1 show that the number of patients in the Se-treated groups, 1 and 2 (9), is significantly less than that in the nontreated groups, 3 and 4 (22) ($\chi^2 = 5.80$, $.01 < P < .025$). The incidence of abnormal changes on ECG and heart roentgenograph was less in the treated groups as well.

Nine patients in Groups 1 and 2 were of the latent type. There were 1 subacute and 13 latent-type patients in Group 3, and 2 chronic and 6 latent-type patient in Group 4.

Hair Se content of the treated groups was low (only the upper limit of the endemic area) because the incidence rate of KD in these areas was decreasing, and the distribution of Se tablets was not taken as seriously by local health workers as it had been before 1978.

From this survey, we can conclude that the incidence of acute damage to the heart as well as insidious changes in the heart were decreased after Se supplementation, though Se may not be able to protect the heart completely from damages due to all harmful factors. Since hair Se content of the treated group was not at a good level (0.2 ppm), perhaps insufficient supplementation is an alternative explanation of this result.

REFERENCE

Chen, X.-S. 1981. Relation of selenium deficiency to the occurrence of Keshan disease. *In* Selenium in Biology and Medicine. J. E. Spallholz, J. L. Martin, and H. E. Ganther (Editors), p. 171. AVI Publishing Co., Westport, CT.

Selenium and Cardiovascular Diseases: Preliminary Results of the WHO/IAEA Joint Research Program

R. Masironi
R. Parr
M. Perry

In recent years the attention of many investigators has been drawn to the beneficial role of selenium in cardiovascular function. Rats and lambs fed Se-deficient diets develop abnormal electrocardiograms accompanied by blood pressure changes. These disorders were prevented by selenium supplementation. It has long been known that in New Zealand animals grazing on selenium-deficient pastures were severely affected by myocardial necrosis and other muscle-wasting diseases. Selenium was therefore added to animal feed to counteract these diseases. Selenium seems to be a favorable element, not only essential for the optimal growth of animals, but perhaps also in preventing cardiovascular disorders in man. No real deficiency is known in man, but there are reports suggesting that such deficiency may be a complicating factor in protein-calorie malnutrition in children. Selenium is also thought to counteract cadmium-induced hypertension in experimental animals. Indeed, as J. Parizek has shown, selenium is an antagonist of cadmium and mercury.

Keshan disease, the now well-known cardiomyopathy found in numerous areas of China and known to be associated with Se deficiency, will not be discussed here.

Shamberger (1) has shown that selenium deficiency in the environment, as indicated by selenium concentration in plants, is inversely associated with death rates from hypertensive heart diseases in the United States. In other words, states where high selenium concentrations are found in plants experience lower death rates from hypertension. A germane association was found between selenium intake in 25 countries and death rates from coronary heart disease. Also the selenium blood levels are inversely associated with death rates from coronary heart disease in the United States. On a European geographic basis, countries such as Sweden and Finland, where cardiovascular death rates are notoriously high, are also characterized by crops that are deficient in selenium, whereas in middle European countries, where death rates from cardiovascular diseases are lower, crops are an adequate source of selenium (2,3).

Certain countries apparently have a problem of marginal selenium deficiency as their soils and crops contain low levels of this element. Typical examples are New Zealand and Finland. In areas of Finland, where the selenium content of vegetables and of men's hair was found to be lower compared to other areas, the frequency of myocardial infarction was found to be twice as high. Other studies in Finland have shown low plasma levels of selenium as well as low glutathione peroxidase activity in subjects under study. These studies have also shown that below a certain level (34 μg/liter) of plasma selenium concentration, the risk of developing myocardial infarction is seven times as high as in those subjects having high (45 μg/liter) selenium concentrations in their blood (4). Supplementation of staple food with imported cereals from high selenium countries, such as those in North America, into Finland has resulted in an increased selenium intake by the Finns and an increase in glutathione peroxidase activity. Indeed, human studies in Canada and in the United States have shown relatively high content of selenium in tissues, i.e., liver, skeletal muscle, kidney cortex, heart, and lungs, whereas tissues from New Zealand and particularly from Finland and Sweden have very low levels. Of all the reported countries, tissues from Sweden have the lowest selenium concentration (Table 1). Geographic differences in selenium tissue levels were also found in a World Health Organization (WHO) study. Tissue from the Philippines and from Israel showed higher selenium concentration than tissues from Czechoslovakia and the United States (5).

WHO and the International Atomic Energy Agency (IAEA) have been collaborating with a number of pathology departments and analytical laboratories in many countries in studying the role possibly played by several trace elements in the pathogenesis of cardiovascular diseases. Selenium is one such element.

TABLE 1. Selenium Content of Adult Tissues in Different Countries

	Selenium content (μg/g dry weight)				
Country	Liver	Kidney cortex	Skeletal muscle	Heart	Lung
Canada	1.48		1.76	0.79	0.95
United States	2.34		1.24	1.00	0.73
	2.14		1.90	1.18	1.36
		5.71			
West Germany	1.81			0.93	0.84
United Kingdom	1.03		0.52		0.45
Sweden	0.24	0.96		0.21	
Finland	0.67	1.95		0.71	
Denmark	1.34				
Norway	1.37				
New Zealand	1.03				
	0.66				
	0.72	3.14	0.29	0.68	0.45

MATERIALS AND METHODS

International collaboration involved collection of autopsy material, namely, heart, liver, kidney cortex, and kidney medulla, from subjects who died with or without myocardial infarction or other cardiovascular diseases. Standardized collection of autopsy samples and pathological diagnoses were done under the coordination of WHO.

Heart samples were taken from the anterior wall of the left ventricle, liver samples were taken from the superior, anterior surface of the right lobe, and kidney sampled from the cortex of the lower pole of the left kidney. Care was taken to keep metal contamination to a minimum by using plastic-coated forceps and titanium knives specially prepared at IAEA, and by avoiding any washing and chemical preservation of the samples. Subjects were males, aged 45–64 years, although females and other age groups were occasionally sampled to see whether any age- or sex-related trend would appear in trace element concentrations. The specimens were quickly frozen and shipped in a frozen state by air to an analytical laboratory where they were analyzed for selenium and other trace elements as well. Analysis was by neutron activation. The analytical procedure is standardized under the general supervision of the International Atomic Energy Agency.

ANALYTICAL METHODS

A variety of analytical methods are available for the determination of selenium in biological materials. The most widely used are atomic absorption spectroscopy and neutron activation analysis. Selenium can also be determined with good sensitivity by fluorometry, and this method now appears to be coming into more widespread use for biological materials.

In this program, selenium was determined exclusively by neutron activation analysis (NAA).

Most of the analyses reported here were carried out in three centers, the Federal Republic of Germany, Norway, and Sweden (Table 2).

It was recognized at the outset of this program that analytical quality assurance would be an important issue, particularly since results were to be reported by several analytical laboratories, thus making it necessary to ensure that the methods used would all be of comparable accuracy.

Fortunately, selenium appears to belong to a small group of trace elements (including Cu, Fe, Mn, and Zn) that can be determined with relatively good reliability by the majority of analytical laboratories engaged in trace element research.

The materials selected for use as quality control materials were bovine liver obtained from the U.S. National Bureau of Standards and two reference materials prepared by the IAEA, animal muscle and animal kidney. These materials were chosen, because of their similarity to the three types of tissue of primary interest in this program, liver, heart muscle, and kidney cortex.

Table 2 records some of the results reported for these materials. As far as can be judged, all the analytical laboratories were producing results of acceptable reliability.

TABLE 2. Analytical Quality Control Data for Selenium[a]

Material	Analytical laboratory	Value found (mean ± SD)	Certified value
Bovine liver NBS-SRM-1577	Federal Republic of Germany	1.26 ± 0.12	1.1 ± 0.1
	Norway	1.02	
	Sweden	1.07 ± 0.18	
Animal muscle IAEA-H-4	Federal Republic of Germany	0.28 ± 0.03	0.28 ± 0.04
	Norway	0.29 ± 0.03	
	Sweden	0.16 ± 0.16	
Animal kidney IAEA-H-7	Federal Republic of Germany	9.5 ± 0.7	9.7 ± 2.4
	Sweden	7.6 ± 1.7	

[a] Values in mg/kg dry weight.

TABLE 3. Selenium Concentration in Kidney Cortex (μg/g, dry weight) in Hypertensive (HT) and Normotensive (N) Subjects

Collection center	HT	N
Hong Kong	0.62	0.90
Malmö	0.61	0.88
Geneva	0.39	0.60
Ibadan	0.39	0.82
Mean	0.50	0.82
$P \leq .001$		

The study consisted of two projects:

1. The study of concentration of Se, Cr, Cu, and Zn in heart, liver, and kidney cortex of subjects who died of myocardial infarction vs control subjects who died of accident and were found at autopsy to be free of disease; and
2. The study of the concentration of the same elements in the same tissues in subjects who died of or with hypertension, as compared to control, normotensive subjects.

Autopsy specimens were collected in various collaborating centers located in Hong Kong, Sweden, Denmark, Switzerland, Nigeria, the Philippines, and the United States. Geographic origin, age, and smoking history of the subjects were taken into account as well as the differences between analytical centers.

From the great number of data obtained, which would not be useful to display here in its entirety, the only meaningful difference found was that the Se concentration in kidney cortex of hypertensive subjects was significantly lower, i.e., by 39%, than in normotensives (0.50 μg/g vs 0.82 μg/g dry weight) (Table 3). This difference is consistent not only across the whole study project, but also within geographic locations. All these analyses were done in one center in Sweden.

No other differences were found in relation to myocardial infarction, tissue, age, and so on.

A few other studies have been conducted by other authors on Se concentrations in tissues of cardiovascular subjects, e.g., Wester in Sweden, Westermark in Finland, and Shamberger in the United States, and no differences were found in the tissues between diseased (both hypertensives and myocardial infarction subjects) and control subjects.

It would be outside the scope of this discussion to try to compare the present data with those of other authors, since the collection and analytical techniques were vastly different.

Kidney cortex seems to be the target tissue where Se, Cd, and Zn concentration vary in relation to disease. The interaction among these three elements could provide a better understanding of the role they play in cardiovascular diseases.

REFERENCES

1. Shamberger, R. J. 1981. Sci. Total Environ. *17*, 59–74.
2. Gissel-Nielsen, G. 1976. Selenium in soils and plants. *In* Selenium and Tellurium in the Environment, pp. 10–25. Industrial Health Foundation, Pittsburgh, PA.
3. Masironi, R. 1979. Philos. Trans. R. Soc. London, Ser. B 288, 193–203.
4. Salonen, J. T. 1982. Lancet *2,* 175–179.
5. Masironi, R. 1976. Selenium and cardiovascular diseases. *In* Selenium and Tellurium in the Environment, pp. 316–325. Industrial Health Foundation, Pittsburgh, PA.

96

Distribution of Selenium in the Microenvironment Related to Kaschin–Beck Disease

Li Ji-Yun   *Liang Shu-Tang*
Ren Shang-Xue   *Zhang Fu-Jing*
Cheng Dai-Zong   *Gao Fen-Min*
Wan Hen-Jun

The distribution of Kaschin–Beck disease (KBD) has a relationship with the microgeomorphological changes in the incidence of KBD zones in the loess plateau in particular. For example, the areas severely attacked by KBD are mostly on the edge of the plateau, the areas lightly attacked are mostly on the surface of the plateau, and the lightest attacked areas are distributed generally in the wide basin or river valley. Furthermore, the areas severely attacked, lightly attacked, and disease free are often interdistributed and adjacent to each other within the range of the narrow microenvironment. The reasons for this are still unknown. In recent years, it has been shown from the results surveyed in wide geographical regions by the authors that KBD was prevalent in the environments containing a very low level of Se (*1–3*). This chapter gives results regarding the relationship between selenium in microenvironments and KBD.

METHODS

Two typical microenvironments were selected for the investigation in the disease region of Shaanxi Province in 1979. One is in the northwestern part of Yunshou County, with an area of only 2.9 km², and contains two production brigades. It is named the first microenvironment. The other is Zhangwu Yuan (Yuan is a high table-like plain with abruptly descending edges) in Zhangwu County. From the center of the edge of the Yuan to the Jinghe River valley is a long sloped belt 12 km in length containing four production brigades. It is named the second microenvironment.

The investigation methods diagnose the incidence of KBD among children in the age group of 5–13 in various production brigades, survey the distribution of soil type in the microenvironment, and study the content and distribution of Se and other essential elements in soils, drinking water, and grains, and the Se amounts in human hair in relation to the incidence of KBD. This chapter discusses only the results of the investigation for Se.

The determination of the amount of Se in water, soil, grains, and children's hair is made through the dissolution of these samples by digestion with nitric–perchloric acid. Se is reduced to the state of selenite, complexed with diaminonaphthalene, extracted with cyclohexane, and analyzed by fluorometry. The following practical method is used.

The determination of the total amount of Se in the soil is made through the digestion of the sample with nitric–perchloric acid, which is then reduced with 6 N HCl. To analyze the content of water-soluble Se in soil, take a portion of soil, add water (water:soil = 5:1), and place it in a mechanical shaker for 2 hr; afterward add 0.5 g NaCl and filter, alkalize 100 ml of filtrate and evaporate to 3 ml, digest with 5 ml of the mixture acid, and then reduce with HCL.

To determine the Se content in drinking water, take 500 ml of the sample, add $CaCl_2$ and HCl, and then oxidize with 0.1 N $KMnO_4$. Evaporate this solution to 150 ml, alkalize, and reduce with NH_4Cl.

To determine the Se content in grains and hair, take the sample after washing with a detergent, dry it at 60°C, and digest it with sulfuric–perchloric acid; then treat it with the same methods as mentioned above.

RESULTS

Environmental Characteristics and Disease Incidence of Residents

The first microenvironment stands on the center of Yunshou Liang (Liang is an elongated loess mound), which belongs to a region severely attacked mid-hill and low-hill of the northwestern part of

Yunshou County and contains two production brigades, Jiangjashan and Beimen. Both villages are adjacent to each other.

Jiangjashan is famous for being a healthy village and lies in the piedmont depression of the south slope on Yunshou Liang, 1276–1442 meters above sea level. In this village there are 26 peasant households, 153 residents, and 2000 mu of land, which includes 820 mu of cultivated land. The purple gravel rock is sparsely distributed near the Yunshou Liang, and the arenaceous shales are exposed in the low-lying land and depression. Because of the long-term effect of microgeomorphology, the various types of soils have been formed by runoff silt and sand deposit, and a part of them has good water and heat regimes. This is one of the main reasons why this area is the lightest attacked by KBD. By X-ray diagnoses, the incidence of KBD in local children was 11.5% in 1979 and 13.0% in 1981 (Table 1).

Beimen production brigade is known as being a village severely attacked by KBD and borders Jiangjashan on the east on the slope of Yunshou Liang at an altitude of 1283–1446 meters. In this village there are 22 peasant households, 124 residents, and more than 2000 mu of land in which the top loess has long been seriously eroded so that the old loess and its red soil layer containing a huge amount of calcareous concretions are extremely exposed. All of these soils make up more than 90% of the total land area, so that water and heat regimes in most of the fields are very bad, and this is one of the main reasons why the disease severely attacks the local residents. By X-ray diagnoses, the incidence of KBD among local children was 92.0% in 1979 and 90.9% in 1981 (Table 1).

TABLE 1. Diagnosis Results on Incidence of Kaschin–Beck Disease among Children in Various Production Brigades of Two Microenvironments

Microenvironment	Locations	Date of diagnoses	Observation by X ray	
			Detected (%)	Detected in metaphyseal region (%)
1	Jiangjashan	October, 1979	11.5	0.0
		May, 1981	13.0	4.35
	Beimen	October, 1979	92.0	76.6
		May, 1981	90.9	81.8
2	Penggong	October, 1979	0.0	0.0
		May, 1983	1.9	1.9
	Langtou	October, 1979	20.0	13.3
		May, 1983	28.3	
	Xianggong	October, 1979	92.5	47.9
		May, 1983	64.6	
	Dongzhui	May, 1983	0.0	0.0

The second microenvironment is in Zhangwu Yuan from the center to the edge of the Yuan and to the Jinghe River valley. Along the long slope belt, soil erosion has been gradually worsening as the relief becomes lower and lower and the gradient gradually increases. From the top area to the bottom, there are four production brigades: Penggong, Langtou, Xianggong, and Dongzhui.

Penggong village is located in the center of the Yuan. In this area the soil is clay black Heilu along the entire profile. There has been less soil erosion in this area. The inhabitants suffer only a light incidence of KBD. By X-ray diagnoses, the incidence in local children was 0 in 1979 and 1.9% in 1983 (Table 1).

Langtou lies near the edge of the Yuan. The soil is an eroded clay black Heilu with loess pan layer, whose upper layer, the mature soil, has been mostly eroded. The incidence of the disease among local children by X-ray diagnoses was 20% in 1979 and 28.3% in 1983.

Xianggong is situated on the edge of the Yuan where the soil is an eroded clay black Heilu, all of whose top loess has been heavily eroded and the deposited layer containing much calcium carbonate is exposed. Most of the local children have suffered from KBD. The incidence by X-ray diagnoses was 92.5% in 1979 and 64.6% in 1983.

Dongzhui village stands on the Jinghe River valley where most of the soil is an eroded black Heilu on the edge of the Yuan, but a small part of the fields is sand soil on the bank of the river. The local children in this village have no KBD.

In the four villages, the staple food of the local inhabitants is wheat and corn taken from their cultivated lands and the living standard of the inhabitants is about the same. The drinking water in the four villages is different. In the center and near the edge of the Yuan drinking water is taken from the well in the deep layer under the loess, while on the edge of the Yuan it is taken from storage pits, and in the gullies near the river it is taken from shallow wells or the river.

Distribution of Selenium in the Environment and Human Hair, and the Relation to the Incidence of KBD

The First Microenvironment. Soil Selenium. It is clear from the results of the surveys of soil distribution in the first microenvironment that the soil types of both brigades are much different. There are various kinds of soils in Jiangjashan which include hard loess distributed on top of Liang, Heilu soil on the platform of the slopes, black benz soil (angular nutty soil) in the depressions under the slope, bicolor soil at the bottom of the slope, and Shan soil (loess-like entisol)

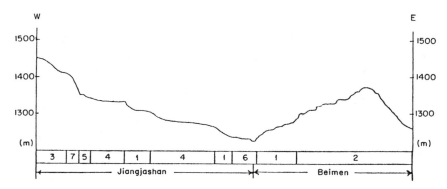

FIG. 1.   Sectional diagram of soil distribution in first microenvironment. (1) Shan soil (loess-like entisol); (2) bicolor soil; (3) hard loess; (4) black benz (angular nutty); (5) Heilu (light mollisol); (6) meadow bog; (7) bare rock; (8) spring.

in the valley bottom. There are a few kinds of soils in the Beimen production brigade, and the major kinds include bicolor and red (old loess), which contains calcareous concretion distribution on the top of Liang and Shan soil on the slope of Liang. Some bicolor soil and silt Shan soil spread only in the valley bottom in a small area. Figure 1 is a sectional diagram of the soil distribution in this environment. It can be seen that the distribution of soil types from the top to the depression of Liang in both production brigades is obviously different.

TABLE 2.   Area and Se Content in Various Soils of Jiangjashan and Beimen Production Brigade

Location	Type of soil	Area (mu)	Soil in total land (%)	Water-soluble Se (ppb)	Total Se (ppb)
Jiangjashan	I-A[a]	449.3	19.0	8.1	340
	I-B	87.0	3.7	3.2	134
	I-C	450.1	19.0	1.6	74
	I-D	571.5	24.1	1.2	71
	I-E	486.9	20.5	2.2	78
	I-F	139.5	5.9	2.0	74
	I-G	186.1	7.8	1.0	70
Beimen	II-A	712.6	33.1	1.6	76
	II-B	916.7	42.6	1.9	71
	II-C	383.3	17.8	2.2	74
	II-D	106.5	5.0	1.8	71
	II-E	30.8	1.5	2.1	76

[a] I-A, Black benz soil; I-B, Heilu soil (light mollisol soil); I-C, Shan soil; I-D, Hard loess; I-E, bicolor soil; I-F, slit Shan soil; I-G, black benz soil with loess pan layer; II-A, Shan soil; II-B, bicolor soil; II-C, red soil (old loess) and bicolor soil; II-D, hard loess; II-E, slit Shan soil.

All of the soil samples collected from both brigades were analyzed for their Se contents, and these results are shown in Table 2 and Fig. 2. The selenium in this map is from surface soil to a depth of 40 cm.

From these results it can be seen that black benz soil makes up 19% of the total land area and the Se content is very high. The total Se is 340 ppb and the water-soluble Se is 8.1 ppb, which is higher than the Se content in soil in most of the disease-free regions of Shaanxi Province. Another soil belonging to the Heilu type comprises 3.7% of

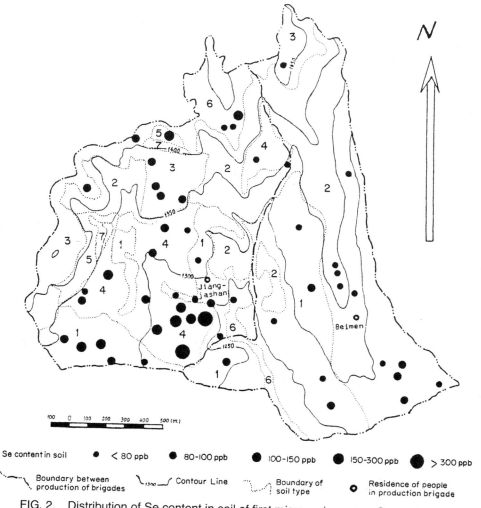

FIG. 2. Distribution of Se content in soil of first microenvironment. See legend to Fig. 1 for soil types.

the total land area. Se content in this soil is much higher than that in any other soils. Total Se is 134 ppb and water-soluble Se is 3.2 ppb, equalling Se contents in the soils of disease-free regions of Guan-Zhong Plain in Shaanxi Province. These two soil types occupy about 24% of the total land area. The remaining 76% of soils contains from 71 to 78 ppb total Se, and from 1.24 to 2.18 ppb water-soluble Se, equal to the Se content in the diseased regions. The Se content in various soils in the Beimen production brigade is very low, where the total Se is 64–76 ppb and the water-soluble Se is 1.6–2.2 ppb. All the Se contents of these soils are equal to the level of Se contents in the disease regions.

Grain Selenium. Grains (wheat and corn) are the staple foods of the local inhabitants. In order to know the effect of Se content in soil on Se in grains and its relationship with KBD, the authors carried out the collection of grains and the analysis of their Se content. Since the production brigades thresh grains harvested in different plots of cultivated lands and give the mixed grains to every peasant household, the wheat and corn grains in 35 and 29 plots of land as well as the flour and corn meal stored in 31 households were collected in both brigades and analyzed for their Se content (Tables 3 and 4; Fig. 3). The results show that the variation of Se content in wheat grains was in accordance with those in the soil. In 25 wheat grain samples of Jiangjashan brigade, 76% contained less than 10 ppb Se, 24% more than 10 ppb (from 10 to 200 ppb), with an average of 16.7 ppb. The Se content of 16 flour samples of Jiangjashan (Table 4) was 75% more than 10 ppb (from 10 to 70 ppb), 25% less than 10 ppb, with an average of 35 ppb. Both the Se content in wheat grains and in flour of Jiangjashan reached the level of Se content in wheat grain in the disease-free regions.

There were 10 samples of wheat grain from Beimen brigade, of which 80% contained less than 10 ppb selenium, 20% more than 10 ppb, from 10 to 30 ppb, with an average of 5.2 ppb. This is equal to the level of Se content of wheat grain in the disease regions. Fifteen flour samples from Beimen were analyzed. As a result, all had Se content less than 10 ppb, with an average of 5.2 ppb, which is such a low level that it is only one-seventh of the average flour Se content of Jiangjashan. There is less variation in the Se content in corn grain from different fields than in wheat because corn is one of the crops which absorbs Se from soil less than others. Moreover, the average Se content of corn grain of Jiangjashan was higher than that of Beimen.

Hair Selenium. Se from the food chain, which contains soil, water, and grains, must ultimately have an effect on the Se level in the human body. This idea has been confirmed from the results in Table 5 in which the Se content in hair of boys aged 5–13 in Jiangjashan was

TABLE 3. Se Content in Wheat and Corn Collected from Various Fields in Jiangjiashan and Beimen Production Brigades

Grade difference of Se content (ppb)	Wheat grain		Grade difference of Se content (ppb)	Corn grain	
	Jiangjiashan	Beimen		Jiangjiashan	Beimen
	No. of samples (%)	No. of samples (%)		No. of samples (%)	No. of samples (%)
Under 10	19 (76)	8 (80)	1–2	7 (44)	1 (8)
11–20	3 (12)	1 (10)	2–3	2 (12.5)	9 (75)
21–30		1 (10)	3–4	4	2 (17)
31–50	1 (4)		4–5		
51–100			5–6	2 (12.5)	
101–150	1 (4)		6–7	1 (6)	
151–200	1 (4)				
$\bar{X} \pm SD$	16.7 ± 31.7	5.2 ± 4.4	$\bar{X} \pm SD$	3.6 ± 1.3	2.6 ± 0.4

TABLE 4. Se Content in Flour and Cornmeal Collected from Various Households in Jiangjiashan and Beimen Production Brigades

	Flour			Cornmeal	
Grade difference of Se content (ppb)	Jiangjiashan No. of samples (%)	Beimen No. of samples (%)	Grade difference of Se content (ppb)	Jiangjiashan No. of samples (%)	Beimen No. of samples (%)
Under 10	4 (25)	15 (100)	1–2		2 (14.3)
11–20	2 (12.5)		2–3		2 (14.3)
21–30	2 (12.5)		3–4	6 (37.5)	7 (50.0)
31–40	1 (6.3)		4–5	4 (25.0)	2 (14.3)
41–50	5 (31.0)		5–6	5 (31.2)	1 (7.1)
51–60	1 (6.3)		6–10	1 (6.3)	
61–70	1 (6.3)		11–20		
$\bar{X} \pm$ SD	35.0 ± 23.1	5.2 ± 1.3	$\bar{X} \pm$ SD	6.3 ± 3.0	4.7 ± 1.7

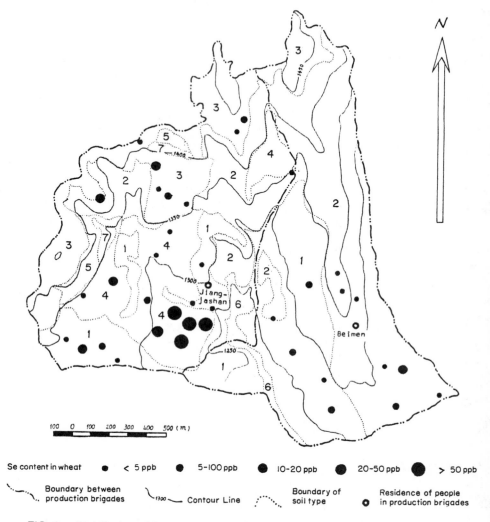

FIG. 3.   Distribution of Se content in wheat of first microenvironment. See legend to Fig. 1 for soil types.

2.2–4.1 times higher than that of Beimen. Furthermore, it can be seen from the variation of hair Se content for various periods of the life of one child who lived in this village that a greater change of Se content in this hair occurred in that of a child who lived in Beimen in the same period (Table 6). The reason for this result may be that the children ate different grains in various periods and the Se content of grains in

TABLE 5. Comparison of Se Content in Children's Hair between Jiangjashan and Beimen Production Brigades

Date of collection of samples	Se content in hair (ppb)		Ratio of Se content in hair, Jiangjashan:Beiman
	Jiangjashan	Beimen	
April, 1979	121 ± 16[a] (6)[b]	32 ± 6.6 (7)	3.8
September, 1979	128 ± 34 (25)	48 ± 26 (21)	2.7
February, 1980	134 ± 50 (10)	33 ± 8.8 (8)	4.1
June, 1980	139 ± 40.5 (10)	63 ± 21.8 (10)	2.2

[a] Average of Se content in hair and standard difference.
[b] Number of samples is in parentheses.

Jiangjashan varied greatly, but that in Beimen varied little since the Se content in grains here was too low. This shows that the Se level in the human body depends on the Se content in grains to a great extent. Based on their investigation in many regions, the authors have put forward a differentiating value of hair Se content between disease and disease-free regions, which is 110 ppb (1). With this value as a standard, 90% of the children of Jiangjashan had hair with a value of more than 110 ppb and nearly 80% of the children in Beimen had hair Se that was not even half of this value. This shows that the incidence of KBD is in close correspondence with the Se level in the human body.

TABLE 6. Se Content in Hair of One Child at Various Periods

Village	Name of children	Se contents in hair (ppb)			
		April, 1979	September, 1979	January, 1980	June, 1980
Jiangjashan	W. X. Ma		112		87
	Z. X. Zhang		131	105	158
	Y. F. Chen		84	164	
	G. X. Yan	143	74	246	175
	G. J. Zao		146	116	173
	W. H. Jiang	104	94		99
	W. H. Jiang		189		124
	G. S. Yan	134	113		153
	X. M. Ma	119	105		203
Beimen	G. Z. Li	28	36	25	49
	P. L. Yan		62	37	78
	S. L. Yan		37	38	62
	Y. L. Yan		43	40	62
	X. P. Ren		38	31	50
	Z. H. Li	24	28	23	
	X. P. Pan	26	57		48
	L. X. Fu	33	97	45	
	Y. F. Zhao	41	34	21	

The Second Microenvironment. The second microenvironment lies in a long slope belt from the top to the bottom and includes four production brigades: Penggong in the center of the Yuan, Langtou near the edge of the Yuan, Xianggong at the edge of the Yuan, and Dongzhui in the valley. All four brigades have different kinds of soils and sources of drinking water.

The variation of Se in this microenvironment shows two tendencies (Table 7): One was the gradual decrease of soil Se with descending altitude from Penggong to Dongzhui; the variation of wheat Se was the same as the soil. The other tendency was that water Se went down gradually only from Penggong to Xianggong, i.e., the center to the edge of the Yuan, while that in Dongzhui located in the river valley became high; the variation of corn Se was the same as that of water owing to the fact that the corn was planted in flooded land and had been irrigated with the river water which contained higher Se.

The change of Se content in children's hair concurred with the Se content in water and grains (Table 8). The Se in drinking water was able to supply the needs of Se in the human body.

TABLE 7. Se Content in Environment and Children's Hair in Various Villages of Zhangwu Yuan

Se content	Lightest attacked village (Penggong)	Lightly attacked village (Langtou)	Severely attacked village (Xianggong)	Disease-free (Dongzhui)
Percentage KBD cases detected by X ray	0.0 (1.9)	20.0 (28.3)	92.5 (64.6)	(0.0)
Altitude (m)	1185	1165	1080	900
Total Se in soil (ppb)	98 ± 7	97 ± 16	85 ± 5.9	70
Soluble Se in soil (ppb)	2.6 ± 0.2	2.2 ± 0.7	1.6 ± 0.6	1.3
Se content in wheat (ppb)	9.4 ± 0.4	7.6 ± 4.3	5.8 ± 1.3	4.4 ± 1.2
Se content in corn (ppb)	4.0 ± 0.6	3.8 ± 0.7	2.8	5.3 ± 2.5
Se content in drinking water (ppb)	0.38	0.24	0.06	2.13
Se daily uptake (μg)[a]	5.03	4.0	2.63	8.73
Se content in hair (ppb)	91.5 ± 14	52 ± 13	49 ± 9.6	93 ± 8.4

[a] The total Se amount provided by the drinking water and grains given to every child in accordance with the proportion of 0.5 kg of grain (70% wheat, 30% corn) and 3 liters of drinking water daily has been calculated.

TABLE 8. Relationship between Se Amounts Provided in Water and Grains with Incidence of KBD through X-Ray Diagnoses

	Relationship with hair Se (r)	Relationship with incidence, detected by X-ray
Drinking water	+0.66	−0.635
Wheat	+0.047	−0.212
Corn	+0.791	−0.846
Se intake from water and grains	+0.812	−0.806
Relationship between hair Se and incidence:	−0.836	

CONCLUSION

The above results indicate the reasons that areas severely and lightly affected by KBD and disease-free areas were interdistributed in the microenvironment of KBD-infected zones. Where Se content in soils and their parent materials are very low, there are different soil types formed by changes of the microgeomophology. There are also great differences caused by Se movement or catchment in soils. Besides, water and heat regimes of soils make Se in soils affect plants differently, resulting in greatly different Se contents in grains in adjacent fields. In addition, hydrogeological and geographic factors can affect differences in the amount of Se in drinking water. On the one hand, all factors may eventually cause the occurrence of KBD in the zones where Se content in grains or drinking water is low. On the other hand, there will be much less KBD in the zones where Se content in drinking water and grains is much higher.

REFERENCES

1. Li, J.-Y. 1982. A study of selenium associated with Kaschin–Beck disease in different environments in Shaanzi. Acta Sci. Circumstantiae 2, 93–101.
2. Li, J.-Y. 1981. A study on the relationship between Kaschin–Beck disease and selenium from Se contents in human hair of various natural environments in Shaanxi Province. J. Environ. Sci. 5, 18.
3. Li, J.-Y. 1979. Relationship between Kaschin–Beck disease and the selenium contents in the environment. Edemic Dis. Newsl. 2.

Pathology and Selenium Deficiency in Kaschin–Beck Disease

Mo Dong-Xu

Kaschin–Beck disease is an osteoarthropathy of unknown etiology with the disturbance of endochondral ossification and deformity of affected joints. It is mainly endemic in eastern Siberia of the Union of Soviet Socialist Republics, northern Korea, and certain areas of China. In 1849, it was first discovered along Urow River in the Transbaikal district of Russia. Kaschin (1859) and Beck and Beck (1906) had separately investigated the epidemiological and clinical aspects.

Pathologically, the disease chiefly affects the hyaline cartilage in the bones of the type in endochondral ossification. Epiphyseal cartilage, articular cartilage, and epiphyseal growth plate at the two ends of tubular bone are the most ordinary sites affected. Thus, the corresponding joints are rather often involved in the pathological process. The injuries of various degree in the cartilage tissue of the whole skeleton including vertebrae and cranial base were described, except for mandibularis articulation.

The cartilage tissue involved shows dystrophic changes, such as atrophy, degeneration, and necrosis, with repair and adaptive changes, which leads to an endochondral ossification disturbance, thus resulting in shortness of fingers, toes, and extremities, even dwarfism and osteoarthrosis deformans.

In the presence of the atrophic process, the chondrocytes are reduced in size and number. The chondrocytic columns become shortened and scanty. The intercolumnar spaces look widened. The affected articular cartilage and epiphyseal plate appear thinner than normal.

The degenerative changes observed in the cartilaginous matrix include red staining, appearance of fibrils, asbestos degeneration, fissure formation, and mucoid degeneration. But these changes also occur in physiological aging, of course, and are not peculiar to the disease.

The more striking finding is chondronecrosis. In the necrotic areas chondrocytes die, lose their stainable nuclei, and become cell ghosts. But the intercellular matrix may stay for a certain period of time. Then the cell ghosts disappear and the necrotic areas gradually become homogenized. We believe that this sort of chondronecrosis, in fact, belongs to the category of coagulation necrosis in pathology.

Chondronecrosis in the disease, according to my observations has six morphological characteristics: (1) In each case the patients cartilage tissue from many bones was involved and the necrotic lesions in articular or epiphyseal cartilage were multiple and localized. (2) Chondronecrotic lesions affected different bones in different cases. The bones of extremities were most often involved, but the affected sites did not strictly show bilateral symmetry. (3) In their distribution, those chondronecrotic areas situated in the deep zone of the cartilage almost always approximated to bone tissue and bone marrow tissue. In severe cases, the chondronecrotic process might expand to the upper zone (or zones). (4) It was the maturating chondrocytes that were mostly affected. Electron microscopy showed two types of cellular damage, namely, cellular vacuolization and condensation, besides some slight degenerative changes in the cytoplasm. (5) Chondronecrosis lesions might be punctate, patchy, or zonal and different in size. (6) Chondrocytes died repeatedly and in batches. The disintegrating processes of necrotic chondrocytes and cartilaginous matrix distinctly assumed the form of phasic changes. In histochemical preparations and in electron micrographs, the degradation and disappearance of proteoglycan and the subsequent decomposition of collagen fibrils were easily confirmed. Newly formed and old lesions could be seen at the same time in the same patient.

After chondrocytes die, secondary changes happen. The surviving chondrocytes adjacent to these necrotic areas soon proliferate and even form chondrocyte clusters. Safranin O staining demonstrates clearly plenty of glycosaminoglycan in the matrix inside the clusters. By electron microscopy, it is found that the chondrocytes inside a cluster have many secretory vesicles and a rough surface endoplasmic reticulum showing the pattern of secretory hyperfunction.

As a result of the reactions from the primordial bone marrow, the processes of absorption, remotion, organization, dystrophic calcification, and ossification gradually take place in the necrotic cartilaginous mass. Eventually, the necrotic areas heal up with the repaired cicatrical bone tissue, which, thereafter, undergoes absorption and adaptive remodeling.

It is impossible to discern the chondronecrotic stage in X rays. Only when the margin of the necrotic area begins to calcify and ossify can the early X-ray signs of Kaschin–Beck disease be seen.

As the normal epiphyseal plate is the growing center of bone, epiphyseal impairment is often followed by developmental disturbance leading to short fingers (toes), short limbs, and even dwarfism. The pathological changes in the articular cartilage may last for a rather long time and progress slowly. Secondary osteoarthrosis with bony enlargement and disfiguration of joints would become prominent.

In addition to the above changes in cartilage and bone, the atrophic process in the striated muscular tissue of extremities is commonly found. No significant change in the patient's viscera is seen.

RELATIONSHIP BETWEEN Se DEFICIENCY AND THE DISEASE

The low selenium content in cereals and drinking water of Kaschin–Beck disease-affected areas was occasionally discovered in experiments feeding rats.

From the spring of 1972 to the summer of 1973, three experiments feeding rats with cereals and drinking water from two endemic regions were performed. In each experiment we unexpectedly found that many rats died of acute massive liver necrosis. In a series of 209 rats fed cereals from the disease-affected area, 83 (39.71%) died of liver necrosis, whereas none died (52 rats) fed cereals from the nonaffected area.

The incidence of muscular dystrophic alteration was even higher than that of liver necrosis. Of 118 rats fed cereals from the disease-affected area, 88 (75.21%) underwent changes in the gastrocnemius.

Reviewing the literature, we suspected that these rats probably suffered from "dietary liver necrosis," which may result from deficiency of cystine, Factor 3 selenium, and vitamin E. For this reason, we estimated the vitamin E content of the feed used and found out that the vitamin E level in the feed was quite low (about 0.006%). The fluorimetric analysis of the Se content of the feed, cereals, and drinking water showed results that were very accurate. The selenium level in the feed and drinking water obtained from the disease-affected area in 1972 was lower than that obtained from the disease-free area, often by 100%. The same results were obtained by analysis in 1973 (Tables 1 and 2).

This is the first discovery of the fact that cereals and drinking water obtained from the Kaschin–Beck disease-affected areas are low in

TABLE 1. Selenium Content in Experimental Feed and Drinking Water (1972)

Origin of feed and water	Selenium in the feed (μg/g)	Selenium in the water (μg/liter)
Disease-affected area	0.051	0.05
Disease-free area	0.080	0.11

selenium and responsible for a selenium-deficiency response in experimental animals.

Later, Dr. Li in Gansu Province began to use selenite to treat the early-stage patients with the disease (1974) and reported that selenite therapy was rather effective (Li 1979).

Geographic researchers proved step by step that the areas affected by both Keshan disease and Kaschin–Beck disease are located mostly in the selenium-deficiency zone of our country. They analyzed quite a number of samples of soil, drinking water, grain, and human hair (1975–1982). They considered that selenium-deficiency in soils may be a main factor of the disease.

In our laboratory, Bai (1983) had made analyses of the selenium content of soil, water, children's hair, and urine once a season in two spots of Kaschin–Beck disease-affected areas in Shaanxi Province. They found that the selenium level in water and the water-soluble selenium in soil were higher in spring and summer than in autumn and winter (Fig. 1 and Table 3).

As the weather in those two spots is very dry in the first half of the year (the total rainfall is 66.7 mm and the total vaporization 724.4 mm), it would promote the accumulation of the water-soluble selenium on the earth and raise the selenium content in the soil and water. However, in the second half of the year, the rainfall is heavy (the total rainfall is 516.7 mm and the total vaporization 408.7 mm). The selenium in the soil and water would be partly washed away and would be greatly decreased.

The selenium content in cereals was influenced by the water-soluble selenium in soil. A positive correlation was found between them. In the spots mentioned above, the selenium content in wheat flour was

TABLE 2. Selenium Content in Experimental Feed (1973)

Origin of feed	Selenium in the feed (μg/g)
Disease-affected area	0.036
Disease-free area	0.061

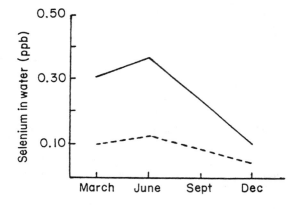

FIG. 1. Seasonal variation of selenium content in drinking water from two areas within Kaschin–Beck disease-affected area. (——), Area 1; (- - -), Area 2.

higher than that in maize flour by about 13.1–51.3% during the 2 years (Table 4). This would be due to the fact that wheat grows and matures in spring and summer, whereas maize grows and matures in autumn and winter in these areas.

The selenium content of children's hair and urine in those two affected areas showed higher selenium in summer and autumn, but lower selenium in winter and spring (Table 5).

Thus, it is obvious from the above results that the variations of selenium in the human body and its environment are not synchronistic. This is to say that the selenium level of the human body is influenced by various kinds of foods rather than completely by cereals and water.

Recently, Bao (1983) reported that the selenium level in human blood and hair in Kaschin–Beck disease-affected and nonaffected areas of Henan Province was also seasonally changeable. It was higher in summer and autumn and lower in winter and spring, just like our findings in human hair and urine. Are the seasonal changes of selenium inside and outside the human body all the same in various endemic areas of the disease each year? The answer is not yet known at the present time.

Another research group at our college studied some selenium-relat-

TABLE 3. Water-Soluble Se Content (ppb) in the Soil from Two Areas within the Disease-Affected Area in Different Seasons

Disease-affected area	March	June	September	December
1	3.80	—	1.98	2.05
2	4.30	—	1.81	1.04

TABLE 4. Selenium Content in Wheat and Maize Flour from the Two Areas within Kaschin–Beck Disease-Affected Area in 1980 and 1981

Area	Flour	Se content (1980)		Se content (1981)	
		(ppm)	(%)	(ppm)	(%)
1	Wheat	0.031	155.0	0.031	147.0
	Maize	0.020	100.0	0.021	100.0
2	Wheat	0.027	117.4	0.025	108.7
	Maize	0.023	100.0	0.023	100.0

ed changes in human blood. The activities of blood glutathione peroxidase (GSHPx) and the content of vitamin E, lipid peroxide (TBA) and free fatty acid (FFA) in children in the Kaschin–Beck disease-endemic areas were determined (Han 1983). They reported that the activities of blood GSHPx and vitamin E levels were lower than those of children in nonendemic areas ($P < .001$). But the levels of blood TBA and FFA were higher than those of children in nonendemic areas ($P < .001$). They suggested that the selenium deficiency state of children's bodies might induce lipid peroxidation.

It was noted that the milieu of the children with Kaschin–Beck disease was in a low selenium nutritional state, but that not all the children with low selenium nutrition could necessarily be patients detected by X ray.

In our laboratory, a study on the relationship between urinary Se levels and X-ray detection rate (incidence) of the disease was made by Jiang (1983). They were able to draw a dose–response curve using the values determined in April and September of 1982 (see Fig. 2). The ED_{50} of these two urinary selenium values were identical, approximately equal to 2.0 μg Se/24 hr.

The preliminary results above show that there may be a causal relationship between the low selenium level in the milieu and the incidence of Kaschin–Beck disease. In addition, the researchers also found that urinary creatinine excretion in the children with a low selenium nutritional state was lower than that in the children without a low selenium state. A significant linear correlation was shown between urinary creatinine and selenium excretion. By analyzing the data obtained in April and September of 1982, they acquired similar results (Figs. 3 and 4).

The authors suggested that the cause of the depression in creatinine metabolism of the children with Kaschin–Beck disease might be due to low selenium levels in the milieu. However, these results still need further confirmation.

TABLE 5. Seasonal Variation of Se Content in Children's Hair and Urinary Se Excretion in Two Areas within the Kaschin–Beck Affected Area

	Area	March $\bar{X} \pm$ SE (N)	June $\bar{X} \pm$ SE (N)	September $\bar{X} \pm$ SE (N)	December $\bar{X} \pm$ SE (N)
Se in hair (ppb)	1	175.0 ± 14.4 (17)	256.9 ± 20.1 (20)	211.3 ± 18.6 (20)	155.9 ± 18.2 (17)
	2	109.7 ± 6.9 (20)	144.0 ± 9.7 (26)	200.4 ± 16.6 (25)	121.6 ± 9.6 (23)
Se in urine (μg/24 hr)	1	4.96 ± 0.49 (16)	4.90 ± 0.41 (24)	5.30 ± 0.55 (18)	4.05 ± 0.30 (17)
	2	2.88 ± 0.22 (21)	3.93 ± 0.40 (29)	3.82 ± 0.37 (24)	3.79 ± 0.28 (21)

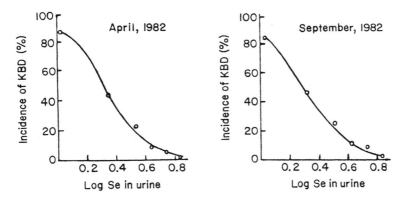

FIG. 2.   Relationship between the incidence of Kaschin–Beck disease and urinary selenium excretion determined in April and September of 1982.

To summarize: Problems worthy of note are as follows: The location of Kaschin–Beck disease-endemic areas are mainly situated in the selenium-deficient zones of our country. The low selenium nutritional state of the population in those areas is affected by selenite in treating the early-stage patients and preventing the disease. The presence of a dose–response relationship between urinary selenium levels and the incidence of the disease in children occurs. All of this tends to support

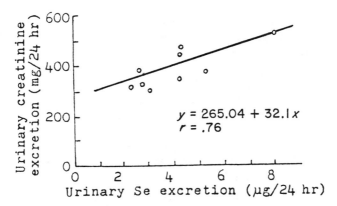

FIG. 3.   Scatter diagram showing creatinine and selenium excretion in the urine of children living in a Kaschin–Beck disease-endemic area (April, 1982).

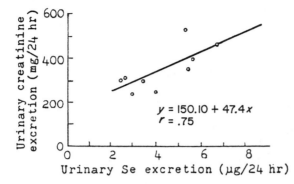

FIG. 4. Scatter diagram showing creatinine and selenium excretion in the urine of children living in a Kaschin–Beck disease-endemic area (September, 1982).

the idea that selenium deficiency probably plays an important part in the etiology and/or the pathogenesis of the disease.

However, the following contradictions merit attention: Some areas with low selenium in the external environment are nonendemic. The hair selenium levels in the population of some endemic areas are higher than those of the generally endemic or nonendemic areas. Feeding rats with cerals low in selenium, which were obtained from the endemic areas, could bring about only atrophy of the epiphyseal growth plate. No typical chondronecrosis in cartilage tissue was found in these rats (1972–1973). Semisynthetic diets, which did not contain cereals from endemic areas but were low in selenium (0.017 ppm), could induce atrophy of epiphyseal growth plate of rats, but not cartilage necrosis (Ren 1982). I would like to emphasize here that multiple focal chondronecrosis is the most important pathologic feature of patients with the disease.

It is well known that selenium deficiency in different species of animals brings about different pathological alterations. To deduce from the results of animal experiments for man is hardly justified. Of course, a negative result from animal experiments cannot be used to deny epidemiological facts either. Feeding tests in primates with selenium-deficient diets or cereals from Kaschin–Beck disease-affected areas may be helpful, and they are being carried out in other laboratories.

It should be emphasized that endemic areas where Keshan disease and Kaschin–Beck disease coexist are not rare. Whether both Keshan disease and Kaschin–Beck disease in human beings are due to the same cause, selenium deficiency, or selenium deficiency as only one common factor of these two diseases, is still a mystery and needs further investigation.

REFERENCES

Bai, C. 1983. A study of selenium variation in the inner and outer environment of Kaschin–Beck disease regions. New Med. Pharm. Shanxi Prov., Suppl. pp. 49–51 (in Chinese).

Bao, B. 1983. A study on the seasonal variation of selenium and vitamin E level of the human body in Kaschin–Beck disease-affected and nonaffected areas (in Chinese). Unpublished.

Han, C. 1983. Determination of blood glutathione peroxidase, vitamin E, lipid peroxide, and free fatty acid of children in the Kaschin–Beck-affected areas. Chin. J. Endemiol. 2 (2), 65 (in Chinese).

Jiang, Y.-F. 1983. A study on urinary creatinine excretion of children with different urinary selenium levels in Kaschin–Beck disease endemic and adjacent nonendemic areas. A discussion about the relationship between Se lowness and incidence of the disease (in Chinese). Unpublished.

Li, C.-Z. 1979. The effect of sodium selenite and vitamin E on 224 cases of patients with Kaschin–Beck disease based upon X-ray examination and discussion about the etiology of the disease. Chin. Med. J. 59 (3), 169 (in Chinese).

Ren, H.-Z. 1982. 1982. A histological study concerning the effects of trace element selenium on the development of metaphyseal cartilage of rat. Chin. J. Pathol. 11 (3), 220 (in Chinese).

Sodium Selenite as a Preventive Measure for Kaschin–Beck Disease as Evaluated in X-Ray Studies

Li Chong-Zheng
Huang Jing-Rong
Li Cai-Xia

Kaschin–Beck disease is an endemic disease which may seriously endanger the health of the inhabitants and affect the normal growth of children living in that area. From 1974 to 1976, we studied the effect of sodium selenite with vitamin E in treating Kaschin–Beck disease and obtained remarkable therapeutic effects. We have suggested that insufficient intake of selenium might be the main cause of the disease. Shortly thereafter, many other medical units repeated our experiment and confirmed the good therapeutic effect of sodium selenite on the metaphyseal lesion of Kaschin–Beck disease. However, the preventive effect of selenium was still to be determined. Therefore, from 1977 to 1983, a series of experiments by oral administration of sodium selenite alone to the children had been carried out in the fifth and sixth production team of Yangzui Brigade, Taichang Commune, Ningxian, Gansu, one of the main endemic areas in Gansu Province, China. This chapter reports some positive results found in the 6 years.

METHODS

Selenium Administration

All children under 10 years of age in the above-mentioned production teams were given sodium selenite tablets with the following dosages: 1 mg/week for children from 6 to 10 years of age; 0.5 mg/week for those below 5; and 2 mg/week for lactating mothers. The number of times and doses of selenium administration were recorded, and only those who had taken selenium more than 35 times in a year were counted.

Roentgenography Examination

Before administration, X-ray films of the right hand were taken of all children from 3 to 10 years of age in the two teams. Thereafter, X-ray examination was made in May of every year. A few individuals dropped out or failed to come for inspection, but over 95% of the children were under observation each year. Those who reached 10 years of age were not included.

Index of Preventive Effect

The percentage of normal X-ray appearances of children 3–10 years of age in each year was used as an index of preventive effects.

RESULTS

The percentage of normal appearances in children for X-ray examinations from 1977 to 1983 is depicted in Tables 1 and 2.

As shown in Tables 1 and 2, the percentage of normal children tended to increase year by year, which was statistically significant.

TABLE 1. Percentage of Normal Children of the 5th–6th Production Teams, Yangzui Brigade, Examined by X Ray from 1977 to 1983[a]

Year	Total number of children observed	Number of normal children	Normal children (%)
1977 (before administration)	47	27	57.5
1978	46	35	76.1
1979	49	38	77.6
1980	51	41	80.4
1981	52	46	88.5
1982	48	44	91.7
1983	49	47	95.9

[a] Calculation of linear regressional significance level for chi-square analysis: $\chi = 27.90$; $P < .001$.

TABLE 2. Number of Normal Children in Different Age Groups

| | Age (yr) | | | | | | | | |
	3	4	5	6	7	8	9	10	Total
1977 Observed	9	8	9	7	6	3	2	3	47
Normal	8	5	5	3	3	1	1	1	27
1978 Observed	4	8	8	9	7	6	3	1	46
Normal	4	8	6	7	5	3	1	1	35
1979 Observed	4	4	9	7	9	5	6	3	49
Normal	4	4	9	5	8	5	2	1	38
1980 Observed	7	4	3	9	6	9	7	6	51
Normal	7	4	3	9	4	8	4	2	41
1981 Observed	7	6	4	4	8	7	9	7	52
Normal	7	6	4	4	8	5	7	5	46
1982 Observed	4	7	6	4	4	8	7	8	48
Normal	4	7	6	4	4	8	5	6	44
1983 Observed	9	4	7	6	4	4	8	7	49
Normal	9	4	7	6	4	4	8	5	47

DISCUSSION

In this intervention study, the percentage of normal children under X-ray study was taken as the index to evaluate the efficacy of selenium treatment. It was found that normal children in nonendemic areas might show X-ray changes in their phalanges during their growth similar to early-stage Kaschin–Beck disease. It was therefore very difficult to distinguish normal and diseased children. If the percentage of Kaschin–Beck disease children was used as the index of the preventive effect, a part of normal children would be included, so the precise assessment on preventive effects could not be drawn. To avoid incorrect diagnosis of normal physiological variations as Kaschin–Beck disease patients, we used the number of children without any X-ray changes in their bones as an index for comparison. All the children with irregular shadows on X-ray films, no matter whether they were due to Kaschin–Beck disease or to normal physiological variations, were both treated as abnormal cases.

It was observed that of the 49 cases X-rayed in 1983, only 2 who were then 10 years old, were abnormal, with both beginning to take selenium when they were 4 years old and already seriously affected. Even though they took the selenium tablets continuously, they failed to respond and did not have normal X-ray pictures. It was shown that preventive selenium treatment must begin in early childhood. In our experiment, none of the children given sodium selenite before 3 years old had developed Kaschin–Beck disease.

CONCLUSIONS

Sodium selenite was used to prevent Kaschin–Beck disease for 6 years in children 3–10 years of age. The percentage of normal children examined by X ray increased from 57.45% before selenium administration in 1977 to 95.5% after continuous selenium administration in 1983.

It was shown that prevention should be started in early childhood to obtain better results.

No side effects of selenium have been found in the children under observation.

Effects of Selenium Supplementation in Prevention and Treatment of Kaschin–Beck Disease

Liang Shu-Tang *Mu Si-Zhang*
Zhang Ju-Chang *Zhang Fu-Jin*
Shang Xuan

Kaschin–Beck disease is one of the severe endemic diseases in China. The cause of the disease remains unknown at present, and there is no absolute method for its prevention or treatment. In order to explore the effects of selenium supplementation in preventing and treating the disease, we have investigated 984 children in an endemic area of Shaanxi Province. By studying radiograms, content of Se in the hair and urine, liver function tests, and other biochemical examinations, we have found that Se supplementation helps in promoting the repair of the lesions in the finger metaphyses and in preventing further disease development (Fig. 1).

THE LEVELS OF Se IN THE HUMAN BODY AND ENVIRONMENT OF THE ENDEMIC AREA

Of all the chemical elements in the environment of the experimental area, we are most concerned with Se and its relation to the occurrence of Kaschin–Beck disease, i.e., the levels of Se in water, soil, grain, and hair of human beings in the endemic area are much lower than those in the nonendemic area. The content of Se in water, soil, and hair was determined with fluorospectrophotometry, and the content of Se in grain was determined with the atomic absorption method (see Table 1).

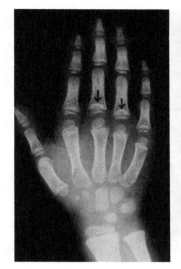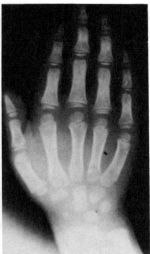

FIG. 1.   Radiograms of typical cases. Left, Before Se supplementation; right, after Se supplementation.

In patients in the early stage of the disease, there is an obvious
metabolic disturbance characterized by a low content of Se in the hair,
decreased activity of the glutathione peroxidase (GSHPx) in whole
blood, and reduced vitamin E in the plasma. The levels of lipid peroxidase (TBA) and free fatty acid (FFA) in the plasma are much higher
than those in the nonendemic area (see Table 2).

METHOD OF THE EXPERIMENT

To study its effects in preventing or treating Kaschin–Beck disease,
Se supplementation was carried out in two periods of time, from June,
1980 to June, 1981 (with Se tablets), and from October, 1981 to May,
1983 (with Se salt). A total of 984 children, aged from 2 to 13 years,
were selected. For the trial using Se tablets, 424 children were selected, and 560 children were selected for the trial in which Se salt was
studied. Children in the experiments were grouped first according to
the development of their bones, location of the lesions on the radiograms, and the nature of the lesions and their extent, and then they
were divided into the Se tablet group and the control group (see
Table   3).

TABLE 1. Content of Se in the Environment of the Endemic Area ($\bar{X} \pm$ SD in ppb)

Area	Water	Soil	Wheat	Corn	Hair
Endemic	0.18 ± 0.16 (90)[a]	2.12 ± 0.67 (92)	4.1 ± 3.0 (92)	3.6 ± 1.9 (83)	44 ± 1.9 (86)
Nonendemic	1.0 ± 1.1 (24)	2.6 ± 0.7 (27)	27.6 ± 19.5 (25)	15.5 ± 9.0 (20)	212 ± 6 (16)

[a] The numbers in parentheses are the number of samples.

TABLE 2. Metabolic Disturbance of Low Se in Patients with Kaschin–Beck Disease

Area	GSHPx (activity unit) ($\bar{X} \pm$ SD)	Vitamin E (μg/ml) ($\bar{X} \pm$ SD)	TBA (nmole/ml) ($\bar{X} \pm$ SD)	FFA (rmm) ($\bar{X} \pm$ SD)
Endemic	105 $\pm$ 11.2	8.51 $\pm$ 1.76	4.18 $\pm$ 0.77	1.05 $\pm$ 0.26
Nonendemic	129.3 $\pm$ 13.8	10.16 $\pm$ 1.99	2.32 $\pm$ 0.8	0.56 $\pm$ 0.22

Standards of Diagnosis and Evaluation of Effectiveness

Radiographic diagnosis was made according to the unified standards in effect in China (for trial implementation). The results of the changes in the metaphysis were described as cured, improved, no change, or worse. Improvement of the metaphyseal lesions was evaluated in three grades: much better, which means that the lesion in the metaphyses has changed from ($++$) to ($\pm$) or from ($+$) to ($-$); better, which means that the lesion in the metaphyses has changed from ($++$) to ($+$) or from ($+$) to ($\pm$); a little better, which means that the lesion had some improvement, but the symbols of ($+$) and ($-$) could not be changed.

The Se tablet group was given tablets of sodium selenite (each containing 1 mg of Se) orally. The control group was given placebos containing starch. Children between 3 and 10 years of age received one table once a week; children between 11 and 13 years of age received two tablets once a week. The Se salt group took the selenized salt daily.

Criteria of Observation

Radiograms of the hands were taken before and after the administration of Se. The results of the group were analyzed and compared

TABLE 3. Categories of Children under Investigation

Category	Se tablet group	Control group	Se salt group	Control group
Number of observed children	220	204	409	151
Number of children with X-ray changes	182	174	266	105
Number of children with metaphyseal lesions	166	159	110	31
Frequency of the X-ray changes (%)	82.7	85.3	65.0	69.5
Frequency of the changes in the metaphyses (%)	75.5	77.9	26.9	205
Number of healthy children	38	30	143	46

with individual changes in the metaphyses. Several criteria for change were frequency of radiographic changes in general (%), frequency of the changes in the metaphyses (%), the index of the activity of the lesion, and the index of the severity of the lesion. Individual results expressed as cured, improved, no change, or worse were compared.

RESULTS OF THE EXPERIMENTAL RESEARCH AND ITS ANALYSIS

In the Se tablet group, the number of patients with changes in their metaphyses (++) was 65, which was reduced to 15 a year later after Se supplementation. Comparison within the group revealed a significant difference ($P < .01$). In the control group, the number of patients was reduced from 62 to 56; no significant difference could be seen from comparison within the group ($P > .05$). One year later, a parallel comparison between these two groups revealed a marked difference ($P < .01$). The index of lesion activity in the Se tablet group was reduced from 115.7 to 81.3, while no obvious reduction could be seen in the control group. In the Se salt group, the number of patients with changes in their metaphyses (++) was 110, which, 1½ years after the experiment, was reduced to 32. The comparison within the group ($P < .01$) revealed a very marked difference; in the control group the cases were reduced from 31 to 6, and no obvious difference could be seen from comparison within the group ($P > .05$). The index of lesion activity in the Se salt group had a greater reduction than that in the control group. Other collective indications before and after the experiment were of no statistical significance (see Table 4).

The metaphyseal lesions responded best to Se and so the improvement of the lesions in the metaphyses served as a main indicator in our experiment. The total improvement rate of the lesions in the metaphyses in the Se tablet group was 81.9% (the acceptable limit of 95% was 75.9–87.9%), showing a significant effectiveness greater than that with other drugs. The total improvement rate in the control group was 39.6%. The comparison between the two groups ($P < .001$) revealed a very marked difference. In the Se salt group, after 1 year and 7 months of supplementary Se, the marked improvement rate of changes in the metaphyses was 11.3%, while in the control group this was only 2.3%. The comparison between the two groups ($P < .01$) revealed a very marked difference. No worsening of the condition was exhibited in the Se tablet group, while in the control group 18.9% of the cases took a turn for the worse ($P < .01$). The difference was very obvious. The cases which became worse were less in number than those

TABLE 4. Changes of the Lesion in Different Groups after Supply of Se

	Se tablet group		Control group		Se salt group		Control group	
	80.6	81.6	80.6	81.6	81.1	83.5	81.1	83.5
Number of observed children	220	220	204	204	409	304	151	86
Number of patients with changes in their metaphyses (++)	65	15	62	56	110	32	31	6
Index of the lesion activity[a]	115.7	81.3	118.2	113.7	103.8	67.0	91.0	68.3

[a] The index of lesion activity $= \left(\dfrac{\text{frequency of changes in metaphyses}}{\text{number of children examined}} \times 100 \right) + \left(\dfrac{\text{number of children with changes } (++) \text{ in metaphyses}}{\text{number of children with changes in metaphyses}} \times 100 \right)$.

TABLE 5. Comparison of Changes in the Metaphyses

Group	Cases	Improvement					No change	(%)	Worse	(%)
		Marked	Some	Slight	Total	(%)				
Se tablet	166	27	86	23	136	81.9	30	18.1	0	0
Control	159	4	38	21	63	39.6	66	41.5	30	10.9
Se salt	152	31	27	43	101	66.4	45	29.6	6	3.9
Control	52	2	15	18	35	67.3	12	23.1	6	11.5

in the control group ($P > .05$), and no obvious difference was observed (see Table 5).

The content of Se in the hair in the Se tablet group increased from the original 40 ppb to 244 ppb, while no notable difference could be seen in the control group (see Table 6).

In both the Se tablet group and the control group, the determinations of aminopolysaccharide, oxyproline, and creatinine in urine, made before and after Se supplementation, showed that Se supplementation can improve and promote the metabolism of polysaccharide and collagen of proteins in the cartilage of the patients and increase the rate of their conversion (see Table 7).

Safety

It has been proved by liver function tests made before and after the experiment that the dosages of Se tablets and Se salt in our experiment were safe in the human body.

MECHANISM OF THE EFFECT OF SELENIUM SUPPLEMENTATION IN PREVENTING AND TREATING KASCHIN–BECK DISEASE

The amount of Se in water, soil, and grain in the endemic area is small. The Se level in the bodies of people in the endemic area is generally low. Se supplementation can greatly improve the metabolism in the cartilage and have a significant preventive and therapeutic effect. All these factors show that Se deficiency leads to metabolic disturbance, decreased activity of GSHPx, and low content of vitamin E. TBA and FFA, on the other hand, are increased. This is unfavorable for removing the harmful factors that can damage the cell membrane and discourage the synthetic process of polysaccharides or proteins in the cartilage. Supplementary Se can increase the activity of GSHPx, which contains Se. This is favorable for removing the "highly active oxide" produced in the process of oxidation in the body and the lipid peroxide, and thus protect the integrity of the cell membrane and the

TABLE 6. Result of Determination of Se in the Hair ($\bar{X} \pm$ SD in ppb)

	Se tablet group	Control group
Before Se supplementation	40 ± 2.5	40 ± 1.5
After Se supplementation	278 ± 46	47 ± 1.9

TABLE 7. Average Amount of Polysaccharide, Oxyproline, and Creatinine in Urine Determined before and after Se Supplementation

Group	Aminopolysaccharide (mg/24 hr)	Oxyproline (mg/24 hr)	Creatinine (mg/24 hr)
Se tablet, before Se supplementation	5.23 ± 0.23[a](36)[b]	97.23 ± 11.23 (32)	363.18 ± 18.75 (34)
Se tablet, after Se supplementation	7.67 ± 0.31 (32)	94.43 ± 0.33 (32)	412.47 ± 17.36 (32)
Control before administration	5.74 ± 0.26 (34)	115.55 ± 6.20 (32)	386.85 ± 23.69 (32)
Control after administration	7.19 ± 0.31 (32)	94.50 ± 5.91 (32)	416.91 ± 22.40 (32)

[a] X̄ ± SD.
[b] Numbers in parentheses are the number of samples.

945

stability of its function. After supplementation, the amount of aminopolysaccharide and oxyproline in urine is increased, which can greatly improve the metabolism of the polysaccharide and collagen of proteins in the cartilage and promote the normal development of the cartilage system in children. Se supplementation has a marked effect in treating the lesion in the metaphyses, promoting the repair of the bone tissue, and preventing it from becoming worse. The possible role of Se in the disease is still not clear. Therefore, to determine the significance of Se in the etiology of the disease and the effectiveness of supplementation of Se, further extensive research is required.

100

Abnormalities of Erythrocyte Membranes from Patients with Kaschin–Beck Disease

F.-Y. Yang   S.-G. Li   W.-H. Wo
F. Huang   S. Sun   R.-Q. Zhang
Z.-H. Lin   J.-W. Chen   Q.-R. Xing
K. Zhang   L.-P. Zhang   W.-W. Chen
B.-S. Shi   B.-G. Guo   J. Zhou
Y.-G. Fu

Kaschin–Beck disease (KBD) is an endemic disease in China which causes serious impairment of health. Many different pathogenic hypotheses for the disease have been suggested, but the cause has not yet been elucidated.

It was reported that the lipid composition of erythrocyte membranes of children suffering from KBD was abnormal. Scanning electron microscopic studies have also shown abnormal membrane morphology (1). It was postulated that KBD might be a kind of membrane disease (1). If true, the structure and function of erythrocyte membranes of infantile patients with KBD would be changed. In this chapter, we present studies of the Na^+,K^+-ATPase activity, membrane fluidity, and membrane skeletal components of erythrocytes of children suffering from KBD.

The endemic areas of the disease seem to be in an Se-deficient belt, and Se deficiency might be a pathogenic factor in the prevalence of KBD. Hence, the Se content of erythrocytes as well as erythrocyte membranes of control children and infantile patients with KBD were determined and compared.

Our preliminary results show that the Se level, Na^+,K^+-ATPase activity, lipid fluidity, and some membrane skeletal components of erythrocytes of infantile patients with KBD differ from those of the control individuals.

EXPERIMENTAL PROCEDURES

The investigation was carried out in the endemic region of Yong Shou (Han Gou) in Shanxi Province in April, 1983.

Sixteen patients with KBD, between the ages of 8 and 12, were chosen for study. According to the diagnosis standard by X ray, there were 11 patients with pathological change in their metaphysis (metaphysis+) and 5 patients with pathological change in bone end (bone end+). Next, 5 children from the endemic region and 20 children from the nonendemic region (Lan Tian), ages 8–12, were chosen as the control.

MATERIALS AND METHODS

Preparation of Human Erythrocyte Membranes (2)

Venous blood (3 ml) was collected from the children of the endemic or nonendemic region and placed into tubes containing 3 ml of ACD, an anticoagulant solution (NaCl, citric acid, glucose solution), and mixed thoroughly by inversion. A suitable amount of normal saline was added and centrifuged at 2000 rpm for 5 min to remove the plasma and buffy coat. Finally, the erythrocytes were washed three times with normal saline, each time centrifuging for 5 min at 1500 rpm. A cold Tris–HCl buffer (10 mM, pH 7.4) was added to the erythrocyte sample to cause hypotonic hemolysis and was centrifuged at 9000 g for 8 min at 0°–4°C: the pellets obtained were the erythrocyte ghosts. The membranes were again washed with Tris–HCl buffer three times. The erythrocyte membranes were stored in the refrigerator for use.

Measurement of Selenium Content of Erythrocyte Membranes (3)

Ten milliliters of a mixture containing 7.5 g sodium molybdate in 150 ml water, 200 ml perchloric acid, and 150 ml Se-free sulfuric acid were added to the erythrocyte membranes previously placed in a covered conical flask and allowed to stand at room temperature for 4 hr. This was then heated on a sand bath until the contents in the flask turned yellowish-green and was removed immediately from the heat source. Digestion was complete when the contents in the flask become colorless, after cooling. Then 20 ml of EDTA–NH$_4$OH mixture containing 0.2 M EDTA:10% NH$_4$OH·HCl:0.02% Cresol Red (=100:10:5) was added and mixed well by shaking. (The pH of the contents was adjusted with NH$_4$OH (NH$_4$OH:H$_2$O = 1:1) and concentrated HCl until the indicator showed a light orange color (pH 1.5–2.0). The flasks were cooled with cold water, and 3 ml of 0.1% DAN (2,3-diaminonaphthalene) reagent were added in subdued light. After mixing well, this was heated in a boiling water bath for 5 min and allowed to cool. Then 1 ml of cyclohexane was added and the tubes were shaken vigorously for 5 min. The contents were transferred into a separatory funnel; after the layers separated, the water layer was discarded and cyclohexane layer was decanted into a covered test tube.

Calculation of the Se content (μg) per milliliter of sample is made as follows:

$$\frac{A - B}{\text{weight of sample}} \times \frac{S}{C - D}$$

where A is the fluorescence reading of the sample; B is the fluorescence reading of blank of sample; C is the fluorescence reading of standard; D is the fluorescence reading of blank of standard; and S is the Se content of standard.

Determination of Na$^+$,K$^+$-ATPase Activity of Erythrocyte Membranes (4,5)

The reaction medium for measuring the activity of Na$^+$,K$^+$-ATPase contained NaCl, 6×10^{-2} M; KCl, 1×10^{-2} M; MgSO$_4$, 1.67×10^{-3} M; phosphoenolpyruvate (PEP), 1×10^{-3} M; pyruvate kinase (PK), 10 μl; ATP, 1×10^{-3} M; and Tris–maleate, 4.2×10^{-2} M, pH 8.0, in a total volume of 2.0 ml. Prior to the determination, we diluted the membranous sample to 400 μg/ml with buffer solution (10 mM Tris–HCl, pH 8.0) and then froze and thawed the samples three times (freezing in dry ice and thawing in 37°C water bath). Then 0.1 ml of the membranous sample (i.e., 40 μg of membrane protein) was added to the reaction medium (without MgSO$_4$ and ATP) in each test tube and preincubated for 10 min at 37°C. The reaction was started by adding MgSO$_4$-ATP solution at 37°C; the decreasing activity was measured for 1 hr at $A_{340\,nm}$. The activity of Na$^+$,K$^+$-ATPase is expressed as the rate of activity decrease at $A_{340\,nm}$ minus the rate of decrease in activity in the same medium with ouabain added.

The activity of Na^+,K^+,Mg^{2+}-ATPase is measured as the rate of decrease in activity at $A_{340\,nm}$ minus the rate of decrease in activity at $A_{340\,nm}$ induced by the spontaneous hydrolysis of ATP in the absence of membranes.

Calculation of Na^+,K^+-ATPase activity is as follows: The specific activity is defined as 1 μmol of ATP hydrolyzed/hr/mg of membrane protein (1 μmol ATP/hr/mg). The ATPase activity can be calculated according to the following formula:

$$\frac{\Delta A_{340\,nm}}{t\ (hr)} \times \frac{2}{6.22} \times \frac{1}{\text{sample protein (mg)}}$$

Discontinuous SDS–Polyacrylamide Gel Electrophoresis of Skeletal Components of Erythrocytes (6)

Discontinuous SDS–polyacrylamide gel electrophoresis of membrane skeletal components of erythrocytes was carried out by the method of Davis (6); a separating gel containing 8.8% (w/v) acrylamide was used with a 3.5% stacking gel. Membrane protein (150 μg) was applied to the gels. Electrophoresis was done in glycine-Tris buffer, pH 8.3, at 3 mA per tube for 4 hr. The gels were stained with 0.25% Coomassie Brilliant Blue R-250 overnight and destained by methanol until the background staining had been suitably reduced. Gels were scanned by Shimadzu CS-910 TLC scanner. The peak area was calculated by microcomputer.

Measurement of Fluidity of Erythrocyte Membranes by Polarization Fluorescence Studies Using DPH (7)

DPH (Sigma) tetrahydrofurane solution ($2 \times 10^{-3}\ M$) was freshly diluted with PBS buffer (NaCl, 171 mM; KCl 34 mM; Na_2HPO_4, 10.2 mM; KH_2PO_4, 1.8 mM, pH 7.4) to $2 \times 10^{-6}\ M$ and mixed on a YKH mixer for 15 min. Then 1.0 ml of the above solution was pipetted out and an equal volume of erythrocyte membrane suspension corresponding to 50 μg protein was added. After vigorously stirring, the solution was incubated at 30°C for 1 hr and assayed by means of a Hitachi 650-60 fluorescence spectrophotometer fitted with a polarization attachment. The wavelength of excitation and emission were 360 nm and 403 nm, respectively. Corrections were made for light scattering and background by using the erythrocyte membrane suspension incubated without DPH. The degree of fluorescence polarization (P), which can reflect the motion and viscosity of lipid molecules, was estimated from the following formula:

$$P = \frac{I_{\parallel} - I_{\perp}}{I_{\parallel} + I_{\perp}}$$

where $I_{\parallel}$ and $I_{\perp}$ are the fluorescence intensities parallel and perpendicular to the polarized excited beam, respectively.

RESULTS AND DISCUSSION

Measurement of Selenium Content of Erythrocytes and Erythrocyte Membranes of the Infantile Patients with KBD

The results shown in Table 1 indicate there were six experimental groups examined including Beijing adults, Beijing children, control subjects from endemic and nonendemic regions, and infantile patients with KBD involving the metaphyses (metaphysis+) or the bone end (bone end+). Regardless of pathological changes having occurred in the infantile patients, the Se content of erythrocytes is found to be 66%

TABLE 1. Se Content of Erythrocytes and Erythrocyte Membranes from Infantile Patients with KBD in Yong Shou

Group	Se content of erythrocytes (μg/mg hemoglobin)	Se content of erythrocyte membranes (μg/mg membrane protein)
Control children in endemic region	0.0293 ± 0.0091 (5)[a]	
Control children in nonendemic region	0.0730 ± 0.0064 (20)	0.933 ± 0.161 (20)
Infantile patients with KBD (metaphysis+)	0.0227 ± 0.0053 (11) $P < .001$[b]	0.562
Infantile patients with KBD (bone end+)	0.0279 ± 0.0091 (5) $P < .001$[c]	
Beijing adults	0.310 ± 0.06 (36)	1.85 ± 0.967 (36)
Beijing children	0.319 (1)	

[a] The numbers in parentheses denote the number of samples measured.

[b] $P < .001$ indicates that the difference between the Se content of erythrocytes of the infantile patients with KBD (metaphysis+) and that of control of the nonendemic region is statistically significant.

[c] $P < .001$ indicates that the difference between the Se content of erythrocytes of the infantile patients with KBD (bone end+) and that of control of the nonendemic region is statistically significant.

less than Se values from control subjects from the nonendemic region. Because of a larger quantity of samples required for determining the Se content, the Se content of erythrocyte membranes of some of the infantile patients suffering from KBD was compared with the control children of the nonendemic region. Table 1 shows that the Se content of erythrocyte membranes of the former is about 50% of the latter. It is interesting that the Se content of erythrocytes and erythrocyte membranes of the children from the nonendemic region (Lan Tian) is invariably lower than that of the adults or children of the Beijing region. Se content of erythrocytes of the former is only 25% of the latter and that of erythrocyte membranes about 50%.

In summary, the Se content of both erythrocytes and erythrocyte membranes from the infantile patients with KBD is significantly lower than the control. It is generally considered that Se content is closely related to glutathione peroxidase activity in the cells, but the physiological significance of Se in erythrocyte membranes is so far not clear. It is worth studying if the significant decrease in the Se content of erythrocyte membranes of the infantile patients with KBD is correlated to the abnormality of erythrocytes. The Se content of human erythrocyte and erythrocyte membranes of the Beijing region is significantly higher than that of the nonendemic region. It is interesting to note that although the Se content of erythrocytes of the controls from the endemic region is significantly lower than of the nonendemic region, the controls from endemic regions do not differ from the infant patients of the endemic region (Table 1). Factors other than Se might also be involved in the pathogenesis of KBD.

Change of Na^+,K^+-ATPase Activity of Erythrocyte Membranes of Infantile Patients with KBD

The activities of Na^+,K^+,Mg^{2+}-ATPase in the presence and absence of ouabain of erythrocyte membranes were determined. The difference is expressed as the Na^+,K^+-ATPase activity (Table 2).

From Table 2 it can be seen that the Na^+,K^+-ATPase activity of erythrocyte membranes of the infantile patients with KBD (metaphysis+) decreased compared to the control values.

Fluidity Change of Erythrocyte Membrane Lipid from Infantile Patients with KBD

From Table 3 it can be seen that the fluorescence polarization (P) values of erythrocyte membranes of infantile patients with KBD (metaphysis+ or bone end+) were higher than those of control children in

TABLE 2. Na$^+$,K$^+$-ATPase Activity of Erythrocyte Membranes of Infantile Patients with KBD

Group	Specific activity of Na$^+$,K$^+$-ATPase	Specific activity of Na$^+$,K$^+$,Mg^{2+}-ATPase	
		(+Ouabain)	(−Ouabain)
Control children in endemic region (5)[a]	0.550 ± 0.205	1.563 ± 0.101	2.113 ± 0.266
Infantile patients with KBD (metaphysis+) (8)	0.316 ± 0.213	1.626 ± 0.308	1.942 ± 0.296
Infantile patients with KBD (bone end+) (3)	0.597	1.474	2.071
Beijing adults (6)	0.758 ± 0.319		
Beijing children (1)	0.675		
Student's t test, comparison of infantile patients with KBD (metaphysis+) with control children in endemic region	$P < .1$	NS	NS

[a] The numbers in parentheses denote the number of samples.

TABLE 3. Fluidity Change of Erythrocyte Membrane Lipid from Children with KBD

Group	Fluorescence polarization (P)	Student's t test	
		Comparison with control children in endemic region	Comparison with control children in nonendemic region
Control children in endemic region	0.294 ± 0.009 (5)[a]		
Control children in nonendemic region	0.294 ± 0.003 (20)		
Infantile patients with KBD (metaphysis+)	0.298 ± 0.004 (11)	NS	$P < .01$
Infantile patients with KBD	0.299 ± 0.005 (5)	NS	$P < .02$

[a] The numbers in parentheses denote the number of samples measured.

the nonendemic region. The difference in fluorescence polarization is quite small, but statistically significant. This indicates that the erythrocyte membrane from children with KBD was less fluid than that of the controls from the nonendemic region.

Change of the Membrane Skeletal Components of Erythrocytes in Infantile Patients with KBD

In order to investigate whether changes occurred in the membrane skeletal components of erythrocytes of infantile patients, discontinuous SDS–polyacrylamide gel electrophoresis was used. Figure 1 shows the SDS–polyacrylamide gel patterns of the Coomassie Brilliant Blue staining proteins of human erythrocyte membranes.

The protein-staining patterns from the erythrocyte membranes of the infantile patients were scanned by a thin-layer chromatography scanner and the peak area was calculated by a microcomputer; the results were compared with controls (Table 4). Membranes from the infantile patients (metaphysis+ or bone end+) contained less actin and more glyceraldehyde-3-phosphate dehydrogenase when compared with nonendemic region control.

From the results it seems likely that some abnormalities in erythrocyte membranes from infantile patients with KBD may exist.

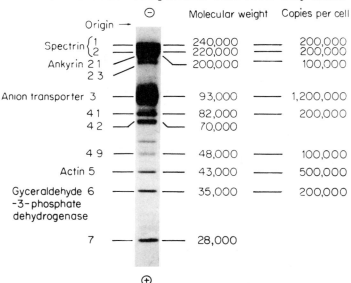

FIG. 1. SDS–polyacrylamide gel electrophoresis of the Coomassie Brilliant Blue staining proteins of the human erythrocyte membrane.
Source: D. J. Anstee et al., Biochem J. 183, 193–205 (1979).

TABLE 4. Membrane Skeletal Components Analysis of Erythrocytes in Infantile Patients with KBD

Protein (band)	Endemic region control (5)[a]	Nonendemic region control (20)	Infantile patients	
			(Metaphysis+) (11)	(Bone end+) (5)
Band 1,2 (spectrin)	29.40 ± 6.34	30.21 ± 4.47	30.17 ± 5.56	30.51 ± 5.60
Band 3	24.19 ± 4.72	24.95 ± 7.56	21.75 ± 3.47	19.26 ± 2.68
Band 4,1;4,2	6.59 ± 2.23	8.02 ± 2.48	8.73 ± 1.86	8.83 ± 1.53
Band 4,3;4,4	8.51 ± 2.22	7.80 ± 2.46	8.35 ± 2.45	8.58 ± 2.25
Band 5 (actin)	9.13 ± 1.70	10.9 ± 2.24	8.07 ± 2.97	7.33 ± 0.91
			$P < .01$[b]	$P < .001$[c]
Band 6 (glyceraldehyde-3-phos- phate dehydrogense)	6.30 ± 0.73	5.77 ± 0.55	6.27 ± 0.80	7.02 ± 0.55
			$P < .05$[b]	$P < .001$[c]

[a] Number of samples.
[b] $P < .01$, $P < .05$.
[c] $P < .001$, as estimated by Student's t test with respect to the control of the nonendemic region.

955

In recent years considerable interest has been focused on the erythrocyte membranes since it was recognized that membrane anomalies may be responsible for numerous disease processes (8). The erythrocytes affected by Duchenne muscular dystrophy are echinocytic in appearance and exhibit a reduced membrane deformability (9). In addition, the membrane fluidity and activities of several membrane-bound enzymes in erythrocytes have been shown to be altered from Duchenne muscular dystrophy (8). The spherocyte of hereditary spherocytosis exhibits the physical measures of increased osmotic fragility, decreased deformability, and an abnormally high sodium ion influx.

It is known that KBD has its main pathological changes in the necrosis of epiphysis cartilage. It is interesting to note that its pathologic manifestations also include changes in the structure and function of erythrocyte membranes.

It has been found that KBD is mainly distributed in low selenium areas, and the populations in the endemic areas of this disease generally have a poor selenium status. Further study is in progress to determine if there is any correlation between the low Se content in erythrocyte membranes of children suffering from KBD and membrane anomalies which appear in this condition. Hopefully, the information accumulated will be helpful in elucidating and diagnosing KBD in its early stages.

ACKNOWLEDGMENTS

We would like to acknowledge X. F. Zhang and F. T. Liu for participating in the preparation of erythrocyte membranes.

REFERENCES

1. Li, F.-S. Unpublished.
2. Dödge, J. T. 1963. Arch. Biochem. Biophys. *100*, 119.
3. Wang, G.-I. 1983. Wei Sheng Yan Jiu (Health Res.) *12*, 1.
4. Lin, Q.-S. 1976. Shengwu Huaxue Yu Shengwu WuLi Jinzhan (Biochem. Biophys.) *4*, 40.
5. Li, S.-G. 1983. Shenywu Huaxue Yu Shengwu WuLi Jinzhan (Biochem. Biophys.) *6*, 70.
6. Davis, B. J. 1964. Ann. N.Y. Acad. Sci. *121*, 404.
7. Shinitzky, M. 1974. J. Biol. Chem. *249*, 2552.
8. Tao, M., and Conway, R. G. 1982. *In* Membrane Abnormalities and Disease M. Tao (Editor), Vol. 1. CRC Press Inc., Boca Raton, FL.
9. Percy, A. K., and Miller, M. F. 1975. Nature (London) *258*, 147.

Approaches to Mechanisms of Prevention of Keshan and Kaschin–Beck Diseases by Selenium

Li Fang-Sheng   Bai Qian-Fu   Zhao Yu-Hua
Wei Feng-Qun   Guan Jin-Yang   Li Li
Fu Zhao-Lin   Duan You-Jin   Jin Qun
Bai Shi-Cheng   Zou Li-Ming

The descriptions of Kaschin–Beck and Keshan diseases appeared a century ago. Although the disease situation undulates, some epidemic areas have had no outbreak of these diseases in recent years. The causes of these diseases remains unclarified, and no adequate methods of prevention and treatment have been available.

In the past 10 years, it has been reported that the endemic area of Keshan disease (KD) was deficient in Se, and patients showed several clinical symptoms similar to animals with white muscle disease. A measure was taken to supplement the diet with sodium selenite in an effort to prevent or treat KD. Favorable results were observed on prevention, but not on treatment of the disease (1,2). A similar effect on the prevention of Kaschin–Beck disease (KBD) with Se-supplemented diets was also reported (3). If the aforementioned observations are correct, it would be interesting to know the mechanism of Se action.

Mammalian cells are thought to be protected from oxidant damage by a variety of defense mechanisms, including enzymes such as superoxide dismutase, glutathione peroxidase (GSHPx), and catalase, as well as free-radical scavengers such as ascorbate, α-tocopherol, and reduced glutathione. These defenses provide protection from toxic free-radical products which cause lipid peroxidation and cellular destruction once they are generated (4,5). The preventive action of Se for KBD and KD may be related to this kind of defense mechanism, particularly in the protection of biomembranes. Since under normal conditions the phospholipid structure in various cellular membranes is rather stable, it often loses its function by the action of highly reactive free radicals. On the other hand, the Se-containing enzyme, GSHPx, may prevent lipid peroxidation (6,7) by catalyzing the decomposition of H_2O_2 and a wide variety of organic peroxides produced by an organism.

MATERIALS AND METHODS

Children with KBD living in Yongshou Prefecture of Shaanxi Province were confirmed with X rays and clinical examinations. Experimental animals used in this study were young dogs and young guinea pigs. The composition of the semipurified basal diet (8) (expressed as a percentage) was corn starch, 64; saccharose, 5; protein mix (protein of corn, soybean, and casein), 15; lard, 5; vitamin mix, 1; mineral mix, 4; cellulose, 8; and it contained less than 0.02 ppm Se. The phospholipids (PL) and their sources in the diet were controlled. All animals were housed in a temperature-controlled (25°C) room. Food and water were supplied *ad libitum* for 2 or 3 months. The exhaled alkanes and the electrocardiogram were examined before being sacrificed.

Total lipids were extracted from erythrocytes and tissues by the method of Folch *et al.* (9). Total PL was determined according to the procedures given by Takemura and Miyazaki (10). Fractionation of PLs was performed by thin-layer chromatography and detected with a dual-wavelength Chromato scanner CS-910 (Shimadzu). The Se content, (11) activity of GSHPx, (12) and fluorescent chromolipids (13) were determined, and the estimation of cytochrome oxidase activity of myocardium (14), cAMP in plasma (15), and exhaled alkanes (16) was made.

RESULTS AND DISCUSSION

Increase in Activity of GSHPx and Decrease in Products of Lipid Peroxidation by Se Supplement

People living in the endemic area of KBD and KD showed deficiency in Se and abnormalities of membranous components due to the ecologi-

TABLE 1. Selenium, Phospholipids (PLs), Sphingomyelin (SM), Phosphatidylcholine (PC), and Phosphatidylethanolamine (PE) of Erythocytes (Plasma Membrane) of Kaschin–Beck Diseased and Control Children[a]

Children	n	Se (ng/10^{10} RBC)[b]	Total PLs	SM	PC	PE	SM/PC
				(μg/10^7 RBC)			
Kaschin–Beck disease	14	76 ± 12	1.86 ± 0.16	0.44 ± 0.07	0.28 ± 0.04	0.70 ± 0.05	1.41
Control in endemic area	13	42 ± 8[**]	2.48 ± 0.17[**]	0.40 ± 0.04[*]	0.47 ± 0.07[**]	0.50 ± 0.09[*]	0.76
Control in nonendemic area	16	148 ± 5[***]	2.73 ± 0.12[***]	0.79 ± 0.14[**]	0.53 ± 0.07[**]	0.80 ± 0.07[*]	1.12

[a] $\bar{X}$ ± SD.
[b] RBC, Red blood cells.
[*] $P > .05$ compared with sick children.
[**] $P < .05$.
[***] $P < .01$.

959

TABLE 2. Effects of Sodium Selenite (0.2 μg Se/g Diet) Supplements for 2 Months on the Activity of Tissue GSHPx and Amount of Lipid Peroxidation Products (Exhaled Alkanes) in Guinea Pigs[a]

Feed	n	Se (ng/10^{10} RBC)[b]	GSHPx activity (U/mg protein)		Exhaled alkanes (mm^2/g)
			Erythrocyte	Myocardium	
Basal diet (deficient in Se and PLs)	5	65 ± 17	174 ± 18	57.6 ± 5.5	80 ± 31
Control	3	136 ± 23**	225 ± 17**	72.3 ± 5.5**	42 ± 9*

[a] $\bar{X}$ ± SD.
[b] RBC, Red blood cells.
* $P > .05$.
** $P < .05$.

TABLE 3.   Effects of Sodium Selenite (0.2 μg/g Diet) Supplements for 2 Months on Content of Chromolipids in Guinea Pigs Fed with Semipurified Diet Deficient in Se[a]

Diet	n	Relative fluorescence	
		Liver	Erythrocytes
Basal	6	45.1 ± 8.8	6.4 ± 3.7
Basal + Se	7	28.8 ± 5.1**	6.5 ± 3.1*
Control	8	14.7 ± 3.6**	4.8 ± 1.9*

[a] $\bar{X}$ ± SD.
* $P > .05$.
** $P < .01$.

cal effects of the particular environment in these areas, with an unbalanced diet defective in green vegetables and protein common. Preliminary work has shown that the PL content of erythrocytes (17) and cartilage from the diseased children was lower than the controls, particularly in phosphatidylcholine (PC). As shown in Table 1, the children in the endemic area were low in Se, but the PLs, especially PC, in erythrocytes (plasma membrane) were decreased only in the diseased children, and sphingomyelin (SM) was increased simultaneously. We have verified this observation with animal experiments, as shown in Table 2. Animals fed a semipurified diet deficient in Se and PL showed lower activity of GSHPx than the control. In addition, the exhaled alkanes increased and the amount of fluorescent chromolipid in several tissues from Se-deficient animals was higher than that of controls (Tables 3 and 4). Animals with Se supplements (0.2 μg of Se as sodium selenite per gram of diet) showed a decrease in these parameters.

TABLE 4.   Content of Chromolipids in Tissues of Young Dogs Fed a Semipurified Diet Deficient in Se[a]

Diet	Relative fluorescence			
	Myocardium	Liver	Erythrocytes	Plasma
Basal	25.1 ± 10.7 (9)[b]	33.6 ± 7.0 (9)	9.1 ± 3.0 (5)	11.0 ± 4.5 (5)
Control	11.4 ± 0.7*(6)	17.3 ± 3.9** (6)	2.9 ± 0.9** (6)	1.4 ± 0.3** (6)

[a] $\bar{X}$ ± SD.
[b] Numbers in parentheses are number of animals.
* $P < .01$.
** $P < .001$.

Phospholipid Level in Tissues (Biomembranes) and the Ratio of Phospholipid Components

The total PL content and PC in cartilage of dogs fed the Se- and PL-deficient diet showed a falling tendency (Table 5), which is generally identical to changes of lipid in cartilage and erythrocytes of KBD children. These changes may be corrected with Se supplement and PLs and have also been observed in guinea pigs (Table 6). The PC level in PLs was increased by the supplement of Se alone. This might be due to the elimination of peroxide free radicals and the deacceleration of the chain reaction of lipid peroxidation by GSHPx. At the same time SM is decreased. Therefore, the mole ratio of several PLs maintained at a certain level was similar to the control. More recent evidence showed that an increase of ratio of SM/PC in membrane bilayer is parallel with aging (*18*). It is interesting that the similar aging changes of cartilage and erythrocytes appeared in KBD children and animals fed a similar diet. These changes may constitute the pathological basis for the phenomenon of early aging in KBD-diseased patients.

The Se content in the erythrocyte membrane (ghost) from KBD children was significantly lower than that from controls in the non-affected area. The membrane fluidity and the activity of Na^+,K^+-ATPase, a membrane-bound enzyme, tended to decrease simultaneously.

The Se supplement in the diet produced a similar effect on PL composition of myocardium (Table 7), but the ratio of different PLs of heart did not improve as in cartilage. It is extremely far from the ratio of PLs in myocardium of animals fed with control diets which had been supplemented sufficiently with all the factors relating to PL metabolism (including soybean PLs, choline, and methionine). The cardiolipid (CL, or diphosphatidylglycerol DPG) was markedly increased, SM decreased, and the ratio of each PL was close to that of control animals fed the normal diet.

Slight Improvement in Function of Membrane

The proportion of PL in membrane, especially that of the boundary lipids, is very important to the functions of biomembrane such as membrane fluidity, permeability, and fragility, the activity of membrane-binding enzymes, cytochrome oxidase in mitochondria, and plasma membrane-bound adenylate cyclase. Figure 1 shows the results of measurement of the activity of cytochrome oxidase (the marker enzyme of mitochondria) in myocardium. The enzymatic activity of groups fed the Se- and PL-deficient basal diet is the lowest. After supplementing

TABLE 5. Effects of Sodium Selenite (0.2 μg/g Diet) Supplements and Phospholipids (20 mg Soybean PL/g of Diet) in Diet on PL Components (μg/mg Tissue and Their % in Total PLs) in Cartilage, 5 Young Dogs per Group[a]

Diet	SM	PC	LL	Total PLs
Deficient in Se and PLs	0.70 ± 0.23 (55.5%)	0.33 ± 0.17 (26.5%)	0.23 ± 0.12 (18.0%)	1.26 ± 0.45
Supplement with Se and PLs	0.78 ± 0.29*(47.0%)	0.50 ± 0.20*(30.0%)	0.39 ± 0.20*(23.0%)	1.67 ± 0.32

[a] $\bar{X}$ ± SD.
* $P > .05$.

TABLE 6. Effects of Se Supplements (0.2 μg/g Diet) for 2 Months on Phospholipid Components (% in Total PLs) in Cartilage of Guinea Pigs[a]

Diet	n	LL	SM	PC	PE
Basal	7	6.1 ± 1.1	43.3 ± 0.1	48.9 ± 0.1	1.7 ± 1.0
Supplement with Se	8	16.6 ± 0.7***	22.3 ± 2.0**	57.8 ± 3.9*	3.4 ± 2.6*
Supplement with Se and PLs	8	9.2 ± 1.5*	19.6 ± 6.0**	67.1 ± 4.3**	4.3 ± 3.5*
Control	8	11.7 ± 4.0*	22.0 ± 4.6**	56.2 ± 8.6*	10.1 ± 1.0**

[a] $\bar{X}$ ± SD.
* $P > .05$.
** $P < .05$.
*** $P < .01$.

TABLE 7. Effect of Se Supplement (0.2 µg/g of Diet) for 3 Months on Phospholipid (PL) Composition: Lysolecithin (LL), Sphingomyelin (SM), Phosphatidylcholine (PC), Phosphatidylethanolamine (PE), and Cardiolipin (CL) % of Guinea Pig Myocardium Fed with Semipurified Diet[a]

Diet	n	LL	SM	PC	PE	CL (DPG)
Basal	5	12.1 ± 4.3	16.9 ± 2.7	43.9 ± 8.2	15.2 ± 7.6	9.8 ± 6.9
Supplement with Se	6	12.2 ± 5.3*	17.9 ± 3.8*	44.7 ± 6.6*	14.9 ± 8.9*	10.3 ± 6.9*
Supplement with Se and PLs	4	7.8 ± 1.8**	17.0 ± 2.9*	35.2 ± 4.8*	27.3 ± 4.4**	12.9 ± 2.3*
Supplement with Se, PLs, and its metabolic factors[b]	5	7.1 ± 1.1***	11.0 ± 2.4***	52.0 ± 15.5*	10.0 ± 5.6*	19.6 ± 10.1*
Control	4	11.2 ± 2.5*	13.2 ± 2.8*	45.6 ± 10.7*	9.7 ± 6.9*	20.4 ± 9.1*

[a] $\bar{X}$ ± SD.
[b] Soybean phospholipids, 20 mg; choline chloride, 2 mg; methionine, 0.5 mg/g of diet.
*$P > .05$.
**$P < .05$.
***$P < .01$.

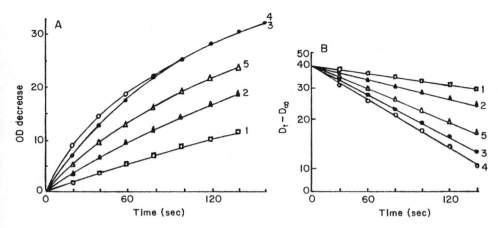

FIG. 1.   Dynamic curves (A) and logarithmic velocity (B) of myocardial cytochrome oxidase of guinea pigs fed for 3 months with (1) basal diet (n = 6); (2) basal diet + Se (0.2 µg/g diet, n = 6); (3) basal diet + Se (0.2 µg/g diet) and phospholipids (20 µg/g diet, n = 7); (4) basal diet + Se (0.2 µg/g), phospholipids (20 mg/g), choline chloride (2 mg/g diet), and methionine (0.5 mg/g diet, n = 8); (5) control diet (n = 8).

with Se alone, the enzymatic activity increased slightly, but the velocity constants were not different from those fed the basal diet. Only when PLs and especially other agents related to the PL metabolism, such as choline and methionine, were supplemented did the enzymatic activity increase significantly ($P < .01$) and sometimes even exceed that of the group fed the control diet. This change in mitochondria enzyme was reflected also in the contents of plasma cAMP (a product of adenylate cyclase in plasma membrane). These data are given in Table 8.

These results as shown in Tables 7 and 8 and Fig. 1 indicate that there is a close relationship between the structure and function of

TABLE 8.   Influence of Se (0.2 µg/g Diet) and Phospholipids (PLs) (20 mg/g Diet) in Diets for 3 Months on Plasma cAMP Content in Guinea Pigs

Diet	n	µmoles ml^{-1}	F test
Basal	3	47.5 ± 23.0	$F < .05$
Basal + Se	3	67.5 ± 27.0	
Basal + Se + PLs	6	102.5 ± 15.0	
Basal + Se + PLs and metabolic factors[a]	5	105.0 ± 42.3	
Control	6	62.5 ± 11.0	

[a] Same as in Fig. 1.

membranes. The results may also illustrate the reason why Se has a preventive effect on myocardium from oxidative damage, but not a therapeutic effect on myocardium already damaged in Keshan disease, because the protective effect of Se on the biomembrane is required only after the structural completion of a relatively completed membrane. If the membrane is defected or damaged, the protective effect of Se will not be effective. The difference between diseased children and nonaffected children in the endemic area or in the Se-deficient nonendemic area may lie in the membrane lipids and their composition.

REFERENCES

1. Keshan Disease Research Group of the Chinese Academy of Medical Science 1979. Epidemic studies on the etiologic relationship of selenium and Keshan disease. Chin. Med. J. *92*, 477–482.
2. Keshan Disease's Laboratory of Xian Medical College 1979. Effect of sodium selenite in prevention of Keshan disease. Chin. J. Med. *59* (8); 457–460.
3. Li, Q.-Z. 1979. Treatment of Kaschin–Beck disease; 224 cases with selenium and vitamin E. Chin. J. Med. *59* (8), 169–171.
4. Frank, L. and Massaro, D. 1980. Oxygen toxicity. Am J. Med. *69*, 177–181.
5. Ramasarma, T. 1982. Generation of H_2O_2 in biomembranes. Biochim. Biophys. Acta *694*, 69–93.
6. Rotruck, J. T., Pope, A. L., and Ganther, H. E. 1973. Selenium, biological role as a component of glutathione peroxidase. Science *179*, 588–590.
7. McCay, P. B., Gibson, D. D., and Fong, K. L. 1976. Effect of glutathione peroxidase activity on lipid peroxidation in biological membranes. Biochim. Biophys. Acta *431*, 459–468.
8. Okano, G., Matsuzaka, H., and Shimojo, I. 1980. A comparative study of the lipid composition of white, intermediate, and red heart muscle in rats. Biochim. Biophys. Acta *619*, 167–175.
9. Folch, J., Lees, M., and Stanley, H. 1957. A simple method for the isolation and purification of total lipids from animal tissues. J. Biol. Chem. *226*, 497–509.
10. Takemura, S., and Miyazaki, M. 1969. Analysis of pancreatic RNase digest of *Torulopsis utilis* tRNA specific for alanine, valine, and several other amino acids. J. Biochem. (Tokyo) *65*, 159–163.
11. Wilkie, J. B., and Young, M. 1970. Improvement in the 2,3-diaminonaphthalene reagent for microfluorescent determination of selenium in biological materials. J. Agric. Food Chem. *18*, 944.
12. Hefeman, D. G., Sunde, R. A., and Hoekstra, W. G. 1974. Effect of dietary selenium on erythrocytes and liver glutathione peroxidase in the rat. J. Nutr. *104*, 580–587.
13. Jain, S. K., and Hochstein, P. 1980. Membrane alteration in phenylhydrazine-induced reticulocytes. Arch. Biochem. Biophys. *201*, 683–687.
14. Mason, T. L., Poyton, R. O., and Wharton, D. C. 1973. Cytochrome oxidase from Baker's yeast. I. Isolation and properties. J. Biol. Chem. *248*, 1346–1354.
15. Gilman, A. G. 1970. Determination of cAMP contents by protein binding assay. Proc. Natl. Acad. Sci. U.S.A. *67*, 305–312.

16. Riley, C. A., and Cohen, G.   1974.   A new index of lipid peroxidation. Science *183*, 208–210.
17. Li, F.-S., and Guan, J.-Y.   1982.   Plasma and erythrocyte levels of Se and lipids in children with Keshan disease and their pathogenic significance. Acta Nutr. Sin. *4*, 221–227.
18. Barenholz, Y., and Thompson, T. E.   1980.   Sphingomyelin in bilayer and biological membranes. Biochim. Biophys. Acta *604*, 129–158.

Selenium Status in Patients with Liver Cirrhosis

B. Åkesson
U. Johansson

Liver damage can be caused by a number of substances. In clinical practice ethanol-induced damage is the most common problem, and several mechanisms for the action of ethanol have been advanced (Van Waes and Lieber 1980). Ethanol many induce peroxidative mechanisms in the liver, as indicated by the finding that acute ethanol intoxication stimulates the production of malondialdehyde (i.e., TBAR, thiobarbituric acid-reactive material) in liver homogenates (Comporti *et al.* 1973). Suematsu *et al.* (1981) found an increased concentration of TBAR in liver biopsies of heavy drinkers. After chronic ethanol feeding to animals, increased hepatic diene conjugation and decreased concentration of glutathione can also occur (Shaw *et al.* 1981). These processes may be influenced by antioxidants such as vitamins E and C and also by the selenoenzyme glutathione peroxidase (GSHPx). Since the liver is an important storage organ for selenium and vitamin E and also has the highest activity of GSHPx (Marklund *et al.* 1982), liver damage may induce losses of these nutrients.

To elucidate some of these problems, we have studied nutritional status with respect to selenium and related nutrients in patients with different degrees of liver damage.

MATERIAL AND METHODS

Three groups of subjects were studied: (1) Twenty-two patients with advanced liver cirrhosis who had previously had at least one episode of bleeding from esophageal varices. Their ages were 33–76 years; they were 6 women and 16 men; (2) 15 men (aged 31–76 years) with chronic alcoholism who had abstained from alcohol at least 1 month prior to study and who had no clinical or biochemical indications of liver cirrhosis; and (3) the control group, consisting of 16 subjects (8 women and 8 men, 29–71 years).

Blood, plasma, and platelets were analyzed for GSHPx using *t*-butyl hydroperoxide as substrate in a coupled assay with glutathione reductase. α-Tocopherol and retinol were quantified by high-performance liquid chromatography (HPLC). Lipid fluorescence in erythrocytes was measured with fluorometry (excitation at 360 nm, emission at 440 nm) after extraction of washed erythrocytes with isopropanol/chloroform (11:7). The fluorescence was related to the amount of lipid phosphorus (Simonsson *et al.* 1982) in the extracts and to a standard of quinine sulfate (Rehncrona *et al.* 1980). TBAR (thiobarbituric acid-reactive material) in plasma was measured by fluorometry.

RESULTS

Plasma selenium was depressed in cirrhotic patients, but was normal in noncirrhotic patients (Table 1). The latter patients and also the controls fell essentially within our reference interval, 0.68–1.28 μmol/liter. Analysis of total blood selenium in a limited number of patients also indicated a decreased concentration in cirrhotic patients compared to controls (Fig. 1).

Since the most well-known form of selenium in blood is GSHPx, the activity of this enzyme was determined using *t*-butyl hydroperoxide as substrate. Figure 2 shows that GSHPx in blood was the same in patients and controls. The enzyme activity in blood probably reflects long-term changes in selenium status, and short-term changes may be reflected in the plasma activity of GSHPx. This activity had the same distribution in the three groups of subjects (Table 1). Also, platelet GSHPx has been proposed as a useful indicator of selenium status. The enzyme activity in platelets was depressed in the cirrhotic patients, but was normal in the noncirrhotic group. There was a significant correlation ($P = .001$) between plasma selenium and plasma GSHPx among cirrhotic patients (Table 2). In the whole material no such correlation was present (Fig. 3). Platelet GSHPx was positively correlated to plasma selenium both within the cirrhotic group and in the whole material. Plasma TBAR and red blood cell lipid fluorescence were used as indicators of lipid peroxidation. No differences among the three groups were observed (Table 1).

TABLE 1. Nutritional Status in Patients with Cirrhotic and Noncirrhotic Liver Disease[a]

Parameter	Controls (I)		Liver cirrhosis (II)		Alcoholic liver disease (III)		Statistical significance		
							I vs II	I vs III	II vs III
Plasma selenium (μmol/liter)	1.01	(0.04)	0.75	(0.03)	0.99	(0.07)	xxx	—	xxx
Blood selenium (μmol/liter)	1.19	(0.06)	0.87	(0.06)	—		xx	—	—
Blood GSHPx (μmol/g Hb × min)	45	(3)	40	(3)	39	(2)			
Plasma GSHPx (μmol/liter × min)	561	(26)	595	(30)	508	(37)			
Platelet GSHPx (nmol/mg × min)	298	(32)	217	(23)	352	(37)	x		xx
Plasma ascorbic acid (μmol/liter)	53	(5)	68	(5)	49	(5)			x
Plasma retinol (μmol/liter)	2.25	(0.10)	1.15	(0.11)	2.89	(0.20)	xxx	xxx	
Plasma α-tocopherol (μmol/liter)	26	(1)	20	(2)	23	(2)	xxx		
Plasma TBAR (μmol/liter)	1.5	(0.1)	2.0	(0.3)	1.3	(0.1)			x
RBC lipid fluorescence (μg quinine sulfate/μmol P)	25	(2)	40	(8)	36	(7)			

[a] Data are expressed as means. Numbers in parentheses represent SE.

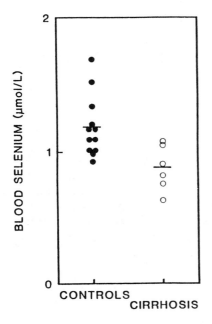

FIG. 1.   Blood selenium in patients with liver cirrhosis.

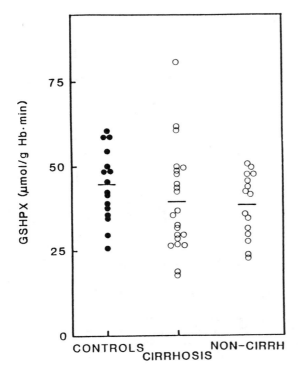

FIG. 2.   Blood GSHPx activity in patients with liver cirrhosis or noncirrhotic alcohol-induced liver disease.

TABLE 2. Relation of Different Measures of Selenium Status in Patients with Liver Cirrhosis[a]

	Plasma GSHPx	Blood GSHPx	Platelet GSHPx
Plasma selenium	0.66[xxx]	−0.14	0.64[xx]
Plasma GSHPx	—	0.26	0.51[x]
Blood GSHPx		—	−0.02

[a] Linear correlation coefficients are given.

Several other nutrients other than selenium which are possibly related to lipid peroxidation and liver damage were measured. The concentration of α-tocopherol and retinol in plasma was significantly depressed in the cirrhotic patients (Table 1). Ascorbic acid may act as an antioxidant in the water phase. Plasma ascorbic acid was therefore

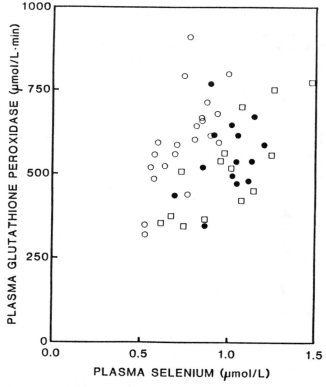

FIG. 3. Relation of plasma selenium to plasma GSHPx activity in patients with liver cirrhosis (○), noncirrhotic alcohol-induced liver disease (□), and in controls (●).

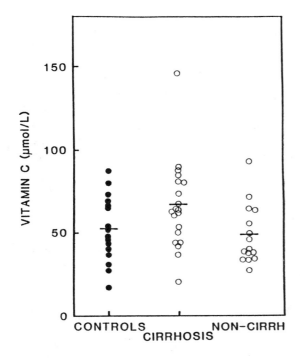

FIG. 4. Plasma ascorbic acid in patients with liver cirrhosis or noncirrhotic alcohol-induced liver disease.

measured to evaluate vitamin C status, but no differences among the three groups were observed (Fig. 4).

DISCUSSION

This study demonstrates several derangements in nutritional status in patients with liver disease. The decrease in plasma selenium may be due to a low selenium intake, deranged selenium metabolism, or may be secondary to the changes in plasma protein concentrations in liver cirrhosis, since much of plasma selenium is protein bound. This study is the first to indicate that blood and plasma GSHPx are unchanged in liver cirrhosis, suggesting that the decrease in plasma selenium affects other selenoproteins or selenium compounds other than GSHPx. Since the platelet activity of GSHPx has been proposed as a sensitive index of selenium status (Levander *et al.*, 1983), the decrease in this enzyme activity and in plasma selenium indicates a deranged selenium status in cirrhotic but not in noncirrhotic patients.

The decrease in plasma retinol and plasma α-tocopherol in cirrhotic patients agrees with several previous studies (Bonjour 1981a,b). The mechanism behind the decrease may be insufficient intake, deranged protein status with low concentration of vitamin-binding proteins, or increased consumption of the vitamins due to increased lipid peroxidation.

In conclusion, the present study demonstrates that patients with liver cirrhosis may have a deranged antioxidant defense, since plasma selenium, platelet GSHPx, and plasma α-tocopherol are depressed. Further studies are necessary to unravel the mechanisms behind these changes and to elucidate whether dietary supplementation will be beneficial.

ACKNOWLEDGMENTS

We thank Ms. B. Mårtensson and Ms. I. Nilsson for skillful assistance. The study was supported by grants from the Swedish Medical Research Council (project 3968) and the Swedish Nutrition Foundation.

REFERENCES

Bonjour, J. P. 1981a. Vitamins and alcoholism. IX. Vitamin A. Int. J. Vit. Nutr. Res. 51, 166–177.

Bonjour, J. P. 1981b. Vitamins and alcoholism. X. Vitamin D. XI. Vitamin E. XII. Vitamin K. Int. J. Vitam. Nutr. Res. 51, 307–318.

Comporti, M., Benedetti, A., and Chieli, E. 1973. Studies on in vitro peroxidation of liver lipids in ethanol-treated rats. Lipids 8, 498–502.

Levander, O. A., Alfthan, G., Arvilommi, H., Gref, C. G., Huttunen, J. K., Kataja, M., Koivistoinen, P., and Pikkarainen, J. 1983. Bioavailability of selenium to Finnish men as assessed by platelet glutathione peroxidase activity and other blood parameters. Am J. Clin. Nutr. 37, 887–897.

Marklund, S. L., Westman, N. G., Lundgren, E., and E., and Roos, G. 1982. Copper- and zinc-containing superoxide dismutase, manganese-containing superoxide dismutase, catalase, and glutathione peroxidase in normal and neoplastic human cell lines and normal human tissues. Cancer Res. 42, 1955–1961.

Rehncrona, S., Smith, D. S., Åkesson, B., Westerberg, E., and Siesjö, B. K. 1980. Peroxidative changes in brain cortical fatty acids and phospholipids, as characterized during Fe^{2+} and ascorbic acid-stimulated lipid peroxidation in vitro. J. Neurochem. 34, 1630–1638.

Shaw, S., Jayatilleke, E., Ross, W. A., Gordon, E. R., and Lieber, C. S. 1981. Ethanol-induced lipid peroxidation: Potentiation by long-term alcohol feeding and attenuation by methionine. J. Lab. Clin. Med. 98, 417–424.

Simonsson, P., Nilsson, A., and Åkesson, B. 1982. Postprandial effects of dietary phosphatidylcholine on plasma lipoproteins in man. Am. J. Clin. Nutr. 35, 36–41.

Suematsu, T., Matsumura, T., Sato, N., Miyamoto, T., Ooka, T., Kamada, T., and Abe,

H. 1981. Lipid peroxidation in alcoholic liver disease in humans. Alcohol.: Clin. Exp. Res. *5*, 427–430.

Van Waes, L., and Lieber, C. S. 1980. Alcohol liver injury. *In* Toxic Injury of the Liver. E. Farber and M. M. Fisher (Editors), pp. 629–653. Marcel Dekker, New York.

Serum Selenium in Patients with Short Bowel Syndrome

Jan Aaseth
Erling Aadland
Yngvar Thomassen

We have previously found a significant lowering of serum selenium in alcohol abusers with liver cirrhosis (Aaseth *et al.* 1980). An accelerated lipoperoxidation associated with the hepatotoxic effect of alcohol has been attributed to a lowering of the selenium-containing enzyme glutathione peroxidase (GSHPx) (Di Luzio 1973; Aaseth *et al.* 1980). Patients subjected to intestinal resection may occasionally develop hepatic lesions quite similar to those present in alcoholics (Peters *et al.* 1975). It has been suggested that deficient nutrition with respect to one or more dietary factors may promote the development of fatty degeneration and liver cirrhosis (Peters *et al.* 1975). Based upon animal experiments, Schwarz (1954) reported that vitamin E, selenium, and cystine could protect against dietary necrotic liver degeneration. To the authors' knowledge, the selenium status of patients with short bowel syndrome has not been systematically studied.

The activity of the selenium-containing enzyme GSHPx depends upon the tissue concentration of the cofactor glutathione. Therefore, it was of interest to study both the selenium status and the glutathione status of the patients with short bowel. In the present work, selenium, glutathione, vitamin E, and other micronutrients have been analyzed in blood or serum from patients subjected to intestinal surgery.

MATERIALS AND METHODS

The clinical material comprised 12 patients who had received intestinal bypass surgery, ad modum Payne and De Wind. The median age of the patients was 50 years (range: 37–63 years), and a median time interval of 8 years (range: 6–9 years) had passed since the surgery. Five additional patients subjected to more extensive intestinal resection, owing to vascular diseases, were also studied. In all cases, histological examination of liver biopsy material was done by an authorized pathologist.

Selenium in sera from the patients was determined by electrothermal atomic absorption after thermal stabilization of selenium compounds by addition of nickel (Saeed *et al.* 1979). Copper and magnesium were determined with conventional atomic absorption technique. Calcium, iron, and inorganic phosphate analyses made use of conventional colorimetric techniques, adapted for the Auto-Analyzer. Glutathione in red blood cells was assayed by using Elman's reagent and the method described by Tietze (1969).

Albumin, prealbumin, ceruloplasmin, and α_2-macroglobulin were determined by single radial immunodiffusion (Mancini *et al.* 1965) using reagents from Behringwerke, Federal Republic of Germany. The vitamin K-dependent coagulation factors were determined with Normotest, the reagents being purchased from Nyegaard and Co., Oslo, Norway. Total iron binding capacity (TIBC) was estimated by the iron method of Caraway (1963). The serum activity of the enzymes alanine aminotransferase (ALAT), aspartate aminotransferase (ASAT), and γ-glutamyl transpeptidase was assayed by recommended methods. Reference values were obtained from healthy blood donors.

RESULTS

The concentration of selenium in serum from patients subjected to intestinal bypass surgery was lowered to around 75% of the mean reference value (Table 1). In the red blood cells, the glutathione concentration in the patients was as low as 54% of the mean control value. It is seen from Table 1 that a marginal lowering was found in the patient group for the serum concentrations of iron, calcium, inorganic phosphate, and potassium.

Of the 12 patients, 8 had histological changes in their liver biopsies, these changes being classified as moderate fatty degeneration in all cases. Only minor alterations were found with regard to serum con-

TABLE 1. Serum Concentrations (Mean ± SD) of Se, Cu, Fe, Mg, Ca, P_i, and K in Patients Subjected to Intestinal Bypass Operation[a]

Substance	Patient group (n = 12)	Reference group (n = 30)
Se (µmol/liter)	1.17 ± 0.32	1.53 ± 0.25
Cu (µmol/liter)	14.5 ± 2.2.	18 ± 3.0
Fe (µmol/liter)	16 ± 5	20 ± 5
Mg (mmol/liter)	0.83 ± 0.20	0.82 ± 0.10
Ca (mmol/liter)	2.2. ± 0.1	2.4 ± 0.1
P (mmol/liter)	1.0 ± 0.2	1.2 ± 0.3
K (mmol/liter)	3.9 ± 0.4	4.3 ± 0.4
GSH (mmol/liter)	1.4 ± 0.3	2.6 ± 0.3 (n = 5)

[a] Glutathione (GSH) concentrations in red blood cells are included.

centrations of liver-synthesized proteins (Table 2). The vitamin K-dependent coagulation factors had slightly lowered activity in the patient group, and the concentration of α_2-macroglobulin was marginally increased in that group. Upon clinical and cardiological examination, one of the patients appeared to suffer from cardiomyopathy. The serum selenium of this patient was 42% of the mean reference group value. In the patient group, serum activities of the enzymes ASAT, ALAT, and γ-glutamyl transpeptidase were, on average, 27 µg/liter, 31 µg/liter, and 31 µg/liter, respectively, with standard deviations of 12, 15, and 33 µg/liter. These activities did not differ significantly from reference values.

Another patient group (n = 5), subjected to more extensive intestinal resection because of vascular disease, had an average concentration of selenium in serum as low as 0.82 µmol/liter, i.e., 54% of the reference group. This latter group was rather heterogenous, including patients with different bowel length, which may explain the rather wide range of selenium concentrations: 0.45–1.08 µmol/liter. Their glutathione

TABLE 2. Serum Concentrations of Proteins in the Same Patients as in Table 1, Expressed as Percentage of Control Values

Albumin	105 ± 25
Prealbumin	103 ± 28
Ceruloplasmin	93 ± 17
α_2-Macroglobulin	129 ± 28
Transferrin[a]	99 ± 22
Coagulation factors[b]	80 ± 15

[a] Estimated as TIBC.
[b] As estimated by Normotest.

concentrations in red blood cells were lowered to an average value of 1.3 mmol/liter. Four of these patients had fatty degeneration of the liver, as diagnosed by histological examination. One of the patients developed severe liver cirrhosis and died 20 months after the intestinal surgery. Two weeks before his death, analyses of blood serum showed low values for selenium (0.45 µmol/liter), copper (6.1 µmol/liter), zinc (10.4 µmol/liter) and vitamin E (3.0 µg/liter), which were as low as 29, 47, 34, and 40%, respectively, of the mean reference group values. Simultaneously, albumin, Normotest, calcium, and magnesium were lowered to 72, 20, 87, and 67% of reference group values, respectively.

DISCUSSION

Previous studies have demonstrated decreased ability to absorb the vitamins E and A in patients with short bowel syndrome (Hove and Baker 1982). The present study shows that patients with short bowel syndrome also have low blood levels of glutathione and selenium. These findings indicate that the antioxidative potential and protective capacity against lipoperoxidation in tissues may be deficient after intestinal resection.

It seems likely that a depletion of selenium, glutathione, and vitamin E from the tissues will lead to tissue damage, at least in the liver (Schwarz 1954; Diplock 1976). In rats, this combined nutritional deficiency leads to liver cell necrosis (Diplock 1976). In patients with short bowel syndrome, other essential nutrients may also be depleted from the body, e.g., zinc, copper, and magnesium (Dyckner et al., 1982), which may further hinder the physiological antioxidative protection of the liver cells (Chvapil 1973; McCord and Fridovich 1978).

Consequently, we propose that the demonstrated lowering of the selenium and glutathione concentrations, combined with other micronutrient deficiencies (vitamins E and A, zinc, and copper) play a role in the development of fatty degeneration and/or liver cirrhosis.

A parallel lowering of selenium and glutathione in the blood is of physiological interest because it has been reported previously that a significant amount of selenium is excreted in rat bile as a glutathione complex (Alexander 1983). Since the glutathione concentration in rat bile is as high as 2 mmol/liter, it is apparent that intestinal reabsorption of glutathione must take place, and it has been suggested that significant amounts of metal–glutathione complexes are reabsorbed in the rat (Alexander and Aaseth 1982). If selenium–glutathione complexes from the bile were reabsorbed in healthy humans as well as in rats, it would be expected that patients with short bowel syndrome could lose complexes into feces, leading to a parallel depletion of selenium and glutathione from the blood. Therefore, our findings indirectly suggest a similar bile/bowel-to-blood circulation of glutathione in humans as in rats.

REFERENCES

Aaseth, J., Thomassen, Y., Alexander, J., and Norheim, G. 1980. Decreased serum selenium in alcoholic cirrhosis. N. Engl. J. Med. *303*, 944–945.

Alexander, J. 1983. Studies on hepatobiliary metabolism of methyl mercury, silver, copper, zinc, and lead—The role of glutathione and interaction with selenium. Thesis, Oslo.

Alexander, J., and Aaseth, J. 1982. Organ distribution and cellular uptake of methyl mercury in the rat as influenced by the intra- and extracellular glutathione concentration. Biochem. Pharmacol. *31*, 685–690.

Caraway, W. T. 1963. Macro- and micromethods for the determination of serum iron and iron binding capacity. Clin. Chem. (Winston-Salem, N.C.) *9*, 188–199.

Chvapil, M. 1973. New aspects in the biological role of zinc. Life Sci. *13*, 1041–1049.

Di Luzio, N. R. 1973. Antioxidants, lipid peroxidation and chemically induced liver injury. Fed. Proc., Fed. Am. Soc. Exp. Biol. *32*, 1875–1881.

Diplock, A. T. 1976. Metabolic aspects of selenium action and toxicity. CRC Crit. Rev. Toxicol. *4*, 271–329.

Dyckner, T., Hallberg, D., Hultman, E., and Wester, P. O. 1982. Magnesium deficiency following jejunoileal bypass operations for obesity. J. Am. Coll. Nutr. *1*, 239–246.

Hove, W., and Baker, H. 1982. Micronutrient deficiency in a case of jejunoileal bypass. J. Am. Coll. Nutr. *1*, 247–253.

Mancini, G., Vaerman, J. P., Carbonara, A. O., and Heremans, J. F. 1965. A single radial diffusion method for the immunological quantitation of proteins. Protides Biol. Fluids *11*, 370–373.

McCord, J. M., and Fridovich, I. 1978. The biology and pathology of oxygen radicals. Ann. Intern. Med. *89*, 122–127.

Peters, R. L., Gay, I., and Reynolds, T. B. 1975. Postjejunoileal bypass hepatic disease. Its similarity to alcoholic hepatic disease. Am. J. Clin. Pathol. *63*, 318–333.

Saeed, K., Thomassen, Y., and Langmyhr, F. J. 1979. Direct electrothermal atomic absorption spectrometric determination of selenium in serum. Anal. Chim. Acta *110*, 285–289.

Schwarz, K. 1954. Factors protecting against dietary necrotic liver degeneration. Ann. N.Y. Acad. Sci. *57*, 878–888.

Tietze, F. 1969. Enzymic method for quantitative determination of nanogram amounts of total and oxidized glutathione: Applications to mammalian blood and other tissues. Anal. Biochem. *27*, 502–521.

Selenium in Plasma, Erythrocytes, and Platelets from Patients with Multiple Sclerosis

B. Ahlrot-Westerlund   *Å. Siden*
L. O. Plantin   *J. Svensson*
I. Savic

As noted more than 100 years ago, the plaques in multiple sclerosis (MS) have an almost ubiquitous perivenular distribution. A possible explanation for that could be hyperaggregability of platelets and formation of microclots in the postcapillary venules of the central nervous system (CNS). Increased adhesiveness of platelets in patients with MS has been observed by many authors in previous studies. Signs of a disseminated intravascular coagulation (DIC) ranging from slight to quite obvious fibrinolytic activity in venous blood of patients in relapses has been reported (*1–3*).

No extensive investigation of ADP-induced platelet aggregation was carried out until Gorbacheva *et al.* (*4*) reported an increased aggregability in 50 MS patients concomitant with or shortly after a relapse. The chronic progressive type of MS showed a continuous hyperaggregability to ADP (*4*). Similar results were reported by Couch and Hassanein (*5*). Neu *et al.* confirmed these results in a study of 30 MS patients who showed an increase in both spontaneous aggregation and aggregation induced by ADP and serotonin (*6*).

The Se content of platelets seems to be crucial to their membrane stability. The platelets have been reported to have the highest level of Se in human tissues (*7*). Thus, we found it worthwhile to investigate the Se content in platelets, erythrocytes, and plasma of MS patients.

MATERIAL AND METHODS

The patient material included 15 subjects (11 females and 4 males) aged from 20 to 57 years (median value 42 years) with clinical definite or probable MS. At the time of blood sampling the duration of the disease ranged from less than 1 to 19 years, with the median value of 7 years. The diagnosis was based on established clinical criteria, and cerebrospinal fluid examinations had been performed in all cases. Twelve patients had exhibited an onset with relapsing-remitting symptoms. Four of these had later entered a progressive course. Three subjects had a progressive course from the beginning of the disease. The age at onset of the disorder ranged from 15 to 50 years, with a median value of 28 years. Of the patients, 3 had an acute bout at the time of blood sampling, 7 were in a progressive phase, and 5 were in a stationary state. The degree of neurological dysfunction was moderate or pronounced in 9 patients, while 6 had a slight or no disability. The controls were 50 healthy blood donors with ages between 19 and 67 years. Platelets were obtained from 13 controls only.

Platelets, red cells, and plasma were separated from 20 ml of blood obtained by venipuncture. Usually we got about 20 mg wet weight of platelets for the neutron activation analysis of Se according to the method by Plantin and co-workers *(8)*. The freeze-dried samples in sealed quartz ampoules were irradiated at a flux of 2×10^{13} n cm^{-1} sec^{-1}. After about 1 week cooling time, the samples were dissolved in a mixture of perchloric and nitric acids together with 50 mg of selenium carrier. The solution was evaporated to near dryness and diluted with hydrochloric acid. The Se was then precipitated in elemental form with a strong reducing agent (hydroxylamine hydrochloride). The gamma spectrum of the sample was measured with a Ge(Li) semiconductor detector coupled to a multichannel analyzer-computer (IN90) which calculated the selenium concentration in the sample.

The megakaryocyte–platelet regeneration time was also measured with a recently described double-isotope method *(9)*. At constant platelet levels the regeneration is a measure of the rate of platelet destruction. In 8 investigated patients, a mean regeneration time of 4.4 days was obtained. In 11 healthy controls, the mean value was 5.2 days (SEM = .15). Platelet numbers for all patients were within the normal range of $150–400 \times 10^9$/liter.

RESULTS

The mean Se concentrations obtained are presented in Table 1. As can be seen, there was no statistical difference between MS patients

TABLE 1. Selenium Concentrations (ng/g) (SD)

Diagnosis	Plasma	Erythrocytes	Platelets[a]	n
MS	77.1 (18.3)	102 (21.1)	242 (44.8)	15
Controls	80.5 (14.2)	99.5 (19.1)	317 (48.0)	50

[a] Platelets were obtained from 13 controls only.

and controls for Se in plasma and erythrocytes. The Se in platelets, however, was significantly lower than in the controls with $t = 4.274$ ($P < .001$).

Our results in this pilot study show that MS patients differ from patients with acute myocardial infarction concerning Se in plasma. They have a decreased Se level not only in platelets, but also in plasma (10).

The shorter half-life of platelets in the MS patients noticed in this study indicates hyperaggregability. The decreased Se content in platelets might be explained by a compensatory thrombocytosis, hiding a thrombocytopenia which is a pathophysiological feature often attributed to chronic DIC.

The reason for the low Se concentration in platelets in MS patients remains to be explained, but it certainly indicates a possible reason for decreased platelet membrane stability.

ACKNOWLEDGMENT

This work was supported by a generous grant from Boliden Metall AB which is gratefully acknowledged.

REFERENCES

1. Sudhakaran-Menon, I., Dewar, H. A., and Newell, D. J. 1969. Fibrinolytic activity of venous blood of patients with multiple sclerosis. Neurology 19, 101–104.
2. Swank, R. L. 1958. Subcutaneous hemorrhages in multiple sclerosis. Neurology 8, 497–499.
3. Sibley, W., Kiernat, J., and Laguna, J. F. 1978. The modification of experimental allergic encephalomyelitis with epsilon-aminocaproic acid. Neurology 28 (2), 102–105.
4. Gorbacheva, F. E., Kvasov, V. T., and Makmudova, M. X. 1976. Platelet aggregation in multiple sclerosis. Zh. Nevropatol. Psikhiatr. 5, 669–672.
5. Couch, J. R., and Hassanein, R. 1977. Platelet hyperaggregability in multiple sclerosis. Trans. Am. Neurol. Assoc. 102, 62–64.
6. Neu, S., Prosiegel, M., and Pfaffenrath, V. 1982. Platelet aggregation and multiple sclerosis. Acta Neurol. Scand. 66, 497–504.

7. Kasperek, K., Iyengar, G. V., Kiem, J., Borberg, H., and Feinendegen, L. E. 1979. Elemental composition of platelets. Part III. Determination of Ag, Au, Cd, Cr, Mo, Rb, Sb, and Se in normal human platelets by neutron activation analysis. Clin. Chem. (Winston-Salem, N.C.) *25*, 711–715.

8. Plantin, L.-O., Meurling, S., and Strandvik, B. 1981. Selenium levels in Swedish children determined with a simple neutron activation analysis method. *In* Mineral Elements '80, Proceedings of a Nordic Symposium on Soil-Plant-Anim.-Man Interrelationships and Implications to Human Health, pp. 453–457. Helsinki.

9. Hamberg, M., Svensson, J., Blombäck, M., and Mettinger, K. L. 1981. Shortened megakaryocyte–platelet regeneration time in patients with ischemic cerebrovascular disease. Thromb. Res. *21*, 675–679.

10. Wang, Y. X., Böcker, K., Reuter, H., Kiem, J., Kasperek, K., Iyengar, G. V., Loogen, F., Gross, R., and Feinendegen, L. E. 1981. Selenium and myocardial infarction: Glutathione peroxidase in platelets. Klin. Wochenschr. *59*, 817–818.

Comparison of Selenium Status in Healthy Patients from a Low-Selenium Area with That of Multiple Sclerosis

Daniel S. Feldman
Diane K. Smith

Multiple sclerosis (MS), a demyelinating disorder of undetermined etiology, exhibits a worldwide incidence gradient that increases with increasing latitude. Nutritional factors have been suggested and eliminated as the cause or an influence on the progression of the disease. During the past decade, a number of studies have suggested an association between MS and selenium status of patients. Selenium status has been measured directly as blood or cellular Se concentration or indirectly as the cellular Se-dependent enzyme, glutathione peroxidase (GSHPx). The initial report of whole-blood Se concentration in MS originated from an Se-poor environment (1). In that study, the low values in MS patients were not significantly different from control subjects residing in the same environment. Other investigators have failed to demonstrate low Se concentrations, but have reported decreased GSHPx in erythrocytes, lymphocytes, and polymorphonuclear leukocytes (2–5).

The present study was undertaken to examine the Se status of MS patients residing in an area of both low MS incidence and bioavailability of that trace mineral. Other nutritional parameters were concomitantly evaluated.

METHODS

Patient Population

Twenty-seven patients from the Multiple Sclerosis Clinic, Department of Neurology, Medical College of Georgia voluntarily agreed to participate and completed an informed consent. Each met the criteria of "clinically definite" ($n = 7$) or "clinically definite with supporting laboratory data" ($n = 20$) MS of Poser et al. (6) in their guidelines for research protocols in MS. All patients received a complete clinical evaluation. Neurological disability was rated on the Kurtzke scale (7). All were evaluated during a stable period of their illness. In 6 patients with remitting-exacerbating disease, serial studies were obtained. Studies were obtained within 5 days of the onset of an exacerbation (mean < 2) and 2 weeks following stabilization. Thirteen patients were receiving adrenocorticosteroids or tropic hormones at the time of sampling. In most cases the drug in use was prednisone; dose varied from 5 mg on alternate days to 120 mg/day.

Anthropometrics

Height, weight, triceps skinfold, and mid-arm circumference were measured. Mid-arm muscle circumference was calculated and compared with standard values. Ideal body weight was calculated from published guidelines (8) and percentage deviation determined.

Biochemical Determinations

Fasting blood samples were promptly centrifuged and plasma separated. Red blood cells (RBC) were washed three times with trace element-free saline. If determinations were immediately carried out samples were stored in trace element-free containers at $-25°C$.

Plasma and erythrocyte Se were determined by the fluorometric method of Whetter and Ullrey (9). Laboratory accuracy was monitored utilizing U.S. National Bureau of Standards bovine liver Se standard. GSHPx was assayed by the coupled method of Paglia and Valentine (10) utilizing hydrogen peroxide as the substrate. Bieri's high-pressure liquid chromatography method (11) was used for vitamin E determination. Atomic absorption photometric plasma zinc analyses of Smith et al. (12) were carried out. Plasma cholesterol and triglycerides and serum albumin determinations used standard automated techniques (Gilford system 103). Fatty acid pattern was determined as methyl esters of extracted lipids on a 15% High Efficiency 2BP on Chromosorb P column (Applied Science Laboratories) isothermally at 180°C by gas-liquid chromatography with flame ionization detection. The hospital clinical laboratory provided total leukocyte count and differential values from which total lymphocyte counts were calculated.

Statistical Methods

Standard descriptive methods (mean, SD, and SEM) were used for the parametric values, and comparisons with normal data from the healthy population utilized t tests. Where parametric values were present, subclasses of the patient population were similarly compared. Data were analyzed using an Apple II+ and "Statpro," an advanced statistical package (Blue Lakes Computing, Madison, WI).

RESULTS

Of the 27 patients reported here, all had clinically definite MS by Poser's criteria (6) and 20 of these had supporting laboratory data (cerebrospinal fluid protein abnormalities). In 11, the disorder was of the progressive variety; when only spinal cord structures were involved myelography had been performed and was negative. Sixteen MS patients had remitting-exacerbating disease; 6 of these developed an acute exacerbation during the course of this study. Patients ranged in age from 19 to 65 years (mean 31.2), and were predominantly women ($n = 17$); 9 were not Caucasian. Thirteen were receiving adrenocorticosteroid or tropic hormones at the time of evaluation. Most patients were receiving prednisone in doses ranging from 5 mg every other day to 120 mg/day in those who were acutely ill. Clinical disability on the Kurtzke scale (7) ranged from 0 to 6 (mean 3.1).

In the MS patients, the plasma Se was 107 ± 4 ng/ml (mean $\pm$ SEM) and the erythrocyte Se was 166 ± 9 ng/ml. The corresponding values in a local healthy population were as follows: plasma Se, 104 ± 1.5 ng/ml; and RBC Se, 158 ± 2.4 ng/ml (13). The MS patients had RBC GSHPx concentrations of 27.8 ± 2 U/g Hb, comparable to values of 26.5 ± 0.5 U/g Hb (13) in the local healthy group. There is no difference between the values found in patients and healthy individuals (t in all cases $< .85$, df $= 231$). Individual observations were normally distributed about the mean (Fig. 1) value in healthy subjects for each of these parameters.

When the Se status of patients with acute exacerbations of MS was compared to the postacute values, mean difference in plasma Se was 1.1, in RBC Se -19 (both ng/ml), and in RBC GSHPx 2.5 U/g Hb. In none of these was the standard error of the difference significant. No correlation was noted between any of the parameters of the Se status

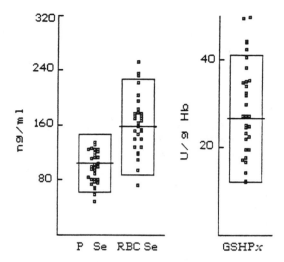

FIG. 1. Selenium in patients with multiple sclerosis. Large rectangles represent 2 SD about mean (horizontal line) in healthy subjects in Georgia. Individual values in the patient population are shown by small squares.

and the disability of the patient as quantified on the Kurtzke scale (7). There was no correlative relationship between the Se values, GSHPx, and the other nutritional parameters, except that between plasma and RBC Se ($r = .62$, $P < .01$).

Table 1 lists the values observed in the MS population for the other nutritional determinations carried out on this group of patients. These too were well within the normal range in the nutrition laboratories at the Medical College of Georgia.

TABLE 1. Other Nutritional Parameters

Parameter	Value[a]
Plasma vitamin E	1055 ± 59 μg/dl
Plasma zinc	98.9 ± 3.7 μg/dl
Serum albumin	4.66 ± 0.10 g/dl
Plasma lipids	
Total cholesterol	237 ± 11.8 mg/dl
Triglycerides	218 ± 30.3 mg/dl
Fatty acid[b]	
14:0	0.78 ± 0.07[c]
16:0	23.16 ± 0.64
18:0	5.36 ± 0.36
18:1	21.64 ± 0.76
18:2	39.73 ± 1.39
20:4	8.41 ± 0.72
Other	0.93 ± 0.56

[a] Expressed as mean ± SEM.
[b] Chain length:double bonds.
[c] Percentage.

DISCUSSION

Previous studies from the nutrition laboratories have shown that low-Se concentrations in plasma and erythrocytes can be found in the Georgia population in patients with protein–calorie malnutrition (14). As can be seen in Fig. 2, the majority of values in such a group of patients falls below the mean of the healthy population in the region. A significant number of observations were below the 95% confidence limits for values seen in good health. If functional Se deficiency is defined as a reduction in the obligatory Se protein, GSHPx, correlations between GSHPx and plasma Se were seen only in those patients most severely Se depleted (lowest plasma Se); these malnourished patients had major deficiencies in other indices of nutrition as well. By comparison the MS patients in this study were adequately nourished and had normal Se and GSHPx values, although their disease was well documented; even acute episodes of MS did not lead to impairment of Se status.

The report of low erythrocyte GSHPx in Danish MS patients did not demonstrate a significant decrease in plasma Se (3). These workers compared then available incidence data for MS in the United States with bioavailability of Se as grain and forage concentration. When the data provided by Kurtzke's study of United States veterans (15) are utilized to attempt correlations between MS incidence and environmental Se (16), no correlation is noted between these data in the 17 states compared by Shukla (3). The random nature of this association is demonstrated in Fig. 3; in this group the correlation coefficient was −0.09, a value virtually identical with 0. None of the studies reporting

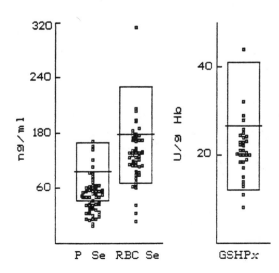

FIG. 2. Selenium in hospitalized patients with protein–calorie malnutrition. Large rectangles represent 2 SD about mean (horizontal line) in healthy subjects in Georgia. Individual values for patients are shown by small squares.

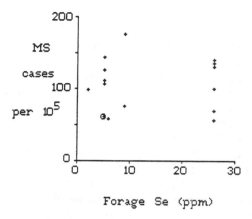

FIG. 3.   Correlation between bioavailability of Se and multiple sclerosis case control incidence in 17 states in the United States. Forage Se reported by Kubota *et al.*, MS incidence by Kurtzke. The correlation coefficient of these data is $r = -.09$, an insignificant value. The value for Georgia is circled.

altered Se status in MS clearly defines the criteria for the diagnosis or the clinical state of the patients in the studies. The present study's findings are similar to those in a small English report (*17*).

Kurtzke has reviewed in detail the epidemiologic data for the disorder (*18*). These studies suggest that climate (latitude) is the major geographic correlate of MS incidence. In addition, there is a suggestion that the geographic factor is acquired early in life and determines the risk after migration to regions of lower MS prevalence.

SUMMARY AND CONCLUSIONS

1. Neither plasma nor erythrocyte Se nor erythrocyte GSHPx concentrations in MS patients differed from values observed in healthy individuals in Georgia. This is a region of low bioavailable Se and low MS incidence.

2. There was no correlation between severity of the disease (degree of disability), recent appearance of symptoms, and Se status.

3. Previous studies in this population have demonstrated impaired Se status only in patients with severe protein–calorie malnutrition. None of the MS patients was malnourished and nutritional status did not correlate with recent onset of symptoms or degree of disability.

4. Treatment of MS with Se supplementation does not seem warranted in the absence of demonstrated deficiency.

ACKNOWLEDGMENTS

This work was supported in part by NIH grant AM30865-03 and by a grant from the Georgia chapter, National Multiple Sclerosis Society.

REFERENCES

1. Wikström, J., Westermarck, T., and Palo, J. 1976. Selenium, vitamin E and copper in multiple sclerosis, Acta Neurol. Scand. *54*, 287–290.
2. Jensen, G. E., Gissel-Neilsen, G., and Clausen, J. 1980. Leukocyte glutathione peroxidase activity and selenium level in multiple sclerosis. J. Neurol. Sci. *48*, 64–67.
3. Shukla, V. K. S., Jensen, G. E., and Clausen, J. 1977. Erythrocyte glutathione peroxidase deficiency in multiple sclerosis. Acta Neurol. Scand. *56*, 542–550.
4. Szeinberg, A., Golan, R., Ben-Ezzer, J., Sarova-Pinhas, I., Sadeh, M., and Braham, J. 1979. Decreased glutathione peroxidase activity in multiple sclerosis. Acta Neurol. Scand. *60*, 265–271.
5. Szeinberg, A., Golan, R., Ben-Ezzer, J., Sarova-Pinhas, I., and Kindler, D. 1981. Glutathione peroxidase activity in various types of blood cells in multiple sclerosis. Acta Neurol. Scand. *63*, 67–75.
6. Poser, C. M., Paty, D. W., and Scheinbert, L. 1983. New diagnostic criteria for multiple sclerosis: Guidelines for research protocols. Ann. Neurol. *13*, 227–231.
7. Kurtzke, J. F. 1965. Further notes on disability evaluation in multiple sclerosis with scale modifications. Neurology *15*, 654–661.
8. Davidson, J. K., and Goldsmith, M. P. 1972. Diabetes Guidebook, Diet Sect. P, pp. 2-3. Litho-Krome, Columbus, GA.
9. Whetter, P. A., and Ullrey, D. E. 1978. Improved fluorometric method for determining selenium. J. Assoc. Off. Anal. Chem. *61*, 927–930.
10. Paglia, D. E., and Valentine, W. N. 1967. Studies on the quantitative and qualitative characterization of erythrocyte glutathione peroxidase. J. Lab. Clin. Med. *70*, 158–169.
11. Bieri, J. G., Tolliver, T. J., and Calignani, G. L. 1979. Simultaneous determination of α-tocopherol and retinol in plasma or red cells by high-pressure liquid chromatography. Am. J. Clin. Nutr. *32*, 2143–2149.
12. Smith, J. C., Butrimovitz, G. P., and Purdy, W. C. 1979. Direct measurement of zinc in plasma by atomic absorption spectroscopy. Clin. Chem. (Winston-Salem, N.C.) *25*, 1487–1491.
13. McAdam, P. A., Smith, D. K., Feldman, E. B., and Hames, C. 1984. Effects of age, sex and race on selenium status of health residents of Augusta, Georgia. Biol. Trace Elem. Res. *6*, 3–9.
14. Smith, D. K., Teague, R. J., and McAdam, P. A. 1984. Selenium status of hospitalized malnourished patients. J. Am. Coll. Nutr. (in press).
15. Kurtzke, J. F., Beebe, G. W., and Norman, J. E., Jr. 1979. Epidemiology of multiple sclerosis in U.S. veterans. 1. Race, sex and geographic distribution. Neurology *29*, 1228–1235.
16. Kubota, J., Allaway, D. L., and Carter, E. E. 1967. Selenium in crops in the United States in relation to selenium responsive diseases of animals. J. Agric. Food Chem. *15*, 448–453.
17. Mehlert, A., Metcalfe, R. A., Diplock, A. T., and Hughes, R. A. C. 1982. Glutathione peroxidase deficiency in multiple sclerosis. Acta Neurol. Scand. *65*, 376–378.
18. Kurtzke, J. F. 1983. Epidemiology of multiple sclerosis. *In* Multiple Sclerosis. J. F. Hallpike, C. W. M. Adams, and W. W. Tourtellotte (Editors), pp. 47–95. Williams & Wilkins, Baltimore, MD.

A Combination Therapy for Chronic Rheumatoid Arthritis with Selenium and Allogeneic Lymphocytes

Masaru Kondo

Previously, I have treated 30 cases of chronic rheumatoid arthritis with allogeneic lymphocyte infusion. Generally speaking, the response to the first infusion seemed very beneficial, but transient. Responses to subsequent infusions were less beneficial and the disease tended to revert to its former severity. However, in 1 case among 30, complete remission was observed, as evidenced by the full normalization of rheumatoid factor titers and disappearance of clinical symptoms. Therefore I started to investigate a new method in the hope of finding a lymphocyte therapy which would be more effective for longer periods of time in more patients.

ASSUMPTIONS

I postulated that the reason why allogeneic lymphocyte infusions were effective only for a short period and why repeated infusions were less effective in comparison to the initial one was the harmful accumulation of oxygen radicals generated in response to the allogeneic lymphocyte treatment. Therefore, to increase the efficacy of the lymphocyte infusion technique, increasing the scavengers of oxygen radicals would be helpful. Selenium and vitamin E were subsequently employed in this study.

CASES

A total of 7 patients (2 males and 5 females, 50 years old or more in 6 cases and 26 years old in 1 case, as shown in Table 1) volunteered for the lymphocyte-selenium-vitamin E treatment. They had become refractory to further infusion of lymphocytes alone. All patients reported severe joint pain for over 4 years, typical of chronic rheumatoid arthritis.

CRITERION

Effectiveness of the method was evaluated through the determination of rheumatoid factor titers before and after treatment; namely, the cases whose titers became normal were only evaluated as effective. Symptomatic remissions were not considered in evaluation of this study.

METHOD

The length of the trial was 4 months. Selenium at the dose of 350 μg/day was administrated orally in the form of "high selenium yeast" (Nutrition 21, San Diego, CA) and simultaneously with vitamin E, 400 IU/day (D-α-tocopherol) for 4 months. Doses of 50 million allogeneic lymphocytes were given at one intravenous infusion. Such infusions were repeated 3 times at 4- to 6-week intervals. The first infusion was given 10–14 days after initiation of the oral administration of selenium and vitamin E. Lymphocytes were isolated from freshly drawn blood of healthy young donors (male and female) by the centrifugation method on Hemaccel-Conray gradient (M. Kondo, unpublished).

RESULTS

The results of this study are summarized in Table 1. Before treatment, rheumatoid factor titers (by the Rose–Waaler test) ranged from 32 to 1024 (normal titer is considered 4 or less). After the 4-month trial, rheumatoid factor titers in all cases dropped significantly and became normal in 4 out of 7 cases. In these 4 cases, joint pain completely disappeared. The condition of the 3 remaining patients improved markedly as evidenced by diminished pain and increased joint

TABLE 1. Results of a Preliminary Rheumatoid Study

No. of case	Sex	Age (yr)	Pretreatment		Posttreatment	
			Rheumatoid factor (normal≤4)	Pain of the joints[a]	Rheumatoid factor (normal≤4)	Pain of the joints
1	M	57	64	++	4	−
2	F	26	32	++	4	−
3	F	50	128	++	4	−
4	M	61	256	++	4	−
5	F	55	1024	++	8 ~ 32	++
6	F	58	128	++	8 ~ 16	++
7	F	53	1024	++	8 ~ 16	+

[a] (++), Severe pain; (+), moderate pain; (−), no pain or occasionally transient slight pain.

994

mobility. Supplementation of selenium and vitamin E was continued in all 7 patients for 5 months after the trial. During that period rheumatoid factor titers remained normal in the 4 patients whose disease had remitted, and only a few slight transient episodes of joint pain were reported by the other patients. In addition, during this time, rheumatoid factor titers had normalized in a fifth patient.

DISCUSSION AND CONCLUSIONS

The following conclusions can be made from this trial.

It seems possible to normalize rheumatoid factor titers with appropriate treatment. This has never been the case for chronic rheumatoid arthritis with any previous therapy, except in very rare cases. While this preliminary study did not include concurrent controls, compared to historical controls, normalization of rheumatoid factor titers in 4 of 7 treated patients within 4 months is a significant effect.

Concurrent supplementation with selenium and vitamin E appears to be necessary for the achievement of these clinical results, since allogeneic lymphocyte treatment alone produced only modest and transient benefits. In our experience, selenium and vitamin E alone may supply some palliative pain relief, but do not seem to influence the course of the disease. Thus, both combined measures seem necessary to achieve an optimal response. Strong stimulation of immunological responses by selenium and vitamin E has been reported in several animal experiments (Spallholz 1980).

REFERENCE

Spallholz, J. E.   1980.   Anti-inflammatory, immunologic and carcinostatic attributes of selenium in experimental animals. *In* Symposium on Diet and Resistance to Disease, pp. 43–62.

Relationship of Serum Selenium and Lipid Peroxidation in Preeclampsia

A Kauppila   L. Viinikka
U. M. Mäkilä   E. Yrjänheikki
H. Korpela

Preeclampsia is a common pregnancy complication characterized by hypertension, proteinuria, and/or edema. Its etiology is unknown and pathogenesis poorly understood. In preeclampsia placental vessels and maternal arterioles in the kidneys and other organs are affected by degenerative changes and disseminated intravascular coagulation. Increased lipid peroxidation (1–3) may contribute to the development of preeclamptic vascular changes as free radicals and lipid peroxides have the potential to cause oxidative damage to cellular membranes (4,5). One factor controlling lipid peroxidation is glutathione peroxidase (GSHPx), which reduces hydrogen peroxides and organic hydroperoxides. The essential trace element selenium is a structural component of this enzyme, and therefore the activity of GSHPx is dependent on the bioavailability of selenium (6). Some previous studies (7–9) have demonstrated that the concentration of selenium in the blood decreases during pregnancy and it is lowest during its last trimester, a manifestation period of preeclampsia. We therefore measured the concentrations of serum selenium and plasma lipid peroxides and their interrelationships in healthy and preeclamptic pregnant Finnish women with suboptimal selenium intake.

MATERIAL AND METHODS

Thirteen healthy pregnant women aged 20–32 years and 21 preeclamptic patients aged 19–37 years participated in this study. Healthy women were followed by collecting serial blood samples at intervals of 1 week from weeks 32–35 of pregnancy until the third to the fifth puerperal day. Preeclamptic patients entered the study upon admission to the hospital from weeks 32–38 of pregnancy, and blood samples were drawn similarly as in the controls. The serum specimens for selenium measurements and plasma specimens for lipid peroxide measurements were stored at −20°C until assayed.

In the determination of serum selenium, acid digestion of samples was followed by the reduction of selenium to hydrogen selenide which was determined by atomic absorption spectrophotometry (10). The plasma lipid peroxides were measured with a malondialdehyde–thiobarbituric acid reaction of the lipids precipitated from plasma with phosphotungstic acid (11). Student's t test and linear regression analysis were used in the statistical treatment of the data.

RESULTS

The concentration of serum selenium in healthy pregnant women decreased significantly during the last trimester of pregnancy (Figs. 1 and 2). The values recorded during weeks 33–35 (0.74 ± 0.02 SEM μmol/liter) and 36–40 of pregnancy (0.66 ± 0.03 μmol/liter) were significantly lower ($P < .01$ and $P < .001$, respectively) than that of 31

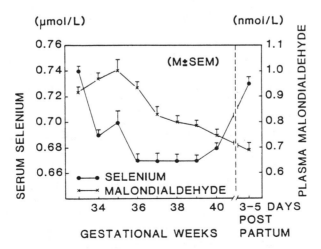

FIG. 1. Concentrations of serum selenium and plasma lipid peroxides in normal and preeclamptic pregnancy during the last trimester of pregnancy and after delivery.

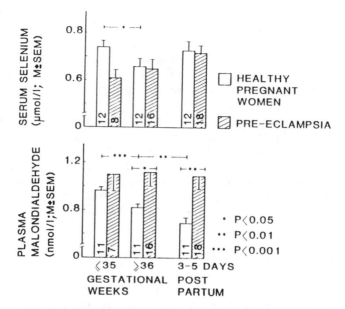

FIG. 2. Concentrations of serum selenium and plasma lipid peroxides in normal and preeclamptic pregnancy during the early and late phases of the last trimester of pregnancy and after delivery. Numbers of determinations are indicated in the columns.

nonpregnant women (0.91 ± 0.04 μmol/liter). The postpartum concentration of selenium was similar to that during weeks 33–35 of pregnancy. Plasma concentration of lipid peroxides also decreased significantly with the advancement of uncomplicated pregnancy (Fig. 1). The mean concentration of lipid peroxides before week 36 of pregnancy was higher ($P < .001$) than during weeks 36–40 (Fig. 2). The postpartum recording decreased further ($P < .01$) in comparison with those values obtained during weeks 36–40.

The serum concentration of selenium did not differ in healthy and preeclamptic women. Plasma concentration of lipid peroxides was significantly higher in preeclampsia than in healthy women both before and after delivery (Fig. 2). There was no correlation between the concentrations of serum selenium and plasma lipid peroxides during normal or preeclamptic pregnancy, but after delivery a positive correlation existed between these parameters in preeclampsia ($r = .66$; $P < .01$).

DISCUSSION

Evidence based on studies of nutritional intake (12,13) and blood concentrations of selenium (14–17) verifies that its bioavailability is suboptimal in Finland. The exact role of deficient selenium intake in the development of diseases is not known, but one study (17) suggested an increased risk of cardiovascular diseases in populations with a very low serum concentration of selenium. Our study evaluated the role of selenium and lipid peroxidation in the development of preeclampsia, a major complication during the last trimester of pregnancy. Consistently with some previous reports (7–9) we found that the serum concentration of selenium was lower in pregnant women than in nonpregnant subjects. The rapid postpartum increase of serum selenium was, however, a new finding, and it may indicate that the decrease in selenium concentration was due partly to the hemodilution occurring during pregnancy. Even after delivery, however, serum selenium was much lower than in healthy, nonpregnant subjects which indicates selenium deficiency during pregnancy and puerperium. It is possible that the increased requirement and decreased availability of selenium during pregnancy may detrimentally affect the balance of oxidative and antioxidative cellular mechanisms. Our finding that the concentration of lipid peroxides in plasma also decreases during pregnancy, however, speaks against this assumption and suggests that the role of selenium in the decomposition of lipid peroxides in humans is not as important as previously thought (18,19). It is worth noticing that we measured lipid peroxides with a thiobarbituric acid reaction in the protein/lipid precipitate of plasma, and this does not necessarily completely reflect the lipid peroxidation at tissue level (20).

We observed that lipid peroxidation was increased in preeclampsia. Together with some previous results (1–3) this suggests that lipid peroxidation is an essential element in the pathogenesis and even in the etiology of this disease. It is likely related to vascular and other tissue damages typical of preeclampsia. A totally new finding is the similar serum concentrations of selenium in healthy pregnant and preeclamptic women, suggesting that selenium deficiency does not play the key role in increased lipid peroxidation. This view is also supported by the lacking correlation between serum selenium and plasma lipid peroxides during pregnancy. Our present study of preeclamptic puerperal women and the previous investigation of nonpregnant women (21) unexpectedly revealed a significant positive correlation between the concentrations of serum selenium and plasma lipid peroxides. This observation is unexplained at present. This, together with the unknown reason of increased lipid peroxidation in preeclampsia, advocates more detailed studies of oxidative and antioxidative mechanisms and the factors controlling these processes during normal and complicated pregnancy.

SUMMARY AND CONCLUSIONS

The concentrations of serum selenium and plasma lipid peroxides were measured in 13 healthy, pregnant and 21 preeclamptic women during the last trimester of pregnancy and 3–5 days after delivery. The serum concentration of selenium in pregnant women was significantly lower than that in nonpregnant women, and it decreased significantly during the course of pregnancy, as did lipid peroxides. In preeclampsia lipid peroxide concentrations were higher than in normal pregnancy both before and after delivery, whereas serum selenium was unchanged. There was no correlation between the concentrations of serum selenium and plasma lipid peroxides during normal or preeclamptic pregnancy, but after delivery a significant positive correlation existed between them in preeclampsia. Our results show that increased lipid peroxidation is a factor in the pathogenesis of preeclampsia, which seems not to be related to the serum concentration of selenium.

ACKNOWLEDGMENTS

This work was supported by grants from the Finnish Academy of Science, the Sigrid Jusélius Foundation, and the Foundation for Nutrition Research.

REFERENCES

1. Ishihara, M. 1978. Studies on lipoperoxide of normal pregnant women and of patients with toxemia of pregnancy. Clin. Chim. Acta *84*, 1–9.
2. Maseki, M., Nishigaki, I., Hagihara, M., Tomoda, Y., and Yagi, K. 1981. Lipid peroxide levels and lipid content of serum lipoportein fractions of pregnant subjects with or without preeclampsia. Clin. Chim. Acta *115*, 155–61.
3. Wickens, D., Wilkins, M. H., Lunec, J., Ball, G., and Dormandy, T. L. 1981. Free radical oxidation (peroxidation) products in plasma in normal and abnormal pregnancy. Ann. Clin. Biochem. *18*, 158–62.
4. Mead, J. F. 1976. Free radical mechanisms of lipid damage and consequences for cellular membranes. *In* Free Radicals in Biology. W. A. Pryor (Editor), Vol. 1., pp. 51–68. Academic Press, New York.
5. Del Maestro, R. F. 1980. An approach to free radicals in medicine and biology. Acta Physiol. Scand., Suppl. *492*, 153–68.

6. Sunde, R. A., and Hoekstra, W. G. 1980. Structure, synthesis and function of glutathione peroxidase. Nutr. Rev. *38*, 265–73.

7. Rudolph, N., and Wong, S. 1978. Selenium and glutathione peroxidase activity in maternal and cord plasma and red cells. Pediatr. Res. *12*, 789–92.

8. Behne, D., and Wolters, W. 1979. Selenium content and glutathione peroxidase activity in the plasma and erythrocytes of nonpregnant and pregnant women. J. Clin. Chem. Clin. Biochem. *17*, 133–135.

9. Butler, J. A., Whanger, P. D., and Tripp, M. J. 1982. Blood selenium and glutathione peroxidase activity in pregnant women: Comparative assays in primates and other animals. Am. J. Clin. Nutr. *36*, 15–23.

10. Hakala, E., and Yrjänheikki, E. 1980. Determination of selenium in urine by hydride generation and atomic absorption spectroscopy. 29. Nordisk yrkeshygieniske möte i Norge, p. 66. Yrkeshygienisk Institutt Publikasjon.

11. Yagi, K. 1976. A simple fluorometric assay for lipoperoxides. Biochem. Med. *15*, 212–216.

12. Mutanen, M., and Koivistoinen, P. 1983. The role of imported grain on the selenium intake of Finnish population in 1941–1981. Int. J. Vitam. Nutr. Res. *53*, 102–108.

13. Levander, O. A. 1982. Clinical consequences of low selenium intake and its relationship to vitamin E. Ann. N.Y. Acad. Sci. *393*, 70–81.

14. Levander, O. A., Alfthan, G., Arvilommi, H., Gref, C. G., Huttunen, J. K., Katama, M., Koivistoinen, P., and Pikkarainen, J. 1983. Bioavailability of selenium to Finnish men as assessed by platelet glutathione peroxidase activity and other blood parameters. Am. J. Clin Nutr. *37*, 887–897.

15. Sundström, H., Yrjänheikki, E., and Kauppila, A. 1984. Low serum selenium concentration in patients with cervical or endometrial cancer. Int. J. Gynaecol. Obstet. *22*, 35–40.

16. Sundström, H., Yrjänheikki, E., and Kauppila, A. 1984. Serum selenium in patients with ovarian cancer during and after therapy. Carcinogenesis (London) *5*, 731–734.

17. Salonen, J. T., Alfthan, G., Huttunen, J. K., Pikkarainen, J., and Puska, P. 1982. Association between cardiovascular death and myocardial infarction and serum selenium in a matched-pair longitudinal study. Lancet *2*, 175–179.

18. Tappel, A. L. 1980. Vitamin E and selenium protection from *in vivo* lipid peroxidation. Ann. N.Y. Acad. Sci. *355*, 18–31.

19. Flohé, L. 1982. Role of GSH peroxidase in lipid peroxide metabolism. *In* Lipid Peroxides in Biology and Medicine. K. Yagi (Editor), pp. 149–159. Academic Press, New York.

20. Capel, I. D., and Thornley, A. C. 1982. Superoxide dismutase activity, caeruloplasmin activity, and lipoperoxide levels in tumour and host tissue of mice bearing the Lewis lung carcinoma. Eur. J. Cancer Clin. Oncol. *18*, 507–513.

21. Sundström, H., Korpela, H., Viinikka, L., and Kauppila, A. 1985. Serum selenium and glutathione peroxidase, and plasma lipid peroxides, and their responses to antioxidants in patients with gynecological cancer. Cancer Lett. (in press).

An Urgent Methodological Need
for Antioxidant Research as Seen
by an Epidemiologist

Curtis G. Hames

Numerous studies have implicated reactive species of oxygen in relation to DNA damage (1), cancer initiation or promotion, aging (2–7), Alzheimer's disease (8), cataract formation, male sterility (9), atherosclerosis (10), hypertension, stroke, congestive heart disease (11,12), Keshan disease (13), pancreatic fibrosis (2), rheumatoid arthritis (14), immunity (15), and hepatic disease (16). If true, they all serve to illustrate the pleiotropic nature of the oxidative process. Could "free radical status" be a new risk factor which can be correlated with the quantitation of antioxidant levels and used with other risk factors such as age, race, sex, socioeconomic status, geographic areas of residence, (13) and dietary patterns (17,18) i.e., as blood pressure with stroke, HDL-LDL lipoproteins and cholesterol with coronary disease, and obesity with longevity?

This rapidly increasing knowledge base suggesting the possible etiological role of oxidative damage to biological systems from free radicals dictates the urgent need to develop analytical methods which can reflect changing antioxidant status in response to diet, medication, disease incidence, or prevalence rates and which can be used for determining the individual's own native baseline endogenous antioxidant capacity. Ames's recent review (19) made a similar plea when he called for the translation of the oxidative DNA damage test used in animal models for use in human epidemiology. However, such a test may not be possible considering the present state and the complexity of the antioxidant systems; i.e., it has been estimated that as many as

50 different components may be involved in protecting the body from oxidative damage. Also, it has not been established which of the currently available tests may truly reflect the pathological process one is attempting to study.

THE ANTIOXIDANT SYSTEM

A free radical has been defined by Freeman and Crapo (20) as any molecule that has unpaired electrons. They are generated *in vivo* as by-products of normal metabolism and are crucial for the normal operation of a wide spectrum of biological processes. How does nature protect against such damage? A number of protective measures against oxidative damage have evolved in the morphology as well as biochemistry of the cell; e.g., morphologically (a) the compartmentalization of the most toxic reactions into a single organelle as a protective measure against damage to the entire cell; (b) utilization of the cell membrane to serve as a barrier. The nucleus and the toxic oxidative products from the mitochondria are kept as far apart physically within the cell as possible. From a biochemical standpoint, the reduction of oxygen is by the monovalent pathway (7), in sequential order: the superoxide radical ($\cdot O_2^-$), hydrogen peroxide (H_2O_2), hydroxyl radical ($\cdot OH$) and possibly singlet oxygen (O_2^1) (16). Defense against these potentially damaging free radicals is achieved by several scavengers, i.e., superoxide dismutases for ($\cdot O_2^-$), and catalases and peroxidases for H_2O_2. Glutathione peroxidase catalyses the reduction of various hydroperoxides to alcohols (7). The antioxidants uric acid, α-tocopherol, and ascorbic acid are also capable of protecting the cell membrane from lipid peroxidation. Cartenoids including β-carotene are known to protect plants against free radical damage during photosynthesis.

WHAT SHOULD WE MEASURE?

To measure the individual's "natural" net antioxidant protective potential would appear to be one approach for use in epidemiological studies, since it would reflect deviations from the norm. Such deviations may represent etiologically related genetic or dietary insufficiency factors.

Cutler (8) has proposed measurement of the extent of the antioxidation protection in serum by determining the amount of malonaldehyde or thiobarbituric acid-reacting material produced during *in vitro* incubations. Because the rate of autoxidation with tissue or serum depends upon the amount of peroxidizable material (e.g., polyunsaturated fatty acids) and the net level of antioxidants present, serum autoxidation is taken as a reflection of the intrinsic oxidant susceptibility of the sample. Cutler has suggested that we would expect to find human tissues and serum to be the most resistant of all the species to autoxidation.

Slater (21) has also pointed out that it is important to remember that the concentrations of observed products of lipid peroxidation (e.g., diene conjugation, lipid hydroperoxides, and malonaldehyde) in tissues at any one instant reflect a balance between the rate of production and the rates of degradation and metabolism. Failure to see significant increases in particular products of lipid peroxidation may hide a stimulation in lipid peroxidation rate above the normal range, but not sufficient to exceed the capacity for degrading or metabolizing of such products. Obviously, the need for better understanding the basis of oxidative–antioxidative homeostatic mechanisms is essential.

Dynamic homeostatic control of the endogenous antioxidant system is evidenced in experiments where a deficiency in one element has been found to cause compensatory changes in one or more of the other antioxidant factors. The importance of better knowledge of these mechanisms has widespread implications for possibly controlling certain human inherited diseases as well as for providing a therapeutic avenue for treating many of the diseases mentioned earlier. Cutler (8) has proposed a model to understand regulation of the oxidative environment within the cell. His proposal is based on the properties of three substances that are unusually sensitive to the effect of oxygen species: (a) arachidonic acid, an essential unsaturated fatty acid (20:4) that is extremely sensitive to lipid peroxidation and is the only precursor of prostaglandins, prostacyclins, thromboxanes, and the leukotrienes; (b) prostaglandin cyclooxygenase, the first enzyme in the conversion of arachidonic acid to prostaglandin endoperoxide; and (c) guanylate cyclase, which catalyzes the synthesis of cGMP and is very sensitive to activation by a large number of active oxygen species.

SUMMARY

Information concerning the possible role(s) of antioxidants in disease processes is increasing. New methodology is needed for use in epidemiological studies for the characterization of the endogenous antioxidant status of individuals. Information so produced may be used with other parameters to develop risk profiles for certain diseases as well as to monitor therapeutic trials. The regulatory mechanisms of control of the multicomponent antioxidant system must be elucidated. This information may provide new approaches for studying the role of oxidative damage in health and disease.

REFERENCES

1. Schrauzer, G. M. 1981. Selenium and cancer: Historical developments and perspectives. *In* Selenium in Biology and Medicine. J. E. Spallholz, J. L. Martin, and H. E. Ganther (Editors), p. 98. AVI Publishing Co., Westport, CT.
2. Combs, G. F., Jr., and Bunk, M. J. 1981. The role of selenium in pancreatic function. *In* Selenium in Biology and Medicine. J. E. Spallholz, J. L. Martin, and H. E. Ganther (Editors), p. 70. AVI Publishing Co., Westport, CT.
3. Csallany, A. S., Zaspel, B. J., and Ayaz, K. L. 1981. Selenium and aging. *In* Selenium in Biology and Medicine. J. E. Spallholz, J. L. Martin, and H. E. Ganther (Editors), p. 118. AVI Publishing Co., Westport, CT.
4. Milner, J. A., Greeder, G. A., and Poirier, K. A. 1981. Selenium and transplantable tumors. *In* Selenium in Biology and Medicine. J. E. Spallholz, J. L. Martin and H. E. Ganther (Editors), p. 146. AVI Publishing Co., Westport, CT.
5. Griffin, A. C., and Lane, H. W. 1981. Selenium chemoprevention of cancer in animals and possible human implications. *In* Selenium in Biology and Medicine. J. E. Spallholz, J. L. Martin, and H. E. Ganther (Editors), p. 160. AVI Publishing Co., Westport, CT.
6. Whiting, R. F., Wei, L., and Stich, H. F., Mutagenic and antimutagenic activities of selenium compounds in mammalian cells. *In* Selenium in Biology and Medicine. J. E. Spallholz, J. L. Martin, and H. E. Ganther (Editors), p. 325. AVI Publishing Co., Westport, CT.
7. Fridovich, I. 1979. Chairman's introduction. Ciba Found. Symp. *65*, 1–4.
8. Cutler, R. G. 1984. Antioxidants, aging and longevity. *In* Free Radicals in Biology. W. A. Pryor (Editor), Vol. 6, pp. 371–428. Academic Press, New York.
9. McConnell, K. P., and Burton, R. M. 1981. Selenium and spermatogenesis. *In* Selenium in Biology and Medicine. J. E. Spallholz, J. L. Martin, and H. E. Ganther (Editors), p. 132. AVI Publishing Co., Westport, CT.
10. Andrews, J. W., Hames, C. G., and Metts, J. C., Jr. 1981. Selenium and cadmium status in blood of residents from low selenium–high cardiovascular disease area of southeastern Georgia. *In* Selenium in Biology and Medicine. J. E. Spallholz, J. L. Martin, and H. E. Ganther (Editors), p. 348. AVI Publishing Co., Westport, CT.
11. Diplock, A. T. 1981. The role of vitamin E and selenium in the prevention of oxygen-induced tissue damage. *In* Selenium in Biology and Medicine. J. E. Spallholz, J. L. Martin, and H. E. Ganther (Editors), p. 303. AVI Publishing Co., Westport, CT.
12. Bryant, R. W., Bailey, J. M., King, J. C., and Levander, O. A. 1981. Altered platelet glutathione peroxidase activity and arachidonic acid metabolism during selenium repletion in a controlled human study. *In* Selenium in Biology and Medicine. J. E. Spallholz, J. L. Martin, and H. E. Ganther (Editors), p. 395. AVI Publishing Co., Westport, CT.
13. Chen, X., Chen, X., Yang, G.-Q., Wen, W., Chen, J., and Ge, K. 1981. Relation of selenium deficiency to the occurrence of Keshan disease. *In* Selenium in Biology and Medicine. J. E. Spallholz, J. L. Martin, and H. E. Ganther (Editors), p. 171. AVI Publishing Co., Westport, CT.

14. Aaseth, J., Thomassen, Y., and Alexander, J.   1981.   Decreased serum selenium in rheumatoid arthritis and in alcoholic cirrhosis. *In* Selenium in Biology and Medicine. J. E. Spallholz, J. L. Martin, and H. E. Ganther (Editors), p. 418. AVI Publishing Co., Westport, CT.

15. Spallholz, J. E.   1981.   Selenium: What role in immunity and immune cytotoxicity? *In* Selenium in Biology and Medicine. J. E. Spallholz, J. L. Martin, and H. E. Ganther (Editors), p. 103. AVI Publishing Co., Westport, CT.

16. Burk, R. F., and Correia, M. A.   1981.   Selenium and hepatic heme metabolism. *In* Selenium in Biology and Medicine. J. E. Spallholz, J. L. Martin, and H. E. Ganther (Editors), p. 86. AVI Publishing Co., Westport, CT.

17. Levander, O. A., Sutherland, B., Morris, V. C., and King, J. C.   1981.   Selenium metabolism in human nutrition. *In* Selenium in Biology and Medicine. J. E. Spallholz, J. L. Martin, and H. E. Ganther (Editors), p. 256. AVI Publishing Co., Westport, CT.

18. Van Vleet, J. F., and Boon, G. D.   1981.   Evaluation of the ability of dietary supplements of silver, copper, cobalt, tellurium, cadmium, zinc, and vanadium to induce lesions of selenium–vitamin E deficiency in ducklings and swine. *In* Selenium in Biology and Medicine. J. E. Spallholz, J. L. Martin, and H. E. Ganther (Editors), p. 366. AVI Publishing Co., Westport, CT.

19. Ames, B. N.   1983.   Dietary carcinogens and anticarcinogens. Oxygen radicals and degenerative diseases. Science *221,* 1256–1265.

20. Freeman, B. A., and Crapo, J. D.   1981.   Biology of disease. Free radicals and tissue injury. U.S.- Can. Div Int. Acad. Pathol. *47* (5), 412–426.

21. Slater, T. F.   1984.   Overview of methods used for detecting lipid peroxidation. Methods Enzymol. *105,* 283–293.

Section IX

Selenium and Cancer

109

Susceptibility of Mammary Carcinogenesis in Response to Dietary Selenium Levels: Modification by Fat and Vitamin Intake

Clement Ip

There is increasing evidence that selenium has a protective effect against tumorigenesis in laboratory animals. Most of the studies reported in the literature involve the use of inorganic selenium supplemented either in the drinking water or in the diet at a concentration ranging from 0.5 to 6 ppm. These levels are considerably higher than the nutritional requirement of about 0.1 ppm established by the National Research Council for animals. The breast cancer models that have been shown to be responsive to selenium chemoprevention include tumors induced by both viruses and chemical carcinogens (*1–6*). Since selenium is effective in suppressing mammary neoplastic development induced by either methylnitrosourea or dimethlybenz[*a*]anthracene, it is unlikely that the primary action of selenium is exerted via changes in carcinogen metabolism. A report from Medina's laboratory (*4*) indicates that selenium markedly inhibits mammary tumorigenesis in BALB/cfC3H mice (MuMTV positive), but has little effect on the incidence of neoplastic transformation in preneoplastic outgrowth lines or the growth rate of primary mammary tumors transplanted subcutaneously in BALB/c mice (MuMTV negative). Thus there seems to be a decreasing sensitivity to selenium-mediated inhibition as cells progress from normal to preneoplastic to neoplastic.

By comparing two sets of worldwide epidemiological data, it becomes apparent that a high ratio of fat to selenium intake is directly proportional to increased mortality rates from breast and colon cancers (7,8). For example, in western European countries and the United States, where breast cancer death rates range from 17 to 25/100,000 population, the per capita fat and selenium intakes average about 160 g and 120 μg/day, respectively. This is in contrast to most Asian and eastern European countries where mortality rates from breast cancer range from 4 to 12/100,000, and their fat and selenium intakes average about 60 g and 300 μg/day, respectively. In other words, high fat and low selenium consumption seems to be associated with an increased incidence of neoplasia.

Spurred by this interesting epidemiological relationship, we have investigated the interaction between dietary fat and selenium in the development of breast cancer in rats. Our work has involved the mammary tumor model induced by 7,12-dimethlybenz[a]anthracene (DMBA) in female Sprague–Dawley rats. The studies described in the present work were designed to address the following questions: (a) What is the effect of selenium deficiency on the susceptibility to carcinogenesis in rats fed different levels and types of fats? (b) Do high levels of dietary selenium protect against tumorigenesis? (c) Can the chemopreventive effectiveness of selenium be modified by other vitamin supplementation? (d) Is the prophylactic action of selenium related to its antioxidant function? Details of the experimental protocol have been published previously (9–13). These references will be alluded to in each of the tables and figures shown in the text. In all our studies, selenium in the form of sodium selenite was added to semipurified synthetic diets.

EFFECT OF SELENIUM DEFICIENCY AND FAT INTAKE ON MAMMARY CARCINOGENESIS

Table 1 shows the effect of selenium deficiency on the incidence and yield of mammary tumors in rats fed diets containing different levels of polyunsaturated fat or a high saturated fat diet. The selenium-deficient and -adequate diets contained <0.02 and 0.1 ppm of selenium, respectively. Animals were sacrificed 22 weeks after DMBA administration. Although tumor incidence increased significantly in proportion to the amount of corn oil in the diet, a deficiency of selenium seemded to have little effect in rats maintained on either the 1% or 5% fat ration. Only in those rats that were fed a high polyunsaturated fat diet (25% corn oil) did selenium deprivation result in a

TABLE 1. Effect of Selenium Deficiency on Mammary Tumor Incidence and Tumor Yield in Rats Fed Different Levels and Types of Fats

Group[a]	Fat in diet		Selenium status[b]	No. of rats	Tumor incidence (%)	Total no. of tumors	Tumors per tumor-bearing rat	Average latency period[c] (days)
	Corn oil	Coconut oil						
1	1%		Deficient	23	17.4	6	1.5	90
2	1%		Adequate	24	12.5	4	1.3	100
3	5%		Deficient	25	44.0	24	2.2	88
4	5%		Adequate	24	33.3	13	1.6	88
5	25%		Deficient	25	96.0	103	4.3	72
6	25%		Adequate	25	60.0	40	2.7	85
7	1% + 24%		Deficient	24	29.2	9	1.3	101
8	1% + 24%		Adequate	25	24.0	8	1.3	97

[a] All rats were given 5 mg of DMBA i.g. at 50 days at age and were killed 22 weeks later.
[b] Selenium-deficient diet, <0.02 ppm; selenium-adequate diet, 0.1 ppm.
[c] Time between DMBA administration and the appearance of the first palpable tumor.
Source: Ip and Sinha (9).

TABLE 2. Effect of Selenium Deficiency and/or Vitamin E Excess on the Incidence and Yield of Mammary Tumors in Rats Fed a 25% Stripped Corn Oil Diet

Group[a]	Selenium status[b]	Vitamin E status[c]	No. of rats	Tumor incidence (%)	Total no. of tumors	Tumors per tumor-bearing rat	Average latency period (wk)
1	Deficient	Adequate	25	88	85	3.8	13.7
2	Adequate	Adequate	25	68	47	2.8	13.9
3	Deficient	Excess	25	80	71	3.6	11.9
4	Adequate	Excess	25	64	41	2.7	14.1

[a] All rats were given 5 mg of DMBA i.g. and were killed 24 weeks later.
[b] Selenium deficient, <0.02 ppm; selenium adequate, 0.1 ppm.
[c] Vitamin E adequate, 30 mg/kg; vitamin E excess, 1000 mg/kg.
Source: Ip (12).

marked enhancement of mammary tumorigenesis. Results in Table 1 also indicate that a diet rich in hydrogenated coconut oil was much less effective in promoting tumor formation. Furthermore, dietary selenium depletion failed to exert any adverse influence on tumor development in rats fed the saturated fat diet.

In view of the sparing effect of selenium and vitamin E on each other, the following experiment was conducted to determine whether vitamin E could abrogate the enhancement in tumorigenesis due to selenium deficiency in rats fed a high polyunsaturated fat diet. The control vitamin E diet contained 30 mg of DL-α-tocopheryl acetate/kg, while the experimental vitamin E diet contained 1000 mg of DL-α-tocopheryl acetate/kg. Table 2 summarizes the effect of selenium deficiency and/or vitamin E excess on mammary tumorigenesis in rats fed a 25% corn oil diet. Results of this experiment indicate that a high level of vitamin E supplementation failed to obliterate the augmentation in tumor development caused by selenium deprivation. Even in those rats that received an adequate supply of selenium, vitamin E by itself provided no protective effect. This study suggests that the effect of selenium is specific on a biological basis and may not be related solely to the antioxidant status of the animal.

ANTICARCINOGENIC EFFECT OF SELENIUM AT HIGH LEVELS OF SUPPLEMENTATION

Our next step was to find out if dietary selenium at levels above recommended requirement can inhibit mammary carcinogenesis, and if so, does the optimal level of supplementation depend on the dose of the carcinogen and the fat intake of the animal? Figure 1 shows the effect of various levels of selenium supplementation in rats that were fed either the 5% or 25% corn oil diet and given either 5 or 10 mg of DMBA. Selenium supplementation of both diets was started from weaning and continued until the end of the experiment 22 weeks after DMBA administration.

We found that dietary selenium had to be raised to 1.5 ppm before its chemopreventive effect became noticeable. The degree of inhibition was proportional to the level of selenium up to 5 ppm. In general, the response was manifested by a combination of a lower tumor incidence, a reduction in tumor yield, and a longer latency period. It should also be noted that the anticarcinogenic efficacy of selenium was diminished by a larger dose of carcinogen. Moreover, selenium was unable to counteract completely the enhancing effect of fat, since rats on the 25% fat diet still developed more tumors than those on the 5% fat diet at comparable levels of selenium supplementation.

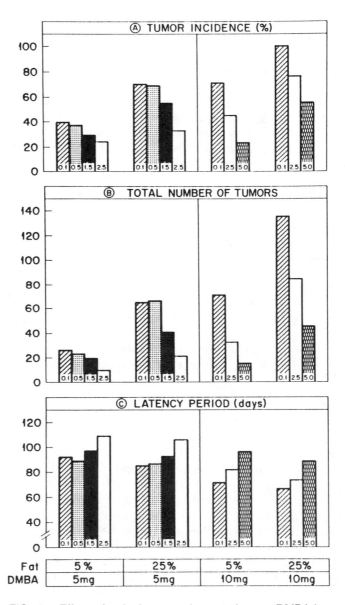

FIG. 1.   Effect of selenium supplementation on DMBA-induced mammary tumorigenesis in rats fed either a 5% or a 25% corn oil diet. The level of selenium in the diet is indicated in each bar.
Source: IP (10).

Supplementation of selenium at 2.5 ppm did not influence the growth rate of rats (10). However, toxicity in the form of a slight suppression in weight gain (about 10%) was evident at 5 ppm. This decrease in growth was most likely due to reduced diet consumption. Periodic measurement of food intake confirmed this assumption. A pair-feeding experiment further indicated that reduced food consumption alone was not sufficient to account for the striking suppression of tumorigenesis in those rats treated with 5 ppm of selenium (10).

INTERACTION OF SELENIUM WITH VITAMINS A AND C

In view of the similarities between selenium and retinoid (vitamin A analog) in their inhibitory effect on carcinogenesis, we have also examined the combined use of both agents. By feeding rats a diet containing 4 ppm of selenium and 250 ppm of retinyl acetate, the yield of mammary tumors was reduced to 8% of control, as compared to 51% and 36%, respectively, for selenium and retinyl acetate alone (Table 3, Experiment A). To our chagrin, this combined regimen, although very effective in inhibiting tumorigenesis, was not well received by the animals, thus resulting in decreases in food consumption and weight gain (approximately 15% reduction compared to the controls). Hence, the significance of this particular treatment cannot be properly evaluated until additional studies on the toxicological and pharmacological effects are clearly differentiated.

The role of vitamin C in cancer prevention still remains controver-

TABLE 3. Interaction of Selenium with Retinyl Acetate (Vitamin A) and Sodium Ascorbate (Vitamin C) on DMBA-Induced Mammary Carcinogenesis

Experiment	Dietary supplement	Tumor incidence (%)	Total no. of tumors	Final body weight (g)
A	None	27/30 (90.0)	112	288
	Se (4 ppm)	16/30 (53.5)	57	269
	Vitamin A (250 ppm)	14/30 (46.7)	40	262
	Se + vitamin A	5/30 (16.6)	9	246
B	None	19/25 (76.0)	59	284
	Se (4 ppm)	10/25 (40.0)	33	270
	Vitamin C (0.5%)	20/25 (80.0)	66	281
	Se + vitamin C	17/25 (68.0)	53	268

Source: Ip and Ip (11).

sial, as publications continue to appear in the literature reporting both positive as well as negative results. In an attempt to understand how selenium may interact with other potential chemopreventive agents, we decided to investigate the effect of supplementing vitamin C and selenium, both individually and collectively, on mammary carcinogenesis. Results are shown in Table 3, experiment B. Vitamin C at a 0.5% level in the diet produced no inhibition of tumor development compared to the controls. However, we were much surprised to find that the protective effect of selenium was nullified by vitamin C supplementation. Studies are now under way to confirm the above observation and to delineate how these nutrients interact with each other when present at excess levels.

IMPROVEMENT OF SELENIUM-MEDIATED PROPHYLAXIS BY VITAMIN E SUPPLEMENTATION

One of the objectives of our research effort in selenium and cancer is to improve the anticarcinogenic efficacy of lower levels of selenium by combining it with another agent. Our experience with vitamin E proved to be most promising. The rationale for selecting vitamin E is twofold. First, both selenium and vitamin E share the role of endogenous antioxidants. Second, there is ample evidence in the literature which shows that they have a sparing effect on each other in the prevention of several nutritional deficiency diseases.

In the present study, rats given DMBA were maintained on a high polyunsaturated fat diet, thus enabling us to evaluate the efficacy of the vitamin E and selenium combination treatment under a more vigorous condition of oxidant stress. Selenium was supplemented in the diet for the entire duration of the experiment (from -2 to $+24$ weeks), while additional vitamin E was present for various lengths of time: -2 to $+24$ weeks, -2 to $+2$ weeks, and $+2$ to $+24$ weeks. The time of DMBA administration was taken as 0. The reason for adopting this protocol is that we have previously found that a continuous intake of selenium is necessary to achieve a maximal inhibitory response. By supplementing vitamin E for a defined period, either around the time of or after DMBA administration, we can examine the effect of vitamin E during the initiation and promotion phases of mammary carcinogenesis.

Earlier in Table 2, it was shown that vitamin E supplementation alone had no prophylactic effect against tumorigenesis. In order to avoid redundancy, we are only presenting here the data in which vi-

TABLE 4. Effect of Selenium and/or Vitamin E Supplementation on DMBA-Induced Mammary Carcinogenesis

Group[a]	Dietary supplement[b]	Duration of vitamin E supplementation[c] (wk)	Tumor incidence (%)	Total no. of tumors	Tumors per tumor-bearing rat
1	None	—	92	113	4.9
2	Selenium	—	72	73	4.1
3	Se + vitamin E	−2 to +24	48	36	3.0
4	Se + vitamin E	−2 to +2	76	66	3.5
5	Se + vitamin E	+2 to +24	56	40	2.9

[a] There were 25 rats per group. All rats received 10 mg DMBA i.g. at 50 days of age and were killed 24 weeks later.

[b] Selenium and/or vitamin E were supplemented in the diet at a concentration of 2.5 mg/kg and 1000 mg/kg, respectively. Additional Se was supplemented starting 2 weeks before DMBA administration and continued until the end of the experiment. Vitamin E was present for different periods of time, as indicated in column 3.

[c] The time of DMBA administration was taken as time 0; minus and plus signs represent the time in weeks before and after DMBA administration, respectively.

Source: Horvath and Ip (13).

tamin E was tested in combination with selenium. Additional selenium and/or vitamin E were present at concentrations of 2.5 mg and 1000 mg/kg of diet, respectively. Rats tolerated these levels of supplementation very well without any obvious undesirable effect. Results in Table 4 show that vitamin E was able to potentiate the prophylactic effect of selenium only when it was present in the postinitiation or promotion phase (Groups 3 and 5). Supplementation with vitamin E around the time of DMBA administration (-2 to $+2$ weeks) produced no beneficial effect (Group 4).

EFFECT OF SELENIUM AND/OR VITAMIN E ON LIPID PEROXIDATION AND GLUTATHIONE PEROXIDASE ACTIVITY

In view of the antioxidant property of both selenium and vitamin E, we proceeded to investigate if there was any correlation between the efficacy of these antioxidants in suppressing peroxidation and their ability to inhibit tumorigenesis. Lipid peroxidation was measured in the liver and the mammary fat pad by the thiobarbituric acid method. As can be seen in Table 5, vitamin E significantly suppressed peroxidation whereas selenium had no effect. A combination of selenium and vitamin E did not result in further inhibition compared to vitamin E alone. Table 5 also describes the activities of several enzymes that are associated with peroxide metabolism. There was no change in the total glutathione peroxidase (GSHPx) activity in any of the supplemented groups. With respect to the selenium-dependent GSHPx activity, selenium supplementation produced a slight but insignificant increase in the mammary fat pad, and only a 25% elevation in the liver. Vitamin E, on other hand, resulted in a 50% increase in the microsomal hydroperoxidase activity. It remains to be determined what role, if any, this enzyme plays in the syner gistic effect of vitamin E and selenium in the inhibition of tumorigenesis.

Several conclusions can be drawn from the effects of vitamin E and selenium on lipid peroxidation in conjunction with their efficacy on cancer prevention. First, although vitamin E is a more potent antioxidant than selenium, it is apparent that systemic suppression of lipid peroxidation by vitamin E alone is not sufficient to inhibit tumor formation. Second, the anticarcinogenic action of high levels of selenium is not related to its biochemical function in the regulation of the selenium-dependent GSHPx. The explanation for this is that the enzyme is already operating at near maximal capacity under normal physiological condition. Additional selenium will not further increase

TABLE 5. Effect of Selenium and/or Vitamin E Supplementation on Lipid Peroxidation and Activities of Glutathione Peroxidase (GSHPx) and Microsomal Hydroperoxidase in Mammary Fat Pad and Liver

Tissue	Dietary supplement[a]	Lipid peroxidation[b]	Total GSHPx[c]	Se-dependent GSHPx[c]	Microsomal hydroperoxidase[d]
Mammary fat pad	None	355 ± 30	99 ± 6	44 ± 3	30 ± 2
	Selenium	320 ± 28	98 ± 6	52 ± 4	27 ± 2
	Vitamin E	142 ± 12	105 ± 8	41 ± 3	46 ± 3
	Se + vitamin E	121 ± 10	112 ± 9	54 ± 4	48 ± 3
Liver	None	228 ± 18	364 ± 28	151 ± 9	92 ± 6
	Selenium	195 ± 16	381 ± 29	186 ± 10	85 ± 6
	Vitamin E	96 ± 8	352 ± 25	160 ± 9	138 ± 9
	Se + vitamin E	87 ± 6	370 ± 30	194 ± 11	131 ± 8

[a] Se and/or vitamin E were supplemented in the diet at a concentration of 2.5 mg/kg and 1000 mg/kg, respectively. Rats were fed these diets for 10 weeks before sacrifice.
[b] Values are expressed as nmoles of malondialdehyde formed/g tissue, mean ± SE.
[c] Values are expressed as nmoles NADPH oxidized/min/mg protein, mean ± SE.
[d] Values are expressed as nmoles TMPD oxidized/min/mg protein, mean ± SE.
Source: Horvath and Ip (13).

its activity, since the enzyme protein becomes the limiting factor. Finally, vitamin E may be able to provide a more favorable climate against oxidant stress, thereby potentiating the action of selenium via some other mechanism.

ROLE OF SELENIUM IN CANCER PREVENTION

The experiments presented in this review are mostly descriptive in nature and do not deal with the mechanism by which selenium deficiency or excess can modify tumorigenesis. Little information is available on this aspect, let alone the formulation of a consensus among investigators in this area of research. Despite the earlier argument that the primary action of selenium is unlikely to be mediated through alterations in carcinogen metabolism, there is some evidence to the contrary. Marshall *et al.* (*14*) have shown that supplementation of selenium increased ring-hydroxylation and decreased N-hydroxylation of 2-acetylaminofluorene, thereby impeding activation and accelerating detoxification of this carcinogen. Grunau and Milner (*15*) have also demonstrated that selenium inhibited the 3,4-oxidation of 7,12-dimethylbenz[*a*]anthracene, but stimulated 12-methyl oxidation. Other evidence that supports a role for selenium in the initiation phase includes protection of liver DNA against single-strand breakage induced by 2-acetylaminofluorene (*16*) and facilitation of the repair process (*17*). Exposure to high concentrations of selenium *in vitro* is known to block the progression of the cell cycle at the S-G2 phase (*18*). Furthermore, pharmacological levels of selenium have been reported to potentiate the primary immune response (*19–21*). It is possible that selenium may be acting through several mechanisms. Any working hypothesis concerning the mode of action of selenium should accommodate the observations that selenium inhibits both virus- and chemical carcinogen-induced tumors and that it is effective during the proliferative or promotional phase of tumorigenesis.

To date, evidence suggestive of an anticarcinogenic effect of selenium is provided largely by animal experiments. Although correlational studies examining selenium intake in relation to cancer rates in different population groups are useful in stimulating research, they are far from conclusive because of the difficulties of circumventing confounding variables. From a nutritional standpoint, the interactions among nutrients cannot be overemphasized. The individual effects of single nutrients can at times be masked or enhanced, depending on the intake and bioavailability of other dietary components as well as the metabolic state of the host. It is this complex combination of biological and biochemical adaptations which modifies the expression of neoplasia. The ultimate test of the effectiveness of selenium in cancer prevention can only be provided by controlled clinical intervention trials. Suffice it to note that before any protocol is designed and implemented, ample consideration should be given to the quantification of a dose–response relationship in conjunction with the nutritional status of the subjects selected for the study.

REFERENCES

1. Schrauzer, G. N., White, D. A., and Schneider, C. J.   1978.   Selenium and cancer. Effects of selenium and of the diet on the genesis of spontaneous mammary tumors in virgin inbred female C3H/St mice. Bioinorg. Chem. 8, 387–396.
2. Thompson, H.J., and Becci, P. J.   1980.   Selenium inhibition of N-methyl-N-nitrosourea-induced mammary carcinogenesis in the rat. JNCI, J. Natl. Cancer Inst. 65, 1299–1301.
3. Thompson, H. J., Meeker, L. D., Becci, P. J., and Kokoska, S.   1982.   Effect of short-term feeding of sodium selenite on 7,12-dimethylbenz[a]anthracene-induced mammary carcinogenesis in the rat. Cancer Res. 42, 4954–4958.
4. Medina, D., and Shepherd, F.   1980.   Selenium-mediated inhibition of mouse mammary tumorigenesis. Cancer Lett. 8, 241–245.
5. Medina, D., and Shepherd, F.   1981.   Selenium-mediated inhibition of 7,12-dimethylbenz[a]anthracene-induced mouse mammary tumorigenesis. Carcinogenesis (London) 2, 451–455.
6. Welsch, C. W., Goodrich-Smith, M., Brown, C. K., Greene, H. D., and Hamel, E. J.   1981.   Selenium and the genesis of murine mammary tumors. Carcinogenesis (London) 2, 519–522.
7. Carroll, K. K., and Khor, H. T.   1975.   Dietary fat in relation to tumorigenesis. Prog. Biochem. Pharmacol. 10, 308–353.
8. Schrauzer, G. N., White, D. A., and Schneider, C. J.   1977.   Cancer mortality correlation studies. III. Statistical associations with dietary selenium intake. Bioinorg. Chem. 7, 23–34.
9. Ip, C., and Sinha, D. K.   1981.   Enhancement of mammary tumorigenesis by dietary selenium deficiency in rats with a high polyunsaturated fat intake. Cancer Res. 41, 31–34.
10. Ip, C.   1981.   Factors influencing the anticarcinogenic efficacy of selenium in dimethylbenz[a]anthracene-induced mammary tumorigenesis in rats. Cancer Res. 41, 2683–2686.
11. Ip, C., and Ip, M. M.   1981.   Chemoprevention of mammary tumorigenesis by a combined regimen of selenium and vitamin A. Carcinogenesis (London) 2, 915–918.
12. Ip, C.   1982.   Dietary vitamin E intake and mammary carcinogenesis in rats. Carcinogenesis (London) 3, 1453–1456.
13. Horvath, P. M., and Ip, C.   1983.   Synergistic effect of vitamin E and selenium in the chemoprevention of mammary carcinogenesis in rats. Cancer Res. 43, 5335–5341.
14. Marshall, M. V., Arnott, M. S., Jacobs, M. M., and Griffin, A. C.   1979.   Selenium effects on the carcinogenicity and metabolism of 2-acetylaminofluorene. Cancer Lett. 7, 331–338.

15. Grunau, J. A., and Milner, J. A. 1983. Selenium modification of 7,12-dimethylbenz[a]anthracene metabolism. Fed. Proc., Fed. Am. Soc. Exp. Biol. *42,* 928 (abstr.).

16. Wortzman, M. S., Besbris, H. J., and Cohen, A. M. 1980. Effect of dietary selenium on the interaction between 2-acetylaminofluorene and rat liver DNA *in vivo.* Cancer Res. *40,* 2670–2676.

17. Lawson, T., and Birt, D. F. 1983. Enhancement of the repair of carcinogen-induced DNA damage in the hamster pancreas by dietary selenium. Chem.-Biol. Interact. *45,* 95–104.

18. Medina, D., Lane, H. W., and Tracy, C. M. 1983. Selenium and mouse mammary tumorigenesis: An investigation of possible mechanisms. Cancer Res. *43,* 2460–2464.

19. Spallholz, J. E., Martin, J. L., Gerlach, M. L., and Heinzerling, R. H. 1973. Immunologic response of mice fed diets supplemented with selenite selenium. Proc. Soc. Exp. Biol. Med. *143,* 685–689.

20. Spallholz, J. E., Martin, J. L., Gerlach, M. L., and Heinzerling, R. H. 1973. Enhanced immunoglobulin M and immunoglobulin G antibody titers in mice fed selenium. Infect. Immun. *8,* 841–842.

21. Spallholz, J. E., Martin, J. L., Gerlach, M. L., and Heinzerling, R. H. 1975. Injectable selenium: Effect on the primary immune response of mice. Proc. Soc. Exp. Biol. Med. *148,* 37–40.

110

Selenium, Lipid Peroxidation, and Murine Mammary Tumorigenesis

Helen W. Lane
Janet S. Butel
Daniel Medina

Selenium is a potent inhibitor of chemical carcinogen-induced tumorigenesis in a variety of epithelial tumor systems. Selenium, given continuously, inhibits tumorigenesis in the skin (*1*), liver (*2,3*), colon (*4–6*), and mammary gland (*7–12*). Figure 1 illustrates the results for the effect of different dietary doses of selenium (0.2, 0.5, 1.0, and 2.0 ppm as selenite) on the appearance of dimethylbenzanthracene (DMBA) -induced mammary tumors in BD2F$_1$ female mice (*13*). Mice were fed the selenium-supplemented diet starting at 7 weeks of age and continued for the length of the experiment (9 months). Mice fed the 0.5 ppm selenium developed 30% mammary tumors as compared to the 56% incidence of mammary tumors seen in mice fed 0.2 ppm selenium. At 2.0 ppm dietary selenium, the mammary tumor incidence was 16%. The initial appearance of tumors in the four groups of mice occurred at the same time (4 months). These results are typical of numerous experiments which have demonstrated that selenium at nontoxic levels is inhibitory to DMBA-induced mouse and rat mammary tumorigenesis (*7–13*).

The mechanism for the chemopreventive action of selenium is unknown, although it is clear from several experiments that increased levels of selenium do not result in increased glutathione peroxidase (GSHPx) activity, the principal selenoprotein in mammalian cells.

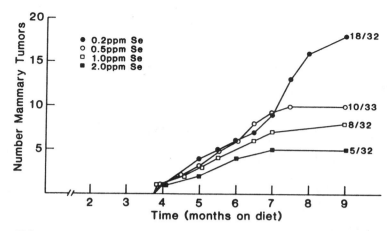

FIG. 1. The effect of different doses of dietary selenium on the appearance of DMBA-induced mammary tumors in BD2F$_1$ female mice. Mice were fed the selenium-supplemented diet starting at 7 weeks of age and continued to length of the experiment. The numbers at the end of each incidence curve represent the number of mice bearing mammary tumors per total mice in each group. DMBA was administered between 8 and 13 weeks.

However, LeBoef and Hoekstra (*14*) demonstrated that this level of selenium altered the enzymatic pathway for glutathione (GSH) metabolism, leading to an increased ratio of GSSG/GSH. Furthermore, the addition of GSH to an *in vitro* ascorbate–iron lipid peroxidation system reduced thiobarbituric acid (TBA) reactive substances (*15*). These results suggest that selenium might inhibit lipid peroxidation and subsequent peroxidative damage of cellular macromolecules which results in enhancement of the tumorigenic process. Since increased levels of dietary fat-enhanced mammary tumorigenesis in rats and of dietary selenium were required to counteract the effect of a high-fat diet, it was possible that the interactions of fat and selenium in mammary tumorigenesis involved lipid peroxidation. The objective of our studies was to determine if dietary selenium could reduce lipid peroxidation at the same dietary levels where selenium reduced mammary gland tumor incidence and if a high-fat diet resulted in increased peroxidation in the mammary gland membranes with a concurrent increase in tumor incidence.

METHODS

Three experiments were completed with BALB/cMed female mice housed in a closed conventional mouse colony in the Department of

Cell Biology, Baylor College of Medicine. Mice were housed 4–6 per cage in temperature- and light-controlled rooms and were free to expressable MTV. The semipurified diet was the AIN-80 casein sucrose diet (16).

In experiment 1, the mice were weaned on a basal diet (5% fat, 0.03 ppm Se) at 3 weeks of age. At 4 weeks of age, a single pituitary isograft was placed under the kidney in order to enhance the number of mammary epithelial cells. At 6 weeks of age, the mice were placed on one of four experimental diets (0.1 ppm Se, 5% fat; 0.1 ppm Se, 20% fat; 2.0 ppm Se, 5% fat; 2.0 ppm Se, 20% fat), with all fat provided as corn oil. At 10 weeks of age, the mammary glands were removed and the glands from 3 mice were pooled to make one sample.

A microsomal–mitochondrial membrane preparation was obtained for each sample (17). Conjugated dienes and TBA reactants (malonyldialdehyde) were measured in these membrane preparations (18). Conjugated dienes were determined by two different methods. Observed diene level was the level found *in vivo* (in the mammary gland preparations). For the second measure, the membrane preparations were incubated at 38°C with a peroxidative reagent which included 25 mm ascorbic acid. This incubation resulted in total peroxidation of all the membranes and thus is a measure of the total potential peroxidation. Difference spectrum was determined by subtracting the *in vivo* observed peroxidation from the *in vitro* peroxidation.

In a separate study, the mammary glands of mice fed 5% corn oil and 0.2 ppm selenium diets were excised. Again glands of 3 mice were pooled and the microsomal–mitochondrial membrane obtained. The membrane fraction was separated into five samples. One of the samples was protected from peroxidation, four were peroxidized *in vitro* with ascorbate, and three of these samples contained various levels of GSH, 0.05, 0.17, or 0.33 mm final GSH concentration.

In experiment 3, the mice were fed a chow diet (Wayne Lab Blox) containing approximately 5% fat and 0.20 ppm Se before and during the DMBA treatments. The mice received a total dose of 1, 3, or 6 mg DMBA, as described previously (13). The dosages were given as 1 mg DMBA once a week for the desired number of weeks, starting at 8 weeks of age. The mice were switched to their respective corn oil diets, 5% or 20% in an AIN-80 diet at 1 week after the last dose of DMBA. The mice were palpated weekly for the presence of mammary tumors and were weighed monthly until the appearance of tumors.

RESULTS

The results of the lipid peroxidation experiments are shown in Figs. 2–4 and Table 1. Figure 2 demonstrates the effect of fat and selenium

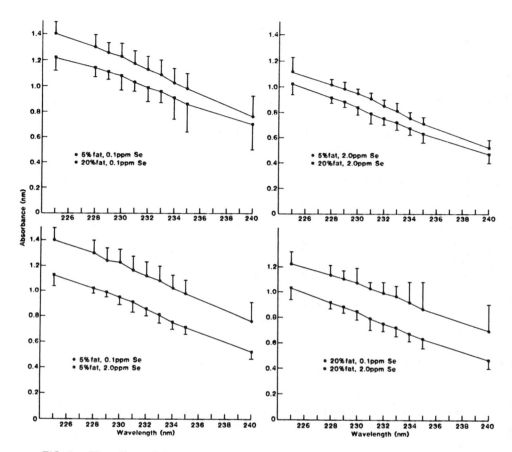

FIG. 2.   The effect of dietary levels of selenium and fat, 0.1 and 2.0 ppm Se and 5
and 20% corn oil, on observed levels of conjugated dienes (*in vivo* in normal
mammary glands of pituitary–isograft-stimulated mice.

on absorbance of the *in vivo* (observed) membranes, and these curves
indicate no detectable levels of conjugated dienes. Figure 3 indicates
the difference spectrum (*in vitro* peroxidation − *in vivo* peroxidation)
for 0.1 ppm Se, 5% fat diet. There were detectable conjugated dienes
for this dietary group, but there were no detectable dienes for any of
the other dietary groups. In Fig. 4, the level of TBA reactants is shown
for the same microsomal–mitochondrial membrane fraction. Increasing dietary selenium to 2.0 ppm resulted in an 18% decrease of TBA
reactants in mice fed the 5% fat diet, while the same increase in
selenium resulted in a 20% increase in TBA reactants from mice fed

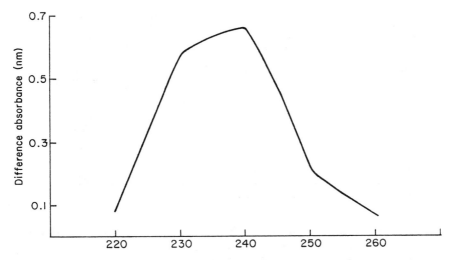

FIG. 3. The difference spectrum for *in vitro* peroxidation minus *in vivo* (observed) peroxidation for the microsomal–mitochondrial membranes from mice fed 5% fat, 0.1 ppm Se diet.

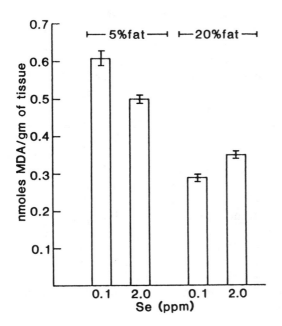

FIG. 4. The level of TBA reactants (malonyldialdehyde) observed in normal mammary glands of pituitary–isograft-stimulated mice fed either 5 or 20% corn oil diets with two levels of dietary selenium (0.1 and 2.0 ppm).

TABLE 1. Effect of Various Levels of Glutathione on MDA Levels after *in Vitro* Peroxidation in the Mammary Gland[a]

Tissue treatment[b]		MDA levels (nmol MDA/g wet weight)
Peroxidized with ascorbic acid	Level of GSH added (nm)	
−	−	1.16
+	−	4.30
+	0.05	1.44
+	0.17	1.33
+	0.33	0.80

[a] AIN-80 diet with 5% fat with 0.1 ppm Se.
[b] Tissues were plasma, microsomal, and mitochondria membranes from mammary glands of female mice receiving a pituitary isograft.

the 20% fat diet. Furthermore, the mice fed the 20% fat diet had decreased levels of TBA reactants compared to the mice fed the 5% fat diet at both levels of selenium.

The effect of the addition of GSH to the *in vitro* peroxidation incubation media is shown in Table 1. The addition of GSH to the peroxidation assay mix resulted in decreased levels of TBA reactants. The decrease was proportional to the amount of GSH added to the assay mix.

The effects of dietary fat on DMBA-induced mammary tumorigenesis are shown in Fig. 5. At each level of DMBA administration (6, 3, and 1 mg), there was a lower mammary tumor incidence in mice consuming the 5% fat diet than in mice consuming the 20% fat diet by 30 weeks after initial DMBA treatment. However, by 50 weeks after DMBA treatment, the mammary tumor incidence in mice treated with 6 or 3 mg DMBA was not significantly different between their respective low (LF) and high (HF) fat diet groups. The decrease in tumor latency period was calculated from the time at which 50% of the tumors appeared in the HF group. Thus, the time of latency period of mice fed 6 and 3 mg DMBA decreased from 23 to 13 weeks and from 40 to 27 weeks, respectively. These results demonstrated that a diet high in corn oil enhanced mouse mammary tumorigenesis primarily by decreasing the latent period for tumor formation. At the lowest dose of DMBA, the corn oil diet (HF) enhanced the tumor incidence as well as decreased the latency period for mammary tumors, although only a few tumors were induced in each group. There was no effect of DMBA dose (3, 1 mg) and dietary fat on weight gain (Table 2). The initial mean body weight was 21–22 g (measured at 10 weeks of age for the treated and untreated groups), and the weight after 8 months on the diets varied from 32 to 34 g.

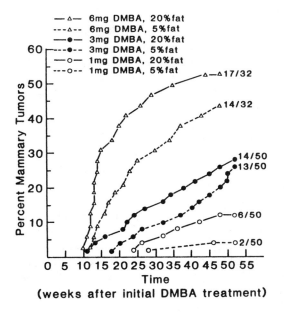

FIG. 5. The effect of different doses of DMBA and dietary fat on number of incidence.

DISCUSSION

The purpose of these experiments was to examine the interrelationships between dietary selenium levels, lipid peroxidation, and DMBA-induced mammary tumorigenesis. The experiments described herein yielded several results which are interesting about these relationships.

First, selenium added at 2 ppm in a semipurified casein, sucrose, 5%

TABLE 2. Effect of Corn Oil and DMBA on Weight[a]

Group			
DMBA (mg)	Fat (%)	Initial weight ($\bar{X} \pm$ SE)	Weight 8 months on diet ($\bar{X} \pm$ SE)	n
0	5	22.96 ± 0.30^{b}	34.32 ± 0.94^{c}	40
0	20	21.50 ± 0.25^{b}	33.60 ± 0.57^{c}	40
1	5	21.22 ± 0.24^{b}	34.13 ± 0.75^{c}	40
1	20	21.20 ± 0.29^{b}	34.40 ± 0.76^{c}	40
3	5	22.40 ± 0.18^{b}	32.79 ± 0.58^{c}	40
3	20	22.49 ± 0.16^{b}	33.60 ± 0.80^{c}	40

[a] Statistics were calculated using multiple analysis of variance. The groups with superscripts of similar letters were not significantly different ($P < .05$).

corn oil diet inhibited DMBA-induced mammary tumorigenesis in mice by decreasing the tumor incidence over an 11-month period. Of interest was the parallel observation that the same level of selenium and fat also decreased mammary gland membrane lipid peroxidation as measured by two different assays. First, a difference spectrum was found for 0.1 ppm Se diet, indicating the presence of conjugated dienes, while no conjugated dienes were detected with the 2 ppm Se diet. These results are similar to those of Recknagel and Ghoshal (*17*) who also found no *in vivo* (observed) -produced conjugated dienes in liver unless the liver was exposed to carbon tetrachloride poisoning. TBA reactants (malonyldialdehyde–MDA) were lower with the 2 ppm Se diet than 0.1 ppm Se diet. These data are consistent with the hypothesis that increased levels of dietary selenium inhibit mammary tumorigenesis by decreasing membrane lipid peroxidation, thereby decreasing peroxide-mediated damage to cellular macromolecules which in turn may lead to cellular alterations favoring neoplastic growth. This idea is further supported by the ability of increased glutathione levels to decrease cellular peroxidation levels. In this scheme (Fig. 6), increased dietary selenium generates an increased level of glutathione, which decreases lipid peroxidation, which in turn decreases cell growth. There are confirmatory data in the literature which support parts of this scheme.

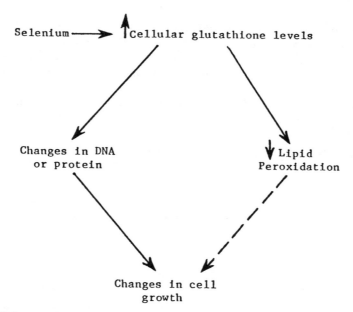

FIG. 6. The proposed relationship of selenium and cell growth.

LeBoeuf and Hoekstra (14 reported that the selenium inhibition of hepatocyte growth *in vitro* correlated with an increase in GSSG/GSH ratios. Lewko *et al.* (19) found a similar correlation between selenium inhibition of mammary tumor cell growth and GSH levels. Furthermore, Lewko *et al.* (19) demonstrated that 2-oxothiazolidine-4-carboxylic acid, a specific stimulator of GSH levels, counteracted the inhibitory effect of high levels of selenium. Additional experiments have demonstrated that increased glutathione levels decrease lipid peroxidation activity in normal liver (15).

Although the above evidence supports the Se–GSH–lipid peroxidation hypothesis, there is little evidence that increased levels of lipid peroxidation are directly involved with enhancement of mammary tumorigenesis. Increased dietary fat, particularly increased levels of PUFA, are purported to enhance membrane lipid peroxidation. It is well documented that HF diets enhance DMBA-induced rats and MMTV-induced mouse mammary tumorigenesis. The experiment described herein demonstrates that an HF diet enhances DMBA-induced mouse mammary tumorigenesis; surprisingly, this is the first experiment demonstrating this fact in the DMBA–mouse mammary tumor system. The enhancement of mammary tumorigenesis was seen at three different dose levels of DMBA and was manifested primarily by a decreased tumor latent period. However, further experiments demonstrated that the HF diet did not enhance membrane lipid peroxidation by the two assays examined herein. Furthermore, the levels of MDA decreased in mice on the HF diet as compared to the LF diet. Similarly, there were no detectable conjugated dienes in membranes of mice fed the HF diet. These results can be explained by the results of Rao and Abraham (20). Rao and Abraham demonstrated that only trace amounts of arachidonate were found in normal mammary glands of virgin, pregnant, and lactating mice fed linoleate-rich diets (high corn oil diets), suggesting that diets high in linoleate inhibited arachidonate formation and thus may affect the level of peroxidation (TBA reactants) that may occur. In summary, with high dietary fat, there was not an increased level of lipid peroxidation, nor was there a consistent pronounced effect of selenium as compared to that found with the LF diet and selenium.

Ip and co-workers (8,9,21) have examined lipid peroxidation in the mammary glands of rats fed HF diets. There are several differences between our results and that of IP (21). Ip demonstrated that an HF diet and not selenium resulted in increased TBA reactants in the homogenates of mammary glands. Besides the differences in species, rat vs mouse, there are also differences in analytical methods between our research and that of Ip (21). Lipid peroxidation changes are generally studied in the membrane factions as opposed to cell homogenates. Second, Rao and Abraham (22) have demonstrated that primarily the phospholipid fraction of these membranes is affected by changing dietary fat, not the cytoplasm fraction of lipids. Thus, we attempted to study the cellular components most sensitive to peroxidation.

In conclusion, the accumulated data so far do not support the general hypothesis that selenium inhibition of tumorigenesis is mediated by increasing levels of GSHPx. The data do support the hypothesis that selenium can decrease levels of lipid peroxidation which occurred in mice fed the 5% corn oil diet, perhaps by affecting glutathione levels. Figure 6 represents two proposed pathways for selenium and tumorigenesis. We have studied one aspect of the lipid peroxidation pathway. However, little is known about the effect of such chronic low-level peroxidation of the cellular events that lead to neoplasms. The glutathione-mediated pathways affecting DNA and protein synthesis (23,24) appear to have some validity, as studies have shown an effect of glutathione on DNA repair, protein synthesis (24), and eventually cell growth itself. Further experiments are necessary to evaluate the interrelationships between selenium, GSH, and cell growth before an understanding emerges on the mechanism of selenium inhibition of tumorigenesis.

ACKNOWLEDGMENTS

We would like to acknowledge the technical support of Frances Shepard, Rekha Halligan, and Cindy Howard. This research was supported by grants from NIH (CA 11944 and CA 33366).

REFERENCES

1. Shamberger, R. J.   1970.   Relationship of selenium to cancer. I. Inhibitory effect of selenium on carcinogenesis. J. Natl. Cancer Inst. (U.S.) *44*, 931–936.
2. Clayton, C. C., and Baumann, C.   1949.   Diet and azo dye tumors; effect of diet during a period when the dye is not fed. Cancer Res. *9*, 575–582.
3. Griffin, A. C., and Jacobs, M. M.   1979.   Effects of selenium on azo dye hepatocarcinogenesis. Cancer Lett. *3*, 177–181.
4. Griffin, A. C.   1979.   Role of selenium in the chemoprevention of cancer. Adv. Cancer Res. *29*, 419–442.
5. Jacobs, M. M., Jansson, B., and Griffin, A. C.   1977.   Inhibitory effects of selenium on 1,2-dimethylhydrazine and methylazoxymethanol acetate induction on colon tumors. Cancer Lett. *2*, 133–137.
6. Soullier, B. K., Wilson, P. S., and Nigro, F. T.   1981.   Effect of selenium on azoxymethane-induced intestinal cancer in rats fed a high fat diet. Cancer Lett. *12*, 343–348.

7. Harr, J. R., Exon, J. H., Weswig, P. H., and Whanger, P. D.   1973.   Relationship of dietary selenium concentration, chemical cancer induction and tissue concentration of selenium in rats. Clin. Toxicol. *6*, 287–293.

8. Ip, C.   1981.   Factors influencing the anticarcinogenic efficacy of selenium on dimethylbenz[a]anthracene-induced mammary tumorigenesis in rats. Cancer Res. *41*, 2683–2686.

9. Ip, C., and Sinha, D.   1981.   Anticarcinogenic effect of selenium in rats treated with dimethylbenz[a]anthracene and fed different levels and types of fat. Carcinogenesis (London) *2*, 435–438.

10. Thompson, H. J. and Becci, P. J.   1980.   Selenium inhibition of *N*-methyl-*N*-nitrosourea-induced mammary carcinogenesis in the rat. JNCI, J. Natl. Cancer Inst. *65*, 1299–1301.

11. Welsh, C. W., Goodrich-Smith, M., Brown, C. K., Greene, H. D., and Hamel, E. J.   1981.   Selenium and the genesis of murine mammary tumors. Carcinogenesis (London) *2*, 519–522.

12. Medina, D., and Shepherd, F.   1981.   Selenium-mediated inhibition of 7,12-dimethylbenz[a]anthracene-induced mouse mammary tumorigenesis. Carcinogenesis (London) *2*, 451–455.

13. Medina, D., Lane, H. W., and Shepherd, F.   1983.   Effect of dietary selenium levels on 7,12-dimethylbenzanthracene-induced mouse mammary tumorigenesis. Carcinogenesis (London) *4*, 1159–1163.

14. LeBoeuf, R. A., and Hoekstra, H. G.   1983.   Adaptive changes in hepatic glutathione metabolism in response to excess selenium in rats. J. Nutr. *113*, 845–854.

15. Burk, R. F.   1983.   Glutathione-dependent protection by rat liver microsomal protein against lipid peroxidation. Biochim. Biophys. Acta *757*, 21–28.

16. American Institute for Nutrition Ad Hoc Committee on Standards for Nutritional Studies   1977.   J. Nutr. *107*, 1340–1348.

17. Recknagel, R. O., and Ghoshal, A. K.   1966.   Quantitative estimate of peroxidative degeneration of rat liver microsomal and mitochondrial lipids after carbon tetrachloride poisoning. Exp. Mol. Pathol. *5*, 413–426.

18. Ohkawa, H., Ohishi, N., and Yagi, K.   1979.   Assay for lipid peroxides in animal tissues by thiobarbituric acid reaction. Anal. Biochem. *95*, 351–358.

19. Lewko, W. M., Winn, D. E., and McConnel, K. P.   1984.   Influence of selenium on cell growth and glutathione levels in cultured breast cancer cells. Fed. Proc., Fed. Am. Soc. Exp. Biol. *43*, 793.

20. Rao, G. A., and Abraham, S.   1976.   Brief communication: Enhanced growth rate of transplanted mammary adenocarcinoma induced in C3H mice by dietary linoleate. J. Natl. Cancer Inst. (U.S.) *56*, 431–432.

21. Ip, C.   1982.   Modification of mammary carcinogenesis and tissue peroxidation by selenium deficiency and dietary fat. Nutr. Cancer *2*, 136–142.

22. Rao, G. A., and Abraham, S.   1976.   Dietary alteration of fatty acid composition of lipid classes in mouse mammary adenocarcinoma. Lipids *10*, 641–643.

23. Kosower, N. S., and Kosower, E. M.   1976.   Functional aspects of glutathione disulfide and hidden forms of glutathione. *In* Glutathione: Metabolism and Function I. M. Arises and W. B. Jakoby (Editors), pp. 159–174. Raven Press, New York.

24. Meister, A.   1983.   Selective modification of glutathione metabolism. Science *220*, 472–477.

111

Selenium and Tumorigenesis

John A. Milner
Maxine E. Fico

Almost 70 years ago selenium was proclaimed as a cancer therapeutic agent (Dalbert 1912; Walker and Klein 1915). In 1956, Weisberger and Surhland reported that selenocystine was effective in reducing leukocyte count and spleen size of patients suffering from acute leukemia. At a dose of 50–200 mg/day symptoms of nausea, vomiting, and diarrhea were observed. Interestingly, these authors indicated that at this dosage hepatic and renal function were normal and the symptoms described were no worse than those occurring with normally employed therapeutic agents.

Supportive evidence for the antitumorigenic role of selenium also comes from the evaluation of the behavior of tumors in humans as a function of circulating blood selenium. Broghamer *et al.* (1976) examined the association between blood selenium concentrations of 110 patients with various types of carcinomas and the biological behavior pattern of the tumor. Although the range of serum selenium concentrations of these 110 patients was wide, the majority had levels lower than healthy controls. Low serum selenium concentrations were associated with a higher frequency of distant metastases, multiple primary tumors, multiple recurrences, and a shortened survival time. Patients with selenium concentrations approaching or exceeding the mean value of all cancer patients had tumors that remained confined to the region of origin, developed less distant metastasis, had fewer primary neoplasms, and had a decreased frequency of recurrences. However, other observations (Broghamer *et al.* 1978) revealed that serum selenium concentration of patients with reticuloendothelial tumors did not correlate with the behavioral patterns of the tumor, i.e., extent of organ involvement, patient survival time, and the incidence of multiple primary neoplasms. Therefore, the efficacy of selenium as an antitumorigenic agent may depend heavily on the tumor cell examined.

Considerable evidence has shown that selenium can inhibit the growth of experimentally transplanted tumors (Abdullaev *et al.* 1973; Greeder and Milner 1980; Ip *et al.* 1981; Watrach *et al.* 1982, 1985; Vernie *et al.* 1981; Poirier and Milner 1984). Milner (1984, 1985) recently reviewed the literature of studies examining parenteral or oral administration of selenium as a means of inhibiting tumor proliferation. Numerous studies have shown that selenium, primarily as sodium selenite, at a dose of approximately 1 μg/g body weight of the host animal, is effective in inhibiting the growth of the transplantable tumors including Guerin carcinomas, sarcomatous M-1 neoplasms, L-1210 leukemic cells, Murphy lymphosarcomas, canine mammary, and human mammary cells. An enhanced effect of tumor inhibition was observed when selenium was given in combination with X-ray (Abdullaev *et al.* 1973) or chemotherapy (Milner and Hsu 1981).

During the last international symposium, our laboratory presented data showing that selenium was effective in inhibiting the growth of two transplantable tumor cells *in vivo* and *in vitro*. Furthermore, our data revealed this response was dependent upon the dose and form of selenium administered. This present chapter addresses the effects of selenium on other tumor-transplantable cells as well as nonneoplastic cells. A possible mechanism by which selenium is inhibiting the growth of neoplastic cells is presented.

TUMOR SPECIFICITY

Previous studies from our laboratory have shown that *in vivo* supplementations of selenium are capable of inhibiting the growth of canine mamary tumor inoculated into mice similar to that previously reported for Ehrlich ascites tumor cells (Watrach *et al.* 1982). Recently, we have examined the effect of selenium upon the growth and metabolism of various canine mammary tumor (CMT) cells in culture. Some canine mammary tumor cell lines were found to be extremely sensitive to selenium supplementation, while others were found to be relatively insensitive to selenium supplementation (Fico *et al.* 1985).

In studies conducted with cultures at near confluency, approximately 24–48 hr were required before a depression in overall viability was detectable (Fico *et al.* 1985). These studies clearly showed that supplementation to twice to five times normal blood concentrations

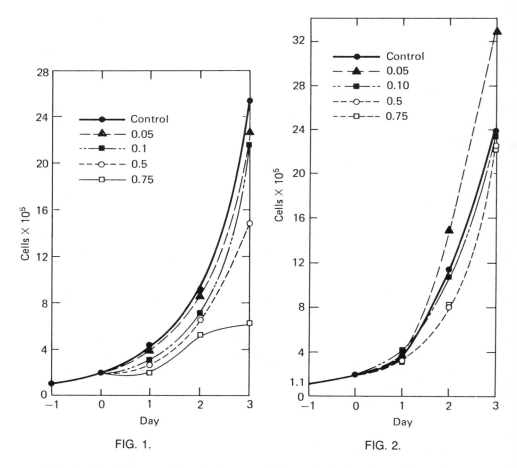

FIG. 1.

FIG. 2.

FIG. 1.   The influence of graded concentrations of selenium as sodium selenite on the growth of CMT-13 cells in culture. Selenium was added 24 hr after plating.
FIG. 2.   The influence of graded concentrations of selenium as sodium selenite on the growth of CMT-11 cells in culture. Selenium was added 24 hr after plating.

was extremely effective in decreasing the viabilities of the canine mammary tumor cells including CMT-13, 14A, and 14B. However, in confluent cultures of CMT-11 very little effect of selenium supplementation was observed. At 1 ppm, selenium addition to the incubation medium only decreased cellular viability by approximately 30%.

Nonneoplastic cells were also examined for their sensitivity to selenium supplementation. Our laboratory has been unable to detect differences in viabilities in nonneoplastic canine mammary tumor cells exposed to selenium up to concentrations of 1 ppm. Similarly, using human nonneoplastic lung cells (MCR-5), no significant differences were detected when selenium was added to the incubation medium at a final concentration of 1 µg/ml (Watrach *et al.* 1985).

Recent studies in our laboratory have examined the growth rate of CMT-11 and CMT-13 cells after exposure to varying concentrations of selenium (Figs. 1 and 2). These studies reveal that growth of CMT-13 is proportionally inhibited as the concentration of selenium increased in the incubation medium. However, in CMT-11 there is initial stimulation of the growth with selenium concentrations of 0.05 µg/ml. Past 0.05 µg/ml, very little effect of selenium was detected in the growth of these cells. Compared to no selenium supplementation, 0.75 µg Se/ml resulted in only a 15% reduction in growth after 3 days of incubation, whereas in CMT-13 approximately 90% inhibition of growth was detectable. Thus, we have classified CMT-13 and CMT-11 cells as relative selenium responsive and nonresponsive, respectively. It is interesting to note that CMT-11 cells may be able to compensate with time to exposure to high selenium, since a reduction growth occurred between days 2 and 3 of incubation, but not with longer incubations. Nevertheless, these two cell lines may be very useful in elucidating the role of the mechanism by which selenium modifies the growth of neoplastic tissue. It is also clear from these observations that not all neoplastic cells are equally sensitive to selenium supplementation.

SELENIUM UPTAKE

The uptake of selenium 75 into the two canine mammary tumor cell lines was completed in an attempt to determine if this could explain the differences in sensitivity of these two cell lines. Poirier (1984) found that selenium supplementation to cultures of CMT-11 and -13 cell lines leads to a rapid accumulation of selenium within the cells. While differences in the total amount of selenium retained were evident between the two cells, they were the opposite of those expected. The CMT-11 cell accumulated greater quantities of selenium as a function of time than did the CMT-13 cell line. The retention was 355 and 144 pmol Se/2 $\times$ 10^6 cells for CMT-11 and CMT-13 cells after a 60-min incubation with 1 µg/ml selenium as sodium selenite (Poirier 1984). Therefore retention, and probably uptake, of selenium is not the decisive factor determining the sensitivity of these cell lines to this trace element.

TABLE 1. Effect of Method of Administration of Selenium on Tumor Incidence in Mice Inoculated with Ehrlich Ascites Tumor Cells[a]

Experiment A	Tumor incidence	First death (hr)	Longevity (hr)
Control	10/10	456	539 ± 19[b]
Gastric gavage	40/40	528	652 ± 20[c]
Intraperitoneal	34/40	768	1447 ± 79[d]

[a] Selenium was administered as sodium selenite on days 0, 1, 3, 5, and 7 following inoculation with 5×10^5 tumor cells. Means ± SEM not sharing a common superscript letter differ $P < .02$.

MODE OF SELENIUM ADMINISTRATION

Ip *et al.* (1981) reported the effect of dietary selenium deficiency and supplementation on the growth of the transplantable MT-W9B mammary tumor in female Wistar–Furth rats. Selenium deficiency had no influence on the growth of this tumor. Supplementation of the diet with 2 ppm of selenium inhibited tumor growth and reduced the final tumor weight by approximately 50% compared to the control rats receiving 0.1 ppm of selenium. The inhibitory response was selective, without inducing any weight loss in the animals. In recent studies by Poirier and Milner (1984), sodium selenite was administered at 2.0 μg selenium six times over a 9-day period by intraperitoneal injection or by gastric gavage. Survival time was significantly increased in Ehrlich ascites-bearing mice by 170 and 20%, respectively, compared to controls (Table 1). These data show that mode of administration of selenium is an additional factor influencing the antitumorigenic properties of this trace element. The reason for these differences in efficacy is unknown, but may relate to differences in the rate or quantity of selenium metabolizing or detoxifying in the liver.

ACTIVE SELENIUM INTERMEDIATE

Our laboratory has been concerned for a number of years as to the active metabolite of selenium responsible for its antitumorigenic properties. In our earlier publication, the typical organic forms of selenium such as selenomethionine and selenocysteine were found to be less effective in inhibiting the growth of neoplastic cells than were sodium selenite, sodium selenate, and selenodioxide (Greeder and Milner 1980; Milner and Hsu 1981; Poirier and Milner 1984). Recent studies in our laboratory have shown that selenodiglutathione is one of the most effective compounds in inhibiting the growth of neoplastic cells (Poirier and Milner 1984). Preincubation of Ehrlich ascites tumor cells with selenium as sodium selenite or selenodiglutathione suppressed the growth of these cells reinoculated into mice. On an equivalent selenium basis, selenodiglutathione is far more effective in inhibiting the growth of these neoplastic cells (Poirier and Milner 1984). Fico *et al.* (1985) has also compared the effects of various forms of selenium on the growth of CMT-13, CMT-11, and nonneoplastic mammary cells. Table 2 reveals that neoplastic cells are, in general, more susceptible to selenium than nonneoplastic cells. Also, these data show that the form of selenium does have a dramatic effect on the efficacy of this trace element. Again, sodium selenite and selenodiglutathione were the most effective forms of selenium examined. Furthermore, differences in the responsiveness of the two neoplastic lines were clearly evident (Table 2).

It appears that the active agent responsible for selenium's antitumorigenic property lies in the pathway for selenium detoxification (Fig. 3). Dimethylselenide, an end product of selenium metabolism, has been examined for its ability to inhibit tumor proliferation. This form of selenium was found to be relatively ineffective in inhibiting the growth of the Ehrlich ascites tumor cells (Table 3). Since the end product of metabolism is less effective and sodium selenite is less effective than selenodiglutathione, we assume that selenodiglutathione or some closely related intermediate is responsible for the antitumorigenic properties of this trace element. Dr. L. Vernie has presented evidence showing that selenodiglutathione does not enter the cell intact. At present we do not know if selenite supplementation increases selenodiglutathione biosynthesis within the cell.

TABLE 2. Effect of Various Forms of Selenium on the Growth of Neoplastic and Nonneoplastic Mammary Cells[a]

	Selenium addition (μg/ml)	Cells (nonneoplastic, × 10^5)	CMT-13 (× 10^5)	CMT-11 (× 10^5)
Control	0	4.40	9.54	15.21
Selenomethionine	0.75	4.46	8.94	13.26
Selenocystine	0.75	4.56	6.59	14.25
Sodium selenite	0.75	4.28	4.55	8.78
Selenodiglutathione	0.75	4.46	0.50	3.60
PSEM		0.21	0.77	0.76

[a] Initially plated at 1.0, 1.2, and 1.0 × 10^5. Nonneoplastic cells were a primary culture obtained from a lactating dog. CMT-13 represents a selenium-responsive cell line and CMT-11 represents a selenium-nonresponsive cell line. Selenium was added 24 hr after plating in fresh medium. Cell counts were determined 48 hr after the addition of selenium.

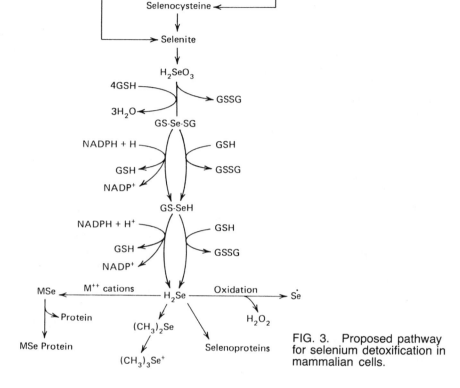

FIG. 3.   Proposed pathway for selenium detoxification in mammalian cells.

TABLE 3.   Longevity of Mice Inoculated with Tumor Cells Preincubated with Various Forms of Selenium[a]

Treatment	Selenium (µg/ml)	Tumor incidence	Time first death (hr)	Mean longevity (hr)
Buffer	—	10/10	480	622 ± 30[b]
Sodium selenite	1.0	10/10	480	626 ± 33[b]
	3.0	10/10	480	607 ± 35[b]
Selenodiglutathione	1.0	9/10	600	789 ± 52[c]
	3.0	7/10	672	1069 ± 145[d]
Dimethylselenide	1.0	10/10	528	605 ± 24[b]
	3.0	10/10	552	607 ± 14[b]

[a] Adapted from Poirier and Milner (1984). All tumor cells were preincubated for 15 min in buffer in the presence or absence of selenium as sodium selenite, selenodiglutathione, or dimethyselenide, before inoculation into Swiss mice. Each mouse received 5×10^5 Ehrlich ascites tumor cells. Vertical means ± SEM not sharing a common superscript letter differ $P < .05$.

TABLE 4. Effect of Selenium Treatment on Macromolecular Biosynthesis in Canine Mammary Tumors[a]

Biosynthesis	CMT-13[b]		CMT-11[b]	
	Se-treated	Control	Se-treated	Control
RNA, dpm ³H-labeled uridine/g RNA	18 ± 8[c]	103 ± 19[d]	107 ± 15[d]	139 ± 18[d]
DNA, dpm ³H-labeled thymidine/g DNA	723 ± 264[d]	865 ± 297[d]	805 ± 129[d]	921 ± 131[d]
Protein, dpm ³H-labeled leucine/g protein	795 ± 234[d]	995 ± 120[d]	1625 ± 576[d]	1554 ± 220[d]

[a] Horizontal paired means ± standard deviation with unlike superscripts differ, $P < .05$.
[b] CMT-13 is selenium responsive, CMT-11 is selenium nonresponsive.

SELENIUM AND MACROMOLECULE BIOSYNTHESIS

Excellent studies by Vernie et al. (1979) have shown that selenodiglutathione is effective in inhibiting protein biosynthesis. We have examined macromolecule biosynthesis after exposure to various forms of selenium in some of our tumor cell lines. Ehrlich and SV40 3T3 cells were incubated with labeled uridine, thymidine, or leucine. After a 2-hr incubation period with 1 μg/ml Se, the only significant reduction in macromolecule biosynthesis occurred in the accumulation of uridine into RNA (Milner 1984). No significant effect of selenium on the incorporation of leucine into protein or thymidine into DNA was observed. We therefore concluded that RNA biosynthesis may be one of the first macromolecules to be modified by selenium supplementation.

Additional studies in our laboratory have examined the synthesis of these macromolecules using labeled precursors in the canine mammary tumor cell lines with different susceptibilities to selenium (Table 4). Again, after short-term incubation with selenium, a marked depression in the rate of RNA biosynthesis was detectable in the sensitive cell line (CMT-13). However, no appreciable alteration in RNA biosynthesis was detected in the insensitive CMT-11 cell line.

CONCLUSION

Selenium is an effective antitumorigenic agent. There appears to be a degree of specificity for neoplastic cells over nonneoplastic cells. However, all neoplastic cells are not equally affected by selenium supplementation. These results clearly indicate that intracellular retention of selenium is not the decisive factor in the sensitivity of tumors. These studies further suggest that selenodiglutathione or some closely related intermediate is responsible for the antineoplastic properties of selenium. Finally, these studies suggest that RNA translation or transcription may be one of the first modifications that occurs following selenium supplementation.

ACKNOWLEDGMENTS

These investigations were supported in part by PHS grant CA33699 awarded by the National Cancer Institute, DHHS, and by an Alpo Cancer Research Grant.

REFERENCES

Abdullaev, G. B., Gasanov, G. G., Ragimov, R. N., Teplyakova, G. V., Mekhutiev, M. A., Dzhafarov, A. I. 1973. Selenium and tumor growth in experiments. Dokl. Akad. Nauk Azerb. SSR 29, 18–24.

Broghamer, W. L., Jr., McConnell, K. P., and Blotcky, A. L. 1976. Relationship between serum selenium levels and patients with carcinoma. Cancer (Amsterdam) 37, 1384–1388.

Broghamer, W. L., Jr., McConnell, K. P., Grimaldi, M., and Blotcky, A. J. 1978. Serum selenium and reticuloendothelial tumors. Cancer (Amsterdam) 41 (4), 1462–1466.

Dalbert, P. 1912. Tentatives de traitement des cancer par selenium. Bull. Assoc. Fr. Etude Cancer 5, 121–125.

Greeder, G. A., and Milner, J. A. 1980. Factors influencing the inhibitory effect of selenium on mice inoculated with Ehrlich ascites tumor cells. Science 209, 825–827.

Ip, C., Ip, M. M., and Kim, U. 1981. Dietary selenium intake and growth of the MT-W9B transplantable rat mammary tumor. Cancer Lett. 14, 101–107.

Milner, J. A. 1984. Selenium and the transplantable tumor. J. Agric. Food Chem. 32, 436–442.

Milner, J. A. 1985. The effect of selenium on virally induced and transplantable tumor models. Fed. Proc. Fed. Am. Soc. Exp. Biol. 44, 2568–2572.

Milner, J. A., and Hsu, C. Y. 1981. Inhibitory effects of selenium on the growth of L1210 leukemic cells. Cancer Res. 41, 1652–1656.

Poirier, K. A. 1984. Selenium: An inhibitor of tumor growth in vitro and in vivo. Ph.D Thesis, Univ. Illinois, Urbana.

Poirier, K. A., and Milner, J. A. 1984. Factors influencing the antitumorigenic properties of selenium in mice. J. Nutr. 113, 2147–2154.

Vernie, L. N., Collard, J. G., Eker, A. P., De Wilt, A., and Wilders, I. T. 1979. Studies on the inhibition of protein synthesis by selenodiglutathione. Biochem. J. 180, 213–218.

Vernie, L. N., Hamburg, C. J., and Bont, W. S. 1981. Inhibition of the growth of malignant mouse lymphoid cells by selenodiglutathione and selenodicystine. Cancer Lett. 14, 303–308.

Walker, C. H., and Klein, F. 1915. Selenium: Its therapeutic value, especially in cancer. Am. Med. 628–633.

Watrach, A. M., Milner, J. A., and Watrach, M. A. 1982. Effect of selenium on growth rate of canine mammary carcinoma cells in athymic nude mice. Cancer Lett. 15, 137–143.

Watrach, A. M., Watrach, M. A., Milner, J. A., and Poirier, K. A. 1984. Inhibition of human breast cancer cells by selenium. Cancer Lett. 25:41–47.

Weisberger, A. S., and Suhrland, L. G. 1956. Studies on analogues of L-cysteine and L-cystine. III. The effect of selenium cystine on leukemia. Blood 11, 19–30.

112

Selective Effects of Selenium on the Function and Structure of Mitochondria Isolated from Hepatoma and Normal Liver

Yu Shu-Yu
Zhu Ya-Jun
Hou Chong
Huang Chang-Zhi

Numerous reports have indicated that Se has inhibitory effects on the growth of various transplanted tumors in laboratory animals including Guerin carcinoma, sarcoma M-1 (*1*), Ehrlich ascites tumor (*1–3*), Murphy lymphosarcoma (*4*), L1210 leukemia (*5*), mammary tumors (*6–7*), ovarian tumors (*8*), malignant lymphoma (*9*), and ascitic hepatoma (*10*). Ao and Pan (*11*) examined the effects of Se on the growth of normal and neoplastic esophageal cells in monolayer cell cultures. They showed that Se as sodium selenite could inhibit the growth of esophageal carcinoma cells Eca 109CIII at concentrations that did not affect the normal esophageal epithelial cells grown in primary cultures. Thus, the data available to date clearly show that Se may have a wide application as a therapeutic agent against tumors. With the purpose to assess the therapeutic potential of Se against cancer, one of our major endeavors has been to study the molecular mechanism of Se-mediated inhibition and to determine if any difference in alteration in metabolic pathways exists between tumor cells and normal cells in response to Se. In the present chapter, the marked differences of Se-mediated effects on mitochondrial energy metabolism between hepatoma cells and normal liver cells are reported.

METHODS

Collection of Tumor Cells and Isolation of Mitochondria

The ascitic hepatoma cells Hep A were transplanted into male Kunming mice weighing 20–22 g. On day 7, ascitic fluid was collected and washed twice in cold 0.25 M sucrose solution by centrifugation (600 g, 10 min). Mitochondria were isolated in ice-cold medium containing 25 mM sucrose and 1 mM EDTA by the procedure of Schneider and Hogeboom (12). The mitochondrial protein concentration in the final suspension was adjusted to about 30 mg/ml.

Methods of Analysis

Measurements of oxidation, respiratory control ratio, and P/O ratio were carried out polarographically at 30°C, using a Clark-type oxygen electrode. ATPase activity was assayed by measurement of the amount of phosphate released (13). Mitochondrial swelling was followed by monitoring changes in optical density at 520 nm. The protein was determined by Biuret reaction (14). The composition of the assay media and the other experimental conditions are specified with the individual experiments.

RESULTS

Oxidative Phosphorylation of Hepatoma and Normal Liver Mitochondria

Table 1 summarizes the respiratory rates, respiratory control ratio (RCR), and P/O ratios. It is evident from the RCR and P/O ratios listed in the table that both the hepatoma and normal mouse liver yielded well-coupled mitochondria.

Effects of Selenium on the Oxidation Function of Mitochondria

As shown in Table 2 and Fig. 1, Se in the form of sodium selenite added to the normal liver motochondria at 1 mM concentration stimulated the oxidation of glutamate, succinate, and ascorbate without affecting the phosphorylative efficiency of their oxidation, as deduced from the P/O values. However, selenite added to hepatoma mitochon-

TABLE 1. Oxidative Phosphorylation Properties of Normal Mouse Liver and Hepatoma Mitochondria

Substrate	Mouse liver			Hepatoma		
	Respiration rate (state III)[a]	P/O	RCR[b]	Respiration rate (state III)	P/O	RCR
L-Glutamate	54–97	1.4–2.8	2.2–6.9	98–100	2.4 –2.7	10
Succinate	94–146	1.0–1.2	1.3–3.6	220–291	1.66–1.71	10

[a] Respiration rates are expressed in nanograms of atoms oxygen min^{-1} mg^{-1} protein. The assay medium 2.0 ml contained 160 mM KCl, 2 mM EDTA, 4 mM MgCl$_2$, and 10 mM potassium phosphate buffer, pH 7.4. Respiratory substrates: L-glutamate, 50 μmol; succinate, 40 μmol (+1 μmol rotenone). The mitochondrial protein, 1.5–4.8 mg, incubation temperature, 30°C.
[b] RCR, Respiratory control ratio.

TABLE 2. Stimulatory Effect of Se (1 mM Na$_2$SeO$_3$) on the Oxidation of Different Substrates by Mouse Liver Mitochondria[a]

Substrate + Se	Stimulation O$_2$ uptake (%)
L-Glutamate	140–225
Succinate	108–246
Ascorbate (+TMPD)	140–150

[a] Assay condition and incubated medium as described in Table 1. Ascorbate, 5 μmol, + TMPD, 0.4 μmol, in the presence of 1.25 μmol antimycin A.

dria at the same concentration inhibited glutamate and pyruvate oxidation, but it did not affect the oxidation of succinate, ascorbate, and malate + NADH, or their associated phosphorylation (Table 3 and Fig. 2). In Fig. 2, it can also be clearly observed that succinate, ascorbate, and malate were readily oxidized in the presence of 1 mM selenite, which completely blocked glutamate and pyruvate oxidation (A–C). Similar results were obtained with Ehrlich ascites tumor mitochondria (3).

Effects of Selenite on Mitochondrial ATPase Activity

Figure 3 shows that the effect of selenite on mitochondrial ATPase of hepatoma was quite different from that on normal liver. It stimulated the mitochondrial ATPase activity of normal liver by a factor up to 3, and the stimulating effects increased with increasing selenite concentrations (10^{-6}–$10^{-3}M$). However, selenite at the same concentrations did not significantly affect the ATPase activity of hepatoma mitochondria.

Effects of Selenite on Swelling of Mitochondria

Figure 4 depicts the effects of Se on swelling of mitochondria. Normal mouse liver mitochondria swelled moderately rapidly in KCl–Tris buffer solution at room temperature. In the presence of selenium, a decrease in optical density occurred immediately. The accelerating effects reached maximum in 5–10 min, and they were proportionally to the increase in selenite concentrations in the range of 10^{-6}–$10^{-3}M$. After this period it approached control level. While under the same condition, the selenite retarded the swelling process of hepatoma mitochondria. The retarding effects increased with increasing concentrations of selenium and with prolonged incubation time.

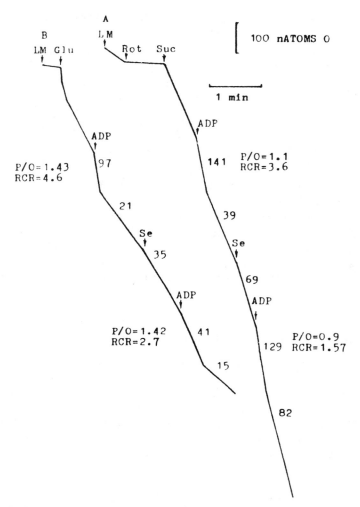

FIG. 1. Action of selenite (1 m*M*) on oxidative processes of normal mouse liver mitochondria (LM). Assay conditions were as described in Table 1. Addition at the points indicated by arrows: 4 mg mitochondrial protein; Glu (glutamate), 50 μmol, Suc (succinate), 40 μmol; Rot (rotenone), 1 μmol; ADP, 0.1 μmol. The numbers shown next to various segments of the oxygraph traces indicate respiration rates in nanograms of atoms of oxygen min^{-1} mg^{-1} protein.

TABLE 3. Effects of Se (1 mM Na$_2$SeO$_3$) on the Oxidation of Different Substrates by Hepatoma Hep A Mitochondria[a]

Substrate + Se	Inhibition O$_2$ uptake (%)	P/O
L-Glutamate	100	—
Pyruvate	100	—
Malate + NADH	0	2.33
Succinate	0	1.66
Ascorbate (+ TMPD)	0	0.44

[a] Assay condition as described in Table 2, L-glutamate, 50 μmol; pyruvate, 20 μmol; malate, 20 μmol; NADH, 5 μmol; succinate, 40 μmol (+rotenone, 1 μmol), ascorbate, 5 μmol, + TMPD, 0.4 μmol (+ antimycin A, 1.25 μmol).

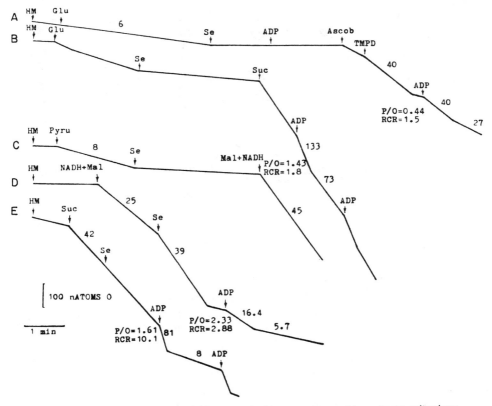

FIG. 2. Action of selenite (1 mM) on oxidative processes of hepatoma mitochondria (HM). Assay conditions were as in Fig. 1. Addition at the points indicated by arrows: Glu (glutamate), 50 μmol; Ascorb (ascorbate), 5 μmol; TMPD, 0.4 μmol; Suc (succinate), 40 μmol in the presence of 1 r rotenone; Pyru (pyruvate), 20 μmol; Mal (malate), 20 μmol; NADH, 5 μmol; ADP, 0.1 μmol and 0.1 μmol (A); mitochondrial protein, 1.5 mg (A,B); and 3 mg (C–E). The respiration rates are as in Fig. 1.

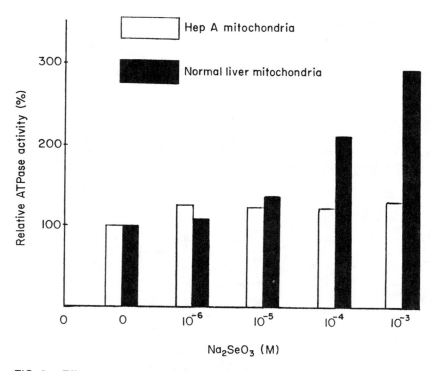

FIG. 3.   Effects of selenite on mitochondrial ATPase. Assay medium: 50 mM KCl; 4 mM MgCl$_2$; 125 mM sucrose; 5 mM ATP; 25 mM Tris–HCl, pH 7.4; 0.2 mg mitochondrial protein, final volume 1 ml, incubation at 30°C for 10 min.

DISCUSSION

Studies of mitochondrial swelling are usually taken to reflect the permeability of the membranes to solutes and the activity of inter-mediate enzymes of electron transport and oxidative phosphorylation. Electron flux together with early stages in energy coupling are neces-sary for the permeability changes leading to water uptake. The results presented here show that Se stimulated the normal mouse liver mito-chondrial ATPase, oxidation, and swelling. The Se-induced ATPase activity, making available additional ADP, can account for the in-crease in oxygen uptake, and it is commonly accepted that respiration is a factor in mitochondrial swelling. Since the stimulatory effect of Se on mitochondrial swelling was seen only for a short period and it stimulated the motochondrial respiration without significantly affect-ing associated phosphorylation, it can be deduced that in the range of concentrations 10^{-6}–10^{-3} M, the effects of Se on normal mouse liver mitochondria are within the bounds of physiologically controllable change in mitochondrial membrane. Experiments on hepatoma mitochondria demonstrated that Se did not affect the ATPase activity and inhibited the swelling of mitochondria. At 1 mM concentration, selenite selectively inhibited the oxidation of glutamate and pyruvate. The observation that malate + NADH oxidation was not inhibited by Se rules out the possibility that selenite can prevent the electron flux through the respiratory chain. The sensitivity of glutamate and pyruvate oxidation to selenite would indicate that the thiol group of glutamate and pyruvate dehydrogenases is more accessible than those of other enzymes to selenite action. Furthermore, in accordance with the swelling data, it is very likely that Se alters more or less the conformation of enzyme molecules and their environment in hepatoma mitochondria.

The most important aspects of the reported results are the selective effects of Se on hepatoma mitochondria. Selenite can inhibit the energy metabolism of hepatoma mitochondria at concentrations that do not adversely affect the normal liver mitochondria. Previous studies in our laboratory have shown that Se administered at 1–2 mg/kg of body weight to hepatoma-bearing mice led to an increase in cAMP levels in hepatoma cells, but not in tumor host or normal liver cells. The biochemical differences between hepatoma and normal liver cells in response to Se provide a suggestion that hepatoma cells may be sensitive to Se supplementation at certain concentrations that can be

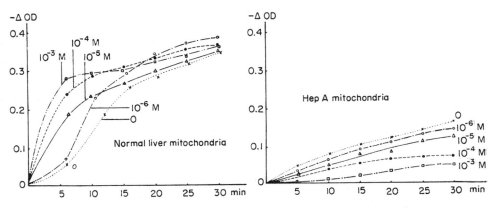

Fig. 4.   Effects of selenite on mitochondrial swelling process. Assay medium: 125 mM KCl; 20 mM Tris–HCl buffer, pH 7.4; mitochondrial protein, 1.5 mg/ml; temperature, 22°C.

tolerated by normal cells. Further studies to prove this hypothesis are now under investigation.

REFERENCES

1. Abdullaev, G. B., Gasanov, G. G., Ragimor, R. N., Teplyakova, G. V., Mekhtiev, M. A., and Dehafarov, A. I. 1973. Dokl. Akad. Nauk Azerb. SSR *29,* 18.
2. Poirier, K. A., and Milner, J. A. 1979. Biol. Trace Elem. Res. *1,* 25.
3. Yu, S.-Y., Zhu, Y.-J., Liu, Q.-Y. and Hou, C. 1983. Chin. J. Oncol. *5,* 8.
4. Weisberger A. S., and Suhrland, C. G. 1956. Blood *11,* 11.
5. Milner, J. A., and Hsu, C.-Y. 1981. Cancer Res. *41,* 1652.
6. Ip, C., Ip, M. M., and Kim, U. 1981. Cancer Lett. *14,* 101.
7. Watrach, A. M., Milner, J. A., and Watrach, M. A. 1982. Cancer Lett. *15,* 137.
8. Randleman, C. Q. 1980. Bios (Madison, N.J.) *51,* 86.
9. Vernie, L. N., Homburg, C. J., and Bont, W. S. 1981. Cancer Lett. *14,* 303.
10. Yu, S. Y., Wang, L. M. and Qian, W. J. 1982. Chin. J. Oncol. *4,* 50.
11. Ao, P., and Pan, J. J. 1983. Cancer Res. Prev. Treat. *10,* 19.
12. Schneider, W. C., and Hogeboom, H. G. 1950. J. Biol. Chem. *183,* 123.
13. Ohmishi, T., Gall, R. S., and Mayer, M. L. 1975. Biochem. *69,* 26.
14. Levin, R., and Braner, R. W. 1951. J. Lab. Clin. Med. *38,* 474.

Effect of Selenium Deficiency on 1,2-Dimethylhydrazine-Induced Colon Cancer in Rats

Barbara C. Pence
Fred Buddingh

It has been shown experimentally that Se supplementation in the diet produces a protective effect against the development of tumors induced by a variety of chemical carcinogens as well as by a virus (*1*). Jacobs *et al.* (*2*) reported that Se added to the drinking water of rats at 4.0 ppm resulted in a 50% reduction in the incidence of colon cancer induced by weekly injection of 1,2-dimethylhydrazine (DMH). Epidemiologic evidence (*3,4*) has shown an inverse relationship between Se levels in forage crops and human cancers of the gastrointestinal tract. It has also been reported (*5*) that there are significant inverse relationships between blood Se levels and cancer of the colon and rectum. Although the Se supplementation studies are numerous (*1*), there are few experimental studies (*6*) which examine the effect of Se deficiency on neoplasia, and none has been reported for cancer of the large bowel. In order to investigate the epidemiologic evidence linking low Se levels to colorectal cancer, the present study was undertaken. Our objective was to examine the effects of an Se-deficient diet and a minimal nutritionally adequate Se (0.1 ppm) diet on DMH-induced colon carcinogenesis in rats.

MATERIALS AND METHODS

Weanling male Sprague–Dawley rats (TTUHSC-bred) were randomized into treatment groups and acclimated for 3 weeks to either an

Se-deficient or Se-adequate diet (Table 1). Animals were housed initially at 5 per cage in suspended wire-mesh cages, and as they grew, they were caged individually. Temperature, humidity, and a 12-hr light/dark cycle were controlled throughout the experiment. Water and food were available *ad libitum*. The drinking water contained very little Se (0.01 ppb). Composition of the diets is shown in Table 1. The diet deficient in Se contained <0.02 ppm Se as determined previously for similar torula yeast diets (6). Selenium was supplemented in the Se-adequate diet to 0.1 ppm as sodium selenite (Na_2SeO_3). The vitamin E level was adequate, being present in the vitamin mix and the 2% corn oil. The diet was made up in large batches and kept refrigerated until use. Food consumption data were taken during week 10 of the experiment.

After 3 weeks on the diets animals were divided into control and DMH-treated groups as shown in Table 2. Carcinogen-treated rats received weekly intraperitoneal injections of 20 mg/kg body weight DMH (*sym*-dimethylhydrazine dihydrochloride, 97%; purchased from Aldrich Chemical Co., Milwaukee, Wl) for 20 weeks. DMH solutions were prepared just prior to injection in 1 m*M* EDTA and brought to pH 6.5 with NaOH to ensure stability of the chemical. Control rats only received the test diets. Body weights were monitored weekly.

At the end of DMH treatment complete necropsies were performed for evaluation of gross and microscopic pathology. Colons were resected at the cecal and rectal junctions and examined for lesions. All lesions were noted, measured, and preserved in 10% buffered Formalin, stained with hematoxylin and eosin, and verified as to tumor type by histological examination. For clinical evaluation, whole blood

TABLE 1. Diet Composition

	g/kg
Se-deficient diet	
Torula yeast[a]	300.0
Sucrose[b]	587.0
AIN-76 mineral salts[c] (Se omitted)	50.0
AIN-76 vitamin mix[c]	10.0
L-Methionine[c]	3.0
Lard[b]	30.0
Corn oil[d]	20.0
Se-adequate diet (Se 0.1 ppm)	
Sodium selenite[c] (Na_2SeO_3) added	0.00021

[a] Lake States Yeast, Rhinelander Paper Co., Rhinelander, WI.
[b] White Swan, Lubbock, TX.
[c] ICN Nutritional Biochemicals, Cleveland, OH.
[d] Best Foods, Englewood Cliffs, NJ.

TABLE 2. Experimental Design

Se status	DMH	20 weeks[a]
Deficient	+	42 (45)[b]
Adequate	+	37 (45)
Control deficient	−	10 (10)
Control adequate	−	10 (10)

[a] 3 + 20 weeks on test diet; 20 weeks of DMH treatment (20 mg/kg).
[b] Original number of animals; decreased survival due to noncancer deaths early in experiment.

(with and without EDTA) was obtained by anesthetizing with ether, severing the axillary vessels, and collecting the blood with a Pasteur pipette. Hematocrit, white blood cell counts, serum total protein, urea nitrogen, and cholesterol determinations were performed. Cholesterol and urea nitrogen assay kits were obtained from Sigma Chemical Co., St. Louis, MO, and total protein was determined spectrophotometrically using the Biuret reagent.

Statistical analysis for tumor incidence was done using the chi square, and all other comparisons were analyzed using the Student's t test (two-tailed, unpaired).

RESULTS

Growth and Food Consumption

Tables 3 and 4 show the effects of Se deficiency on body weights of DMH-treated and control rats at elected intervals through week 20 of the experiment. The body weight-to-spleen weight ratio is also shown, taken at necropsy (week 20). By week 10 there was a highly significant ($P < .01$) difference in the body weights of the deficient vs adequate animals. This difference, however, is only seen in the animals treated

TABLE 3. Effects of Selenium Deficiency on Body Weights of Rats Treated with DMH

Se status	Week 1[a]	Week 10	Week 20	Body/spleen ratio
Deficient	136 ± 4[b]	333 ± 6	414 ± 6	584 ± 17
Adequate	134 ± 3	355 ± 5	443 ± 9	541 ± 18
P value	>.05	<.01	<.01	.05 > P < .10

[a] Week 1, start of DMH treatment.
[b] Weight (g) ± SE.

TABLE 4. Effects of Selenium Deficiency on Body Weights of Control Rats

Se status	Week 1[a]	Week 10	Week 20	Body/spleen ratio
Deficient	180 ± 6[b]	412 ± 14	489 ± 18	726 ± 34
Adequate	181 ± 3	425 ± 13	495 ± 19	701 ± 29

[a] Week 1, corresponds to start of DMH treatment in treated rats.
[b] Weight (g) ± SE.

with DMH. Control animals (Table 4) show no significant weight differences during the entire course of the study. Control animals started out as weanlings at the beginning of the study at the same approximate weight as those treated with DMH (45 g). The weight differences between control and DMH-treated rats seen at week 1 reflect differences in caging during the 3-week acclimatization period. Controls were housed individually instead of 5 per cage for the first few weeks. Food consumption data (Table 5) do not show any differences which would account for the weight disparity seen in the DMH-treated rats. There is a trend seen in the difference between body/spleen weight ratios in DMH-treated animals, not seen in controls, which may indicate a possibility of decreased spleen sizes in the deficient animals treated with the carcinogen.

Hematology

Hematologic parameters and serum chemistries for DMH-treated rats are shown in Table 6. There were no differences in any of these parameters in control rats (no DMH) due to Se status. In DMH-treated animals (Table 6), the following significant differences were noted: Blood urea nitrogen was lower in Se-deficient animals, 9.6 ± 0.7 vs 13.7 ± 0.9 ($P < .01$) and serum cholesterol was higher in Se-deficient rats, 69.0 ± 5.5 vs 50.7 ± 3.9 ($P < .05$).

TABLE 5. Food Consumption[a]

Food	g/rat/day
Se deficient + DMH	17.7 ± 1.1
Se adequate + DMH	17.2 ± 0.9
Se-deficient control	17.3 ± 1.0
Se-adequate control	16.9 ± 1.2

[a] Mean ± SE taken at week 10 of experiment.

TABLE 6. Effect of Selenium Deficiency on Hematologic and Serum Parameters in DMH-Treated Rats[a]

Parameter	Se deficient	Se adequate
WBC ($\times$ $10^3/mm^3$)	10.9 ± 1.0^b	12.9 ± 1.5
Hematocrit (%)	41.7 ± 2.0	39.7 ± 2.3
Total protein (g/dl)	7.7 ± 0.3	7.4 ± 0.4
BUN (mg/dl)	$9.6 \pm 0.7^{**}$	$13.7 \pm 0.9^{**}$
Cholesterol (mg/d)	$69.0 \pm 5.5^*$	$50.7 \pm 3.9^*$

[a] Control rats (no DMH) showed no differences in these parameters due to Se status.
[b] Mean $\pm$ SE.
** Significant difference ($P < .01$)
* Significant difference ($P < .05$)

Colon Tumors

Table 7 shows the incidences of colon adenocarcinomas, neoplastic changes, and total tumors in the DMH-treated animals. Control animals receiving no DMH showed no evidence of any colon lesions. Chisquare analysis showed no difference in incidence of tumors attributable to Se status. Tumor incidence was identical in both deficient and adequate dietary groups.

DISCUSSION

The most striking finding in this study was that Se deficiency before and during DMH administration did not result in increased tumor incidence. Ip and Sinha (6) reported similar results in that they did not find an enhancing effect of Se deficiency on 7,12-dimethyl-benz[a]anthracene (DMBA) mammary tumorigenesis at the 5% dietary fat level. This was similar to fat levels used in our study. These two studies [(6) and this report] do not agree with the epidemiologic

TABLE 7. DMH-Induced Colon Tumor Incidence with and without Selenium

Se status	Adenocarcinomas[a]	Neoplastic change[a]	Total
Deficient	62 (26/42)[b]	14 (6/42)	76 (32/42)
Adequate	57 (21/37)	19 (7/37)	76 (28/37)

[a] After 20 weeks of DMH treatment.
[b] Incidence (%), actual numbers in parentheses.

data linking low Se availability to increases in cancer of the large bowel (3–5).

Another notable result in the present study was the highly significant difference in body weights in animals treated with DMH. This effect of Se deficiency was not seen in control (no DMH) rats. Usually, Se deficiency in the presence of an adequate intake of vitamin E does not affect the growth of animals (7). The diets used in this study contained adequate levels of vitamin E, and control (diet only) animals showed no significant body weight differences. Ip and Sinha (6) in their study of mammary tumorigenesis found that Se deficiency did not affect the growth of animals treated with DMBA. These previous studies (6,7) suggest that the growth difference seen in our study was due to a toxic effect of DMH, induced by Se deficiency. This suggests that Se deficiency may interfere with the normal detoxification of DMH.

The differential effects of Se deficiency on serum urea nitrogen and cholesterol levels in DMH-treated animals are not seen in control animals. The role of Se deficiency in controlling these serum parameters again appears to be linked to DMH-induced toxicity and may be a response to generalized liver damage. Control Se-deficient rats had a urea nitrogen value of 12.9 ± 1.1 and cholesterol of 55.3 ± 2.4, and control Se-adequate rats had respective values of 13.4 ± 0.9 and 50.6 ± 3.7. Normal values for serum urea nitrogen in rats range from 15 to 22 mg/dl (8) and for total serum cholesterol from 28 to 76 mg/dl, with a mean value of 52 (9). Thus, Se deficiency coupled with DMH treatment produces a synergistic toxic response affecting weight gain, protein utilization, and cholesterol metabolism. What role Se plays in these physiologic responses to xenobiotic stress is unknown at this time and warrants further investigation.

REFERENCES

1. National Academy of Sciences 1982. Diet, Nutrition, and Cancer, pp. 10-2–10-8. National Academy Press, Washington, DC.
2. Jacobs, M. M., Jansson, B., and Griffin, A. C. 1977. Inhibitory effects of selenium on 1,2-dimethylhydrazine and methylazoxymethanol acetate induction of colon tumors. Cancer Lett. 2, 133–138.
3. Schrauzer, G. N., White, D. A., and Schneider, C. J. 1977. Cancer mortality correlation studies. IV. Associations with dietary intakes and blood levels of certain trace elements, notably Se antagonists. Bioinorg. Chem. 7, 35–56.
4. Shamberger, R. J., and Willis, E. C. 1971. Selenium distribution and human cancer mortality. CRC Crit. Rev. Clin. Lab. Sci. 2, 211–221.
5. Robinson, M. F., Godfrey, P. J., Thomson, C. D., Rea, H. M., and van Rij, A. M. 1979. Blood selenium and glutathione peroxidase activity in normal subjects and

in surgical patients with and without cancer in New Zealand. Am. J. Clin. Nutr. *32,* 1477–1485.

6. Ip, C., and Sinha, D. K.   1981.   Enhancement of mammary tumorigenesis by dietary selenium deficiency in rats with a high polyunsaturated fat intake. Cancer Res. *41,* 31–34.

7. McCoy, K. E. M., and Weswig, P. H.   1969.   Some selenium responses in the rat not related to vitamin E. J. Nutr. *98,* 383–389.

8. Baker, H. J., Lindsey, J. R., and Weisbroth, S. H. (Editors)   1979.   The Laboratory Rat, Vol. 1, p. 114. Academic Press, New York.

9. Altman, P. L., and Dittmer, D. S.   1971.   Blood and Other Body Fluids, p. 81. FASEB, Bethesda, MD.

Effect of Sodium Selenite on MNNG- and MNU-Induced Chromosomal Aberrations of V79 Cells in Vitro

Chen Quan-Guang   *Liu Jun*
Hu Guo-Gang   *Ke Yang*
Gao Fu-Zheng   *Luo Xian-Mao*
Liu Xiu

Epidemiological studies have shown an inverse relationship between dietary selenium and cancer mortality (*1,2*). Selenium exhibited anticarcinogenic properties *in vivo* (*3–7*) and *in vitro* (*8–9*) when used in combination with carcinogens. Yet, when selenium is used alone, it is a clastogenic agent, inducing sister-chromatid exchanges (SCEs) (*10*), chromosomal aberrations (*11*), and causing other types of DNA damage (*12*). In order to clarify these problems, more work should be done. In the present study we tested the effect of sodium selenite (Na_2SeO_3) on N-methyl-N'-nitro-N-nitrosoguanidine (MNNG) and methlynitrosourea (MNU) induced chromosomal aberrations of V79 cells *in vitro*.

MATERIALS AND METHODS

Chemicals

MNNG (Sigma) and MNU (Sigma) were freshly prepared in Eagle's minimal essential medium (MEM). Sodium selenite (Na_2SeO_3) was dissolved in MEM, and stock solution was filtered by 0.45-μm metrical membrane filter and stored in 4°C. These chemicals were added to the cells from a 10-fold higher concentration to give the required doses.

Cell Culture

V79 cells, Chinese hamster lung fibroblasts, were cultured in MEM supplemented with calf serum, penicillin, and streptomycin to a concentration of 15%, 100 IU/ml and 100 μg/ml, respectively. A constant exponentially growing population of cells was obtained by seeding $2.5–3.0 \times 10^6$ cells per 75-cm^2 flask and subculturing every 3–4 days. For these experiments $7–8 \times 10^5$ cells were seeded per 25-cm^2 flask.

Mode of Chemical Treatment of Cells

1. Cells were pretreated with various doses of selenium ($10^{-7}–10^{-4}$ M) for 4 hr, the medium was removed and replaced by medium containing MNNG or MNU to a final concentration of 10^{-4} M and 0.1 mg/ml, respectively, and then cells were incubated at 37°C for 44 hr.

2. Simultaneous cell incubation took place with Na_2SeO_3 ($10^{-7}–10^{-4}$ M) and MNNG (10^{-4} M) or MNU (0.1 mg/ml) for 48 hr.

3. Incubation with Na_2SeO_3 ($10^{-7}–10^{-4}$ M) for 48 hr was performed.

4. Cells were cultured with MNNG (10^{-7} M) or MNU (0.1 mg/ml) for 48 hr.

5. Cells were treated with MEM as a control.

Chromosome Preparations

Cells ($7–8 \times 10^5$) were cultured in a flask containing 5 ml of medium with Na_2SeO_3 and carcinogens described above. Before harvesting, cells were blocked for 4 hr in 0.06 μg/ml colchicine, swollen with 0.075 M KCl, fixed with methanol:acetic acid (3:1), and dropped onto slides. After air drying, the preparations were stained with Giemsa and the number of cells containing chromosomal aberrations in 100 cells of each sample were counted under the microscope with 1500 × magnification.

RESULTS

Na_2SeO_3 alone at $10^{-7}–10^{-4}$ M increased the frequencies of chromosomal aberrations, which increased with a rising concentration of

TABLE 1. Effect of Sodium Selenite on MNNG-Induced Chromosomal Aberrations

Na_2SeO_3 (M)	Mode of selenium treatment of cells containing chromosomal aberrations		
	Preincubated with cells for 4 hr and then mixed with MNNG	Mixed with MNNG simultaneously	Selenium control (no MNNG)
10^{-4}	63	—	35
10^{-5}	30	61	19
10^{-6}	46	63	16
10^{-7}	51	61	12
0	60	64	5

Na_2SeO_3. Sodium selenite at 10^{-4} M increased 7-fold cells containing chromosomal aberrations above that of control. Preincubated with cells for 4 hr, Na_2SeO_3 at 10^{-7}–10^{-5} M, but not 10^{-4} M, reduced the number of cells with chromosomal aberrations induced by MNNG. No protective effect was observed if the cells were coincubated with Na_2SeO_3 and MNNG simultaneously (Table 1). Preincubated with cells or coincubated with MNU simultaneously, Na_2SeO_3 at the concentration of 10^{-7}–10^{-4} M decreased the frequencies of chromosomal aberrations (Table 2).

DISCUSSION

Does selenium cause chromosome damage? There are some contradictory reports. For example, Nakamura *et al.* (11) proved that Na_2SeO_3 induced chromosomal aberrations in cultured human lym-

TABLE 2. Effect of Sodium Selenite on MNU-Induced Chromosomal Aberrations

Na_2SeO_3 (M)	Mode of selenium treatment of cells containing chromosomal aberrations		
	Preincubated with cells for 4 hr and then mixed with MNU	Mixed with MNU simultaneously	Selenium control (no MNU)
10^{-4}	26	22	32
10^{-5}	27	23	21
10^{-6}	24	22	15
10^{-7}	26	25	12
0	40	42	3

phocytes, but no detectable chromosome damage in selenium-treated V79 cells *in vitro* was found by Whiting (9). Our results supported the former report, i.e., Na_2SeO_3 alone at $10^{-7}-10^{-4}$ M does increase the frequencies of cells containing chromosomal aberrations.

On the one hand, Na_2SeO_3 is a clastogenic agent when it is used alone. On the other hand, sodium selenite combined with carcinogens can function as an anticarcinogenic agent. Whiting (9) reported that sodium selnite at $10^{-7}-10^{-5}$ M with the exception of 10^{-4} M reduced the frequencies of cells containing chromosomal aberrations induced by MNNG when cells were preincubated with Na_2SeO_3 before addition of this carcinogen. When Na_2SeO_3 was mixed with MNNG at 37°C in test tubes for 30 min before addition of the mixture to cells, there was no protective effect. Our results are similar to Whiting's in this case.

Selenium has anticarcinogenic activity, but the mechanism of reduction of tumor incidence and chemical mutagenicity by selenium is still unknown. Banner *et al.* (13) suggested that the protective effects of selenium are probably due to a mechanism other than influence with carcinogen activation and interaction with cellular macromolecules. Whiting (9) proposed that Na_2SeO_3 and its metabolite might react competitively with MNNG or change the structure of the membrane or the chromatin and decrease the rate of reaction of the carcinogen with DNA. According to these reports, the mechanism of antimutagenicity by selenium remains to be elucidated.

Na_2SeO_3 can reduce the number of cells with chromosomal aberrations induced by MNNG only when preincubated with those cells. Na_2SeO_3 has protective effects whether preincubated with cells or mixed with MNU simultaneously.

In order to demonstrate the mechanism of protective effects by Na_2SeO_3 *in vitro,* other methods such as SCE as well as UDS should be used and more carcinogens should be tested for comparison.

REFERENCES

1. Schrauzer, G. N., White, D. A., and Schneider, C. J. 1977. Cancer mortality correlation studies. III. Statistical association with dietary selenium intakes. Bioinorg. Chem. 7, 23–24.
2. Shamberger, R. J., Tytko, S. A., and Willis, C. E. 1976. Antioxidants and cancer. VI. Selenium and age-adjusted human cancer mortality. Arch. Environ. Health *31,* 231–235.
3. Schrauzer, G. N., and Ishmael, D. 1974. Effects of selenium and of arsenic on the genesis of spontaneous mammary tumors in inbred C₃H mice. Ann. Clin. Lab. Sci. *4,* 441–447.
4. Jacob, M. M., Jansson, B., and Griffin, A. C. 1977. Inhibitory effects of selenium

of 1,2-dimethylhydrazine methylazoxymethanol acetate induction of colon tumors. Cancer Lett. *2*, 133–138.

5. Ip, Clement, Ip, Margot, M., and Untake, K.   1981.   Dietary selenium intake and growth of the MT-W9B transplantable rat mammary tumor. Cancer Lett. *14*, 101–107.

6. Marshall, M. V., Arnott, M. S., Jacobs, M. M., and Griffin, A. C.   1979.   Selenium effects on the carcinogenicity and metabolism of 2-acetylaminofluorene. Cancer Lett. *7*, 331–338.

7. Thompson, H. J., and Beccl, P. J.   1980.   Selenium inhibition of *N*-methyl-*N*-nitrosourea-induced mammary carcinogenesis in the rat. JNCI, J. Natl. Cancer Inst. *65*, 1299–1301.

8. Shamberger, R. J., Baughman, F. F., Kalchert, S. L., Millis, C. E., and Hoffman, G. C.   1973.   Carcinogen-induced chromosomal breakage decreased by antioxidants. Proc. Natl. Acad. Sci. U.S.A. *70*, 1461.

9. Whiting, R. F., Wei, L., and Stich, H. F.   1981.   Mutagenic and antimutagenic activities of selenium compounds in mammalian cells. *In* Selenium in Biology and Medicine J. E. Spallholz, M. L. Martin, and H. E. Ganther pp. 325–330. AVI Publishing Co., Westport, CT.

10. Ray, T. H., Altenburg, L. C., and Jacobs, M. M.   1978.   Effect of sodium selenite and methyl methanesulfonate or *N*-hydroxy-2-acetylaminofluorene coexposure on sister chromatid exchange production in human whole blood cultures. Mutat. Res. *57*, 359–368.

11. Nakamura, K., Yoshikawa, K., Sayato, Y., Kurata, H., Tonomura, M., and Tonomura, A.   1976.   Studies on selenium-related compounds. V. Cytogenetic effect and reactivity with DNA. Mutat. Res. *40*, 177.

12. Lo, L. W., Koropanick, J., and Stich, H. F.   1978.   The mutagenicity and cytotoxicity of selenite, "activated" selenite, and selenate for normal and DNA repair-deficient human fibroblasts. Mutat. Res. *49*, 305–312.

13. Banner, W. P., Queng, H. T., and Zedeck, M. S.   1982.   Selenium and the acute effects of the carcinogens, 2-acetylaminofluorene and methylazoxymethanol acetate. Cancer Res. *42*, 2985–2989.

115

Effect of Sodium Selenite on MNNG- and MNU-Induced Sister Chromatid Exchanges of V79 Cells

Hu Guo-Gang   Ke Yang
Chen Quan-Guang Liu Jun
Gao Fu-Zheng   Luo Xian-Mao
Liu Xiu

Epidemiological surveys have shown that there was an inverse relationship between dietary selenium intake, selenium level of serum, and mortality of type A tumors (1). Shamberger et al. (2) indicated that serum selenium level in patients suffering from certain types of tumors was lower than that of normal human serum. It is known that selenium exerts an anticarcinogenic effect on spontaneous, chemical carcinogen-induced and transplantable tumors in animal experiments (3–7). Selenium also can function as an antimutagenic agent in in vitro systems (8–10). Hence, it was of interest to examine the effect of selenium on mammalian chromosomes exposed to agents known to cause DNA damage, as evidenced by increased sister-chromatid exchange (SCE) frequency, which has been shown to be a sensitive, effective, and reliable indicator of DNA damage induced by mutagens/carcinogens.

MATERALS AND METHODS

V79 Chinese hamster lung fibroblast cells were routinely cultured in Eagle's minimal essential medium (MEM) supplemented with 15% fetal calf serum and antibiotics (100 U/ml of penicillin and 100 μg/ml of streptomycin). Cells were subcultured by transferring $2.5-3.0 \times 10^6$ cells per flask at 3- to 4-day intervals.

N-Methyl-N'-nitro-N-nitrosoguanidine (MNNG) and methylnitrosourea (MNU) (Sigma) were dissolved in Eagle's MEM. A stock solution of the chemicals were freshly prepared before each experiment. A stock solution of Na_2SeO_3 was filtered by 0.45-μm metrical membrane filter and stored at 4°C.

Incubation of Na_2SeO_3 before the Addition of Mutagens

Cells ($7-8 \times 10^5$) seeding in flask (25 cm²) were treated for 4 hr with various doses of Na_2SeO_3 ($10^{-7}-10^{-4}$ M) in Eagle's MEM. After the treatment periods, the medium was discarded. The cells were reincubated with fresh medium containing MNNG or MNU (final concentration was 10^{-6} M and 0.01 mg/ml, respectively).

Coincubation of Na_2SeO_3 with Mutagens

Exposed to various doses of Na_2SeO_3, cells were treated with MNNG or MNU immediately. Bromodeoxyuridine (BUdR) was added at a final concentration of 6 μg/ml at 37°C for 48 hr in total darkness. Colchicine (final concentration 0.04 μg/ml) was present during the final 4 hr of incubation. Cells were harvested by centrifugation (615 g, 10 min), treated with 0.075 M KCl at 37°C for 20 min (hypotonic treatment), and fixed 3 times in methanol-glacial acetic acid (3:1). Preparations were made on ice-cold slides and air dried. The procedure of differential Giemsa staining of sister chromatids described by Goto *et al.* (*11*) and Wu and Wang (*12*) was slightly modified for analysis of SCE. After 2–3 days, the slides were treated with 0.14 M NaCl, 0.004 M KCl, and 0.01 M PBS for 5 min, respectively, stained in Hoechst 33258 solution (1 μg/ml) for 20 min, transferred to 0.01 M PBS for 10 min, and washed thoroughly with distilled water. The slides were soaked in 2 $\times$ SSC (sodium chloride and sodium citrate) solution (pH 7.0) at 50°C and exposed to two 20-W blacklight lamps (5 cm above slides) for 30 min. The specimens were washed with distilled water and stained by 10% Giemsa for 10 min. SCE frequency of each sample was determined by scoring 25–50 metaphases.

RESULTS

SCE frequency of V79 cells treated with MNNG or MNU was increased. This increase depended on the concentration of these muta-

TABLE 1. MNNG- and MNU-Induced SCEs in V79 Cells

MNNG and MNU levels	Total cells scored	SCE/cell (mean ± SD)	SCE/chromosome (mean ± SD)	P
Control	70	5.20 ± 2.90	0.25 ± 0.14	
MNU (0.001 mg)	25	10.44 ± 4.26	0.52 ± 0.22	<.001
MNU (0.01 mg)	25	28.12 ± 7.92	1.40 ± 0.42	<.001
MNU (0.1 mg)	25	46.84 ± 8.72	2.29 ± 0.43	<.001
MNNG (10^{-7} M)	44	6.07 ± 3.23	0.29 ± 0.15	>.05
MNNG (10^{-6} M)	45	18.02 ± 7.52	0.85 ± 0.34	<.001
MNNG (10^{-5} M)	36	37.69 ± 12.02	1.78 ± 0.60	<.001
MNNG (10^{-4} M)	7	64.71 ± 18.92	3.17 ± 0.84	<.001

gens (Table 1). High levels of Na_2SeO_3 (10^{-4} M) alone could induce a 5-fold SCE frequency (25.96 ± 17.50/cell) above that of background (4.78 ± 1.99/cell). There was no effect of 10^{-7}–10^{-5} M Na_2SeO_3 on SCE frequency.

Preincubation of Na_2SeO_3 at the concentration of 10^{-6}–10^{-5} M significantly decreased the SCE frequency induced by MNNG or MNU (Tables 2 and 3). The SCE frequency of V79 cells incubated with 10^{-6}–10^{-5} M of Na_2SeO_3 before the addition of 10^{-6} M of MNNG was 8.78 ± 4.78 and 10.36 ± 5.87 compared with the value of treatment of MNNG alone (13.72 ± 7.48/cell). Preincubation with 10^{-6}–10^{-5} M of Na_2SeO_3 also resulted in the decrease of SCE frequency induced by 0.01 mg/ml MNU. Meanwhile, Na_2SeO_3 had no effect on these mutagen-induced SCE frequencies treated simultaneously with Na_2SeO_3 (Table 4).

DISCUSSION

Although evidence from a number of epidemiological studies and experiments *in vivo* and *in vitro* indicated that selenium can function as an anticarcinogenic or antimutagenic agent, the mechanisms of reduction of tumor incidence and chemical mutagenicity by selenium are still unknown. SCE technique is a sensitive, simple, and reliable method for determination of DNA damage induced by mutagens/carcinogens in the environment. In these experiments, a study of the effect of Na_2SeO_3 alone or with the addition of MNNG or MNU on SCE frequency in V79 cells was carried out.

MNU, a potent mutagen/carcinogen, can significantly increase the SCE frequency in mammalian cells. Goth-Goldstein (*13*) found that O⁶-meGua was reduced 28% of the initial amount 48 hr after MNU ex-

TABLE 2. Effect of Preincubation of Na_2SeO_3 on MNNG-Induced SCEs in V79 Cells

	Total cells scored	SCE/cell (mean ± SD)	SCE/chromosome (mean ± SD)	P
Control	50	4.78 ± 1.99	0.23 ± 0.10	
Na_2SeO_3 (10^{-7} M)	50	4.78 ± 5.41	0.23 ± 0.27	>.05
Na_2SeO_3 (10^{-6} M)	50	4.58 ± 2.05	0.22 ± 0.09	>.05
Na_2SeO_3 (10^{-5} M)	50	5.76 ± 2.85	0.28 ± 0.14	>.05
Na_2SeO_3 (10^{-4} M)	50	25.96 ± 17.50	1.23 ± 0.85	<.001
MNNG (10^{-6} M)	47	13.72 ± 7.48	0.65 ± 0.36	
Na_2SeO_3 (10^{-7} M) + MNNG (10^{-6} M)	29	11.93 ± 6.41	0.56 ± 0.30	>.05
Na_2SeO_3 (10^{-6} M) + MNNG (10^{-6} M)	50	10.36 ± 5.87	0.49 ± 0.28	<.05
Na_2SeO_3 (10^{-5} M) + MNNG (10^{-6} M)	32	8.78 ± 4.87	0.42 ± 0.23	<.01
Na_2SeO_3 (10^{-4} M) + MNNG (10^{-6} M)	33	12.21 ± 5.20	0.58 ± 0.25	>.05

TABLE 3. Effect of Preincubation of Na_2SeO_3 on MNU-Induced SCEs in V79 Cells

	Total cells scored	SCE/cell (mean ± SD)	SCE/chromosome (mean ± SD)	P
Control	50	5.34 ± 3.29	0.27 ± 016	
Na_2SeO_3 (10^{-7} M)	42	7.00 ± 2.78	0.33 ± 0.13	<.05
Na_2SeO_3 (10^{-6} M)	50	4.50 ± 3.09	0.22 ± 0.17	>.05
Na_2SeO_3 (10^{-5} M)	50	5.82 ± 2.72	0.28 ± 0.13	>.05
Na_2SeO_3 (10^{-4} M)	—	—	—	
MNU (0.01 mg)	55	21.27 ± 4.51	1.08 ± 0.32	
Na_2SeO_3 (10^{-7} M) + MNU (0.01 mg)	36	19.61 ± 7.47	0.94 ± 0.36	>.05
Na_2SeO_3 (10^{-6} M) + MNU (0.01 mg)	50	15.42 ± 6.74	0.75 ± 0.32	<.001
Na_2SeO_3 (10^{-5} M) + MNU (0.01 mg)	50	16.46 ± 5.53	0.80 ± 0.28	<.001
Na_2SeO_3 (10^{-4} M) + MNU (0.01 mg)	55	21.31 ± 8.16	1.04 ± 0.39	>.05

TABLE 4. Effect of Coexposure of Na_2SeO_3 on MNNG- and MNU-Induced SCEs in V79 Cells

	Total cells scored	SCE/cell (mean ± SD)	SCE/chromosome (mean ± SD)	P
Control	50	4.86 ± 2.20	0.23 ± 0.10	
Na_2SeO_3 (10^{-7} M)	50	4.74 ± 2.06	0.23 ± 0.10	>.05
Na_2SeO_3 (10^{-6} M)	50	5.74 ± 2.37	0.29 ± 0.12	>.05
Na_2SeO_3 (10^{-5} M)	50	5.84 ± 2.51	0.28 ± 0.13	>.05
Na_2SeO_3 (10^{-4} M)	—	—	—	>.05
MNNG (10^{-6} M)	50	14.84 ± 5.57	0.70 ± 0.26	
Na_2SeO_3 (10^{-7} M) + MNNG (10^{-6} M)	50	14.88 ± 6.61	0.70 ± 0.30	>.05
Na_2SeO_3 (10^{-6} M) + MNNG (10^{-6} M)	50	15.68 ± 6.41	0.75 ± 0.30	>.05
Na_2SeO_3 (10^{-5} M) + MNNG (10^{-6} M)	38	17.29 ± 7.55	0.81 ± 0.36	>.05
MNU (0.01 mg)	50	17.20 ± 6.58	0.83 ± 0.31	
Na_2SeO_3 (10^{-7} M) + MNU (0.01 mg)	50	14.80 ± 6.03	0.71 ± 0.28	>.05
Na_2SeO_3 (10^{-6} M) + MNU (0.01 mg)	50	15.00 ± 6.64	0.70 ± 0.31	>.05

posure of normal human fibroblasts compared with 63% of the initial O^6-meGua in xeroderma pigmentosum-derived fibroblasts in which SCE was much higher than normal cells after treatment with mutagen. He proposed that there might be a causal relation between the unexcised O^6-meGua and the increased SCEs because O^6-alkylguanine is chemically stable in DNA and must be enzymatically excised. However, Connell and Metcalf (14) recently could not prove the result, and they demonstrated that the number of SCEs (in 50 cells) of V79 cells treated with 0.08 mM of ^{14}C-labeled MNU for 0–24 hr was 2015–1225, the 7-meGua level was 17.5–5.9, the 3-meAde was 1.0–0.1, while there was no change of O^6-meGua. Moreover, they suggested that formation of 7-meGua 3-meAde may be responsible for MNU-induced SCEs in V79 cells. The discrepancy of these results perhaps was related to the difference of cell types. In our experiments, preincubation of 10^{-6}, 10^{-5} M Na$_2$SeO$_3$ to protect DNA methylation induced by MNU might be responsible for the decrease of SCE frequency in V79 cells.

Popescu et al. (15) indicated that MNNG increased 4- to 5-fold in SCE of V79 cells. Whiting et al. (16) demonstrated that preexposure to 10^{-7}–10^{-5} M Na$_2$SeO$_3$ significantly decreased chromosome aberrations induced by MNNG in Chinese hamster cells, while neither exposure simultaneously to 10^{-7}–10^{-5} M Na$_2$SeO$_3$ with MNNG at 37°C for 30 min nor preexposure to a high level of Na$_2$SeO$_3$ (10^{-4} M) has any effect on MNNG-induced chromosome damages. We agree with Whiting's view that Na$_2$SeO$_3$ might react competitively with MNNG or modify some modes of cellular metabolism, change the structure of membrane or chromatin, and result in the reduction of DNA damage. Although Na$_2$SeO$_3$ acts as an antimutagen if applied together with some mutagens in mammalian cells, the mutagenic potential of selenium has been demonstrated in human lymphocytes. Norppa et al. (17) found that hamster treated with Na$_2$SeO$_3$ (3–6 mg/kg body weight, intraperitoneally) showed a very significant increase in SCEs of bone marrow cells. They postulated that increases in chromosome damage at very high doses of selenium may be a reflection of the general toxicity of selenite. Ray and Altenburg (18) proposed that RBC Na$_2$SeO$_3$ metabolism such as selenodiglutathione (GSSeSG) and glutathione selenopersulfide (GSSeH), whose formation require reduced glutathione and glutathione reductase, respectively, may represent an active SCE-inducing form of Na$_2$SeO$_3$ at high doses (32 μg/ml), creating a significant increase in SCE in V79 cells. Also, S9 enhanced the SCE rate produced by various doses of the selenium, just as the red blood cell lysate correspondingly elevated the SCE frequency induced by Na$_2$SeO$_3$. The result of our experiments has shown that SCE inducing capacity of Na$_2$SeO$_3$ may be due to a direct stimulation to the cells rather than the presence of red blood cell lysate. More work is needed to elucidate the mechanism of the protective effect of selenium on DNA damage induced by mutagens/carcinogens.

REFERENCES

1. Schrauzer, G. N., White, D. A., and Schneider, C. J. 1977. Cancer mortality correlation studies. III. Statistical associations with dietary selenium intakes. Bioinorg. Chem. *7*, 23–24.
2. Shamberger, R. J., Rukovena, E., Longfield, A. K., Tytko, S. A., Deodhar, S., and Willis, C. E. 1973. Antioxidants and cancer, I. Selenium in the blood of normals and cancer patients. J. Natl. Cancer Inst. (U.S.) *50*, 863–870.
3. Jacobs, M. M., Jansson, B., and Griffin, A. C. 1977. Inhibitory effects of selenium on 1,2-dimethylhydrazine and methylazomethanol acetate induction of colon tumors. Cancer Lett. *2*, 133–138.
4. Schrauzer, G. N., and Ishamael, D. 1974. Effects of selenium and of arsenic on the genesis of spontaneous mammary tumors in inbred C_3H mice. Ann. Clin. Lab. Sci. *4*, 441–447.
5. Clement, I. P., Margot, M. I. P., and Untake, K. 1981. Dietary selenium intake and growth of the MT-W9B transplantable rat mammary tumor. Cancer Lett. *14*, 101–107.
6. Marshall, M. V., Arnott, M. S., Jacobs, M. M., and Griffin, A. C. 1979. Selenium effects on the carcinogenicity and metabolism of 2-acetylaminofluorene. Cancer Lett. *7*, 331–338.
7. Thompson, H. J., and Beccl, P. J. 1980. Selenium inhibition of *N*-methyl-*N*-nitrosourea-induced mammary carcinogenesis in the rat. JNCI, J. Natl. Cancer Inst. *65*, 1299–1301.
8. Martin, S. E., Afams, G. H., Schillaci, M., and Milner, J. A. 1981. Antimutagenic effect of selenium on acridine orange and 7,12-dimethylbenz[a]anthracene in the Ames salmonella/microsomal system. Mutat. Res. *82*, 41–46.
9. Shamberger, R. J., Baughman, F. F., Kalchert, S. L., Millis, C. E., and Hoffman, G. C. 1973. Carcinogen-induced chromosomal breakage decreased by antioxidants. Proc. Natl. Acad. Sci. U.S.A. *70*, 1461–1463.
10. Whiting, R. F., Wei, L., and Stich, H. F. 1980. Unscheduled DNA synthesis and chromosome aberrations induced by inorganic and organic selenium compounds in the presence of glutathione. Mutat. Res. *78*, 159–169.
11. Goto, K., Maeda, S., Kano, Y., and Sugiyama, T. 1978. Factors involved in differential Giemsa staining of sister chromatids. Chromosoma *66*, 351–359.
12. Wu, M., and Wang, X.-Q. 1980. An improved method for differential staining of sister chromatids. Acad. Sin. *25*, 239–240.
13. Goth-Goldstein, R. 1977. Repair of DNA damaged by alkylating carcinogens in defective xeroderma pigmentosom-derived fibroblasts. Nature (London) *267*, 81–82.
14. Connell, J. R., and Metcalf, A. S. C. 1982. The induction of SCE and chromosomal aberrations with relation to specific base methylation of DNA in Chinese hamster cells by *N*-methyl-*N*-nitrosourea and dimethyl sulfate. Carcinogenesis (London) *3*, 385–390.
15. Popescu, N. C., Amsbaugh, S. A., and Dipalol, J. A. 1980. Reduced *N*-methyl-*N'*-nitro-*N*-nitrosoguanidine sister chromatid to 5-bromodeoxyuridine. Chromosoma *76*, 329–338.

16. Whiting, R. F., Wei, L., and Stich, H. F. 1981. Mutagenic and antimutagenic activities of selenium compounds in mammalian cells. *In* Selenium in Biology and Medicine, J. E. Spallholz, J. L. Martin, and H. E. Ganther (Editors). AVI Publishing Co., Westport, CT.
17. Norppa, H., Westermarck, T., and Knuutila, S. 1980. Chromosomal effects of sodium selenite *in vivo*. III. Aberrations and sister chromatid exchanges in Chinese hamster bone marrow. Hereditas *93*, 101–105.
18. Ray, J. H., and Altenburg, I. W. 1982. Sister-chromatid exchange induction by sodium selenite plasma protein-bound selenium is not the active SCE-inducing metabolite of Na_2SeO_3. Mutat. Res. *102*, 285–296.

Inhibition of Protein Synthesis and Anticarcinogenicity of Selenium Compounds

L. N. Vernie
A. van Leewenhoekhuis

For centuries selenium, named after the moon, has left its footprints on earth. It followed the path from a very toxic, although at first not recognized as such, carcinogenic element and thus an ungraceful substance, to an essential trace element with antioxidant and anticarcinogenic properties and other potential health benefits. In view of its beneficial properties, we attempted to counteract, with selenium, peroxidation of the unsaturated fatty acids of liver endoplasmic reticulum membranes and reduction of amino acid incorporation observed in the livers of rats treated with hepatocarcinogens (1). Besides the effect of addition of sodium selenite, via food or drinking water, on the carcinogen-treated animals, the effect of selenite on an *in vitro* amino acid-incorporating system was tested as well. In the cell-free system consisting of free polyribosomes and a 150,000 g supernatant from rat liver fortified with cofactors, a complete loss of amino acid incorporation was observed at nanomole quantities of sodium selenite per milliliter of incubation mixture (1). To our knowledge, Everett and Holley (2) were the first who noted an inhibition of amino acid incorporation in a rat liver system by sodium selenite.

CELL-FREE SYSTEM

In a cell-free system inhibition of amino acid incorporation by selenite only became apparent if a thiol, such as glutathione (GSH), was present in the incubation mixture. In a more purified system, selenite was not inhibitory at all (1). When GSH reacts with selenite, two products are formed which have been identified by Ganther as oxidized glutathione, GSSG, and selenodiglutathione, GSSeSG (3). Separation of these two compounds showed GSSeSG to be inhibitory at very low concentrations, 20 nmol/ml (Fig. 1). In the system used, it was shown that elongation factor 2, EF-2, was blocked by GSSeSG (4); neither the peptidyltransferase reaction, elongation factor-1 (EF-1), nor the polyribosomes or the cofactors were affected (4). This indicates that by a reaction between GSSeSG and EF-2 this enzyme is inactivated. The inactivation could be reversed by addition of glutathione reductase and NADPH (4). Additional experiments demonstrated that GSSeSG inhibited the ribosome-dependent GTPase activity of EF-2.

In a cell-free system derived from *Escherichia coli,* amino acid incorporation was not changed by GSSeSG, indicating that the corresponding bacterial elongation factor G (EF-G) is not blocked by GSSeSG (5). All experiments point to a specific inactivation of EF-2 by GSSeSG. Glutathione is not the only thiol that reacts with selenite.

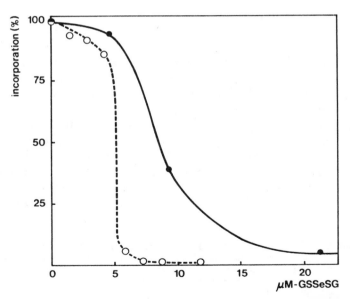

FIG. 1. Inhibition of amino acid incorporation in a cell-free system (●———●) and inhibition of protein synthesis in 3T3-f cells (○---○) as a function of the GSSeSG concentration.

In general, monothiols, formula RSH, react with selenite according to the equation

$$4\text{ RSH} + \text{H}_2\text{SeO}_3 \rightarrow \text{RSSeSR} + \text{RSSR} + 3\text{H}_2\text{O}$$

This reaction, originally proposed by Painter (6), shows the formation of selenotrisulfides, RSSeSR, and disulfides, RSSR (7,8). Also with vicinal dithiols a reaction with selenite has been described. In this case it is stated that more complex and stable reaction products are formed with a linkage of one selenium atom to three (9) or four (9,10) sulfur atoms and hydroxyl groups (9). Several reaction products of selenite with monothiols (glutathione, cysteine, sodium thioglycolate, 2-mer-captoethanol, thiophenol, cysteamine, ethanethiol, 2-propanethiol, and 1-butanethiol) and one dithiol (2,3-dimercapto-1-propanol) were prepared by us; they all proved to be very potent inhibitors of amino acid incorporation in the cell-free system. It was concluded, therefore, that the inhibition of amino acid incorporation depends on the Se moiety of the compounds irrespective of the side group. The several compounds tested differed slightly in their dose–response curves; the concentrations for maximal inhibition ranged from 6 to 20 nmol/ml (11). A summary of the results is presented in Table 1.

INTACT CELLS

The observation that all reaction products of selenite with thiols tested so far are inhibitory in a cell-free system prompted a study of their effect on cultured cells. First, the effect of GSSeSG on 3T3-f cells was examined. It was observed that GSSeSG inhibits the incorporation of [³H]leucine into protein by 3T3-f cells (Fig. 1). This inhibition could not be reversed by removing GSSeSG (refreshing the culture medium), was correlated with the uptake of GSSeSG, and led to cell death (5). In the next experiments with P815 mastocytoma and L1210 leukemic cells, various unseparated reaction products of selenite with thiols were used. In Table 2, the results with intact cells are summarized. In

TABLE 1. Effects of Sodium Selenite and Its Reaction Products with Glutathione and Other Thiols on the Amino Acid Incorporation in a Cell-Free System Derived from Rat Liver

Selenite itself is not inhibitory
Reaction products of selenite with thiols are inhibitory
GSSeSG specifically inactivates elongation factor 2, EF-2
The ribosome-dependent GTPase activity of EF-2 is blocked
Inactivated EF-2 can be reactivated by glutathione reductase and NADPH

TABLE 2. Effects of Sodium Selenite and Its Reaction Products with Thiols
on Intact Cells

Leucine incorporation is blocked
The same reaction product has different effects on different cell lines
Different reaction products have different effects on the same cell type
Incubation of the cells with the reaction products leads to cell death
Selenite itself is only inhibitory after a longer incubation period and at higher concentra-
  tions than the reaction products
In a bacterial system (*E. coli*), no effect has been found

---

contrast to the cell-free system, a more pronounced difference in inhibitory effect was observed with the various compounds; moreover, a
variance in effect of the same product tested on several cell lines was
noted (*11*). If the incorporation of leucine into proteins was taken as a
parameter, the inhibitory action of GSSeSG, for example, was highest
on L1210 cells, followed by P815 and the 3T3-f cells. The differences in
dose–response curves of the various products on L1210 cells were explained by a difference in uptake of the compounds; the nonpolar reaction products showed the highest inhibitory effect.

ASCITES TUMOR CELLS IN MICE

The reaction products of selenite and each of various thiols were
tested next on *in vivo* ascites tumor cells. The purified reaction products, GSSeSG and selenodicysteine. CySSeSCy, were studied first. Intraperitoneal injections of these two compounds in mice, 1 hr after the
inoculation of the tumor cells, inhibited tumor growth and increased
the life span of treated as compared with untreated control mice up to
6 days (*12*). As the doubling time of the tumor cells in mice is about 18
hr, such a survival suggests a considerable tumor cell death. This
assumption could be confirmed by cell countings in the ascites fluid of
the mice (*12*).

In mice injected with a relatively high dose of GSSeSG (0.83 μmol
per mouse), the quantity of tumor cells was also counted at the time of
death. It was observed that in this case death did not correlate with the
amount of cells in the abdomen. This suggested that the higher doses
of GSSeSG, together with a restricted number of tumor cells or lytic
products of the killed cells, were lethal to the mice (*12*). The percentage of tumor cell kill was also demonstrated by staining with Trypan
Blue. Following inoculation of the tumor cells, GSSeSG was injected 1
hr or 24 hr later, and the percentage dead cells was determined 1 day
thereafter. This resulted in a high percentage, 97 and 89%, respectively, of cells which could be stained by Trypan Blue. The outcome

showed that both the increased survival time of the treated over the control mice and the lower amount of tumor cells in the ascites fluid of treated mice was due to cell kill and not to a reduction in proliferation of the tumor cells.

In one experiment set up to compare the survival time of mice injected with a single high dose of GSSeSG (1.0 μmol) with that of mice who received the same amount of GSSeSG, but divided over 3 successive days, the high single dose turned out to be most toxic; 4 out of 5 mice died even before the control mice. However, one mouse lived for 387 days. In the second group, 4 mice survived the control mice for from 1 to 4 days, but 1 mouse lived for a considerably longer period of time: 189 days. Eventually, the long-lived survivors died of the ascites tumor. Such remarkably long survival times have also been observed in another experiment with DBA-II mice inoculated with L1210 leukemic cells. In this system, reaction products of selenite with either one of several thiols, viz., glutathione, 2,3-dimercapto-1-propane sulfonic acid, ethanethiol, and β-mercaptoethanol, were tested. Since the reaction products of these compounds showed considerable differences in effect on tissue culture cells (supra), we hoped that some of them would preferentially kill tumor cells *in vivo*. The results of these experiments were as follows:

Control mice inoculated with L1210 tumor cells died between day 12 and 18, averaging 14 days. The various reaction products gave rise to different survival times (Table 3). The reaction product of selenite with β-mercaptoethanol turned out to be very toxic, and the mice all died before the control mice. With the other three reaction products some mice lived for considerably longer periods of time, up to 1 year. All these mice died eventually from the ascites tumor, and when ascites fluid from these mice was inoculated into other mice, the latter quickly developed ascites tumor cell growth and died from the tumor. The results show that certain selenium compounds very effectively counteract tumor cell growth *in vivo*. The spread in individual survival

TABLE 3. Longevity of DBA-II Mice Inoculated with L1210 Ascites Tumor Cells and Injected, 1 Hour Later, with Unseparated Reaction Products of Selenite and Thiols[a]

Thiol	Survival (days)				
Control	12	12	14	14	18
Glutathione	14	32	326	344	354
2,3-Dimercapto-1-propane sulfonic acid	17	21	317	323	337
Ethanethiol	17	17	18	20	273
β-Mercaptoethanol	3	3	3	7	7

[a] An amount corresponding to 0.45 μmol Se per mouse was given.

times remains puzzling. When GSSeSG was injected after more than one day after the inoculation of the tumor cells, it had no effect on survival. This might be due to a smaller susceptibility of the tumor cells at this stage, but might also be due to the invation of blood cells in the abdomen of mice, which can be observed some days after the inoculation of the tumor cells (infra). This raises an important question. Do the results discussed so far (inhibition of amino acid incorporation in a cell-free system and in intact cells, and the inhibition of proliferation of ascites tumor cells *in vivo*) bear on an anticarcinogenic effect of the selenium compounds, or is the effect due to the antioxidant action of selenium per se via the enzyme glutathione peroxidase (*13,14*), and is it as a radical scavenger (*15*) or are other mechanisms involved? Recent results obtained by others suggested that the anticarcinogenic action of selenium is not mediated by its antioxidant function in lipid peroxidation (*16,17*), and the possibility that selenium decreased the ability of cells to proliferate through other mechanisms has been put forward (*18,19*). Inhibition of protein synthesis might be involved.

Poirier and Milner (*20*) and Milner and Hsu (*21*) tested the inhibitory effects of several selnium compounds on the growth of L1210 and Ehrlich ascites tumor cells *in vitro* and *in vivo*. Sodium selenite and selenium dioxide were most effective and it is probable that these compounds reacted with thiols such as glutathione or cysteine present in the biological system, and that the corresponding selenotrisulfides were the ultimate inhibitors. This hypothesis is supported by the results in a subsequent paper of Poirier and Milner (*22*) in which they compared the antitumorigenic action of Na_2SeO_3 and GSSeSG. In some experiments, the effect of GSSeSG was more outstanding than that of Na_2SeO_3. They also reported that Na_2SeO_3 when administered in the diet increased the survival time of mice inoculated with Ehrlich ascites tumor cells and reduced solid Ehrlich tumor growth (*22*). Previously, Ip *et al.* (*23*) have reported that selenium supplementation inhibited the growth of the transplantable MT-W9B rat mammary tumor.

RED BLOOD CELLS

Under physiological conditions, it is very probable that if selenite is converted into GSSeSG, a further reduction to GSSeH will occur (*3*). It has been shown that intravenously injected selenite is rapidly taken up by the erythrocytes (*24*). From *in vitro* experiments it was originally postulated that after the uptake of selenite by the red blood cells, selenium was expelled as GSSeSG (*25*). Afterward Gasiewicz and

Smith showed that not GSSeSG, but another reaction product of selenium metabolism is released (26).

In our experiments, red blood cells from humans or rats were separated from plasma and buffy coat, washed twice, and, after addition of plasma, preincubated with glucose, according to the method of Gasiewicz and Smith (26). GSSeSG was added and, following incubation, the red blood cells and plasma were again separated. After addition of this plasma to intact L1210 cells, no inhibition of protein synthesis was observed. However, after incubation of GSSeSG in plasma alone, a strong reduction by addition of plasma on protein synthesis in intact L1210 cells was still found (Table 4).

Next, the uptake of ^{75}Se-labeled GSSeSG by erythrocytes was studied, and it was found that there was a partition between the amounts of selenium recovered in the plasma fraction and in the erythrocyte fraction (Fig. 2). By increasing the amount of GSSeSG added, a plateau value of the quantity of selenium in plasma was observed, and selenium became mainly associated with the red blood cells. Besides ^{75}Se-labeled GSSeSG, ^{35}S-labeled GSSeSG was studied. In the latter case, all ^{35}S radioactivity was recovered in the plasma. This underlines the remark of Gasiewicz and Smith that "the term uptake and release as used are not to be interpreted as indicating active transport, but simply describe the alteration in distribution of selenium between erythrocytes and plasma" (26). From these experiments our conclusion must be that GSSeSG is metabolized by the erythrocytes and that the resulting product is not inhibitory to protein synthesis in intact cells. In a control experiment, the fate of GSSeSG after incubation in plasma alone was examined. To this end, both [75]Se- and ^{35}S-labeled GSSeSG were incubated with plasma, and the incubation mixture was

TABLE 4. Effect of Incubation of GSSeSG in Plasma with or without Erythrocytes on the Inhibition of Plasma on [14C]Leucine Incorporation in Intact L1210 Cells

	Incorporation (%)	
GSSeSG (μM)[a]	With erythrocytes	Without erythrocytes
0	100	100
40	100	3
50	100	3
100	93	3

[a] GSSeSG, in the concentration indicated, was incubated for 7 min in plasma alone or in plasma with erythrocytes (human), according to the method of Gasiewicz and Smith (26). After incubation, plasma and erythrocytes were separated, and from the plasma fraction 0.1 ml was added to 1 ml culture medium with 1×10^6 intact L1210 cells. The 100% incorporation in the control experiment was 1935 dpm.

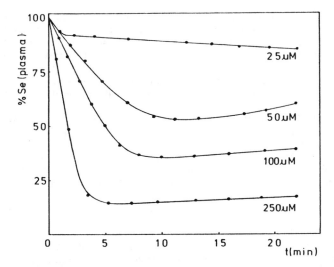

FIG. 2.   Percentage of original Se from 75Se-labeled GSSeSG in plasma. GSSeSG, in the concentrations indicated, was incubated in rat plasma with erythrocytes according to the method of Gasiewicz and Smith (26). After incubation for various times, plasma and erythrocytes were separated, and the percentage of radioactivity in plasma was determined.

fractionated over Sephadex G-25. The selenium label was eluted with the protein peak, in the void volume, while the sulfur label was retarded. Therefore the inhibitory action by plasma, incubated with GSSeSG, on protein synthesis in intact L1210 cells was not due to GSSeSG still present, since the latter compound was degraded. After testing the variously eluted fractions for an inhibitory effect, it was shown that the section of the protein peak was inhibitory. It could have been possible that elemental selenium was formed and was the blocking agent, but we have previously shown that elemental selenium was not inhibitory (11). Therefore, we are left with an unknown product that reduces protein synthesis in intact L1210 cells, and we wonder whether the anticarcinogenic action of selenium is related to this product.

REFERENCES

1. Vernie, L. N., Bont, W. S., and Emmelot, P.   1974.   Inhibition of *in vitro* amino acid incorporation by sodium selenite. Biochemistry *13*, 337–341.

2. Everett, G. A., and Holley, R. W.   1961.   Effect of minerals on amino acid incorporation by a rat liver preparation. Biochim. Biophys. Acta *46*, 390–391.
3. Ganther, H. E.   1971.   Reduction of the selenotrisulfide derivative of glutathione to a persulfide analog by glutathione reductase. Biochemistry *10*, 4089–4098.
4. Vernie, L. N., Bont, W. S., Ginjaar, H. B., and Emmelot, P.   1975.   Elongation factor 2 as the target of the reaction product between sodium selenite and glutathione (GSSeSG) in the inhibiting of amino acid incorporation *in vitro*. Biochim. Biophys. Acta *414*, 283–292.
5. Vernie, L. N., Collard, J. G., Eker, A. P. M., de Wildt, A., and Wilders, I. T.   1979.   Studies on the inhibition of protein synthesis by selenodiglutathione. Biochem. J. *180*, 213–218.
6. Painter, E. P.   1941.   The chemistry and toxicity of selenium compounds, with special reference to the selenium problem. Chem. Rev. *28*, 179–213.
7. Ganther, H. E.   1968.   Selenotrisulfides. Formation by the reaction of thiols with selenious acid. Biochemistry *7*, 2898–2905.
8. Kice, J. L., Lee, T. W. S., and Pan, S.   1980.   Mechanism of the reaction of thiols with selenite. J. Am. Chem. Soc. *102*, 4448–4455.
9. Czauderna, M., and Samochocka, K. 1982.   Studies on the reactions of selenite ion with 1,2-dimercaptoethane or thioacetic acid. Tetrahedron *38*, 2421–2423.
10. Friedheim, E. A. H.   1970.   Selenium compounds. U.S. Pat. 3,544,593.
11. Vernie, L. N., de Vries, M., Karreman, L., Topp, R. J., and Bont, W. S.   1983.   Inhibition of amino acid incorporation in a cell-free system and inhibition of protein synthesis in cultured cells by reaction products of selenite and thiols. Biochim. Biophys. Acta *739*, 1–7.
12. Vernie, L. N., Homburg, C. J., and Bont, W. S.   1981.   Inhibition of the growth of malignant mouse lymphoid cells by selenodiglutathione and selenodicysteine. Cancer Lett. *14*, 303–308.
13. Rotruck, J. T., Pope, A. L., Ganther, H. E., Swanson, A. B., Hafeman, D. G., and Hoekstra, W. G.   1973.   Selenium: Biochemical role as a component of glutathione peroxidase. Science *179*, 588–590.
14. Flohé, L., Günzler, W. A., and Schock, H. H.   1973.   Glutathione peroxidase: A selenoenzyme. FEBS Lett. *32*, 132–134.
15. Tappel, A. L.   1965.   Free-radical lipid peroxidation damage and its inhibition by vitamin E and selenium. Fed. Proc., Fed. Am. Soc. Exp. Biol. *24*, 73–78.
16. Medina, D., Lane, H. W., and Tracey, C. M.   1983.   Selenium and mouse mammary tumorigenesis: An investigation of possible mechanisms. Cancer Res. *43*, Suppl., 2460s–2464s.
17. Ip, C.   1983.   Selenium-mediated inhibition of mammary carcinogenesis. Biol. Trace Elem. Res. *5*, 317–330.
18. Ankerst, J., and Sjögren, H. O.   1982.   Effect of selenium on the induction of breast fibroadenomas by adenovirus type 9 and 1,2-dimethylhydrazine-induced bowel carcinogenesis in rats. Int. J. Cancer *29*, 707–710.
19. LeBoeuf, R. A., and Hoekstra, W. G.   1983.   Adaptive changes in hepatic glutathione metabolism in response to excess selenium in rats. J. Nutr. *113*, 845–854.
20. Poirier, K. A., and Milner, J. A.   1979.   The effect of various selenocompounds on Ehrlich ascites tumor cells. Biol. Trace Elem. Res. *1*, 25–34.
21. Milner, J. A., and Hsu, C. Y.   1981.   Inhibitory effects of selenium on the growth of L1210 leukemic cells. Cancer Res. *41*, 1652–1656.
22. Poirier, K. A., and Milner, J. A.   1983.   Factors influencing the antitumorigenic properties of selenium in mice. J. Nutr. *113*, 2147–2154.
23. Ip, C., Ip, M. M., and Kim, U.   1981.   Dietary selenium intake and growth of the MT-W9B transplantable rat mammary tumor. Cancer Lett. *14*, 101–107.

24. Sandholm, M.   1973.   The initial fate of a trace amount of intravenously adminis-
    tered selenite. Acta Pharmacol. Toxicol. *33,* 1–5.
25. Jenkins, K. J., and Hidiroglou, M.   1972.   Comparative metabolism of ^{75}Se-sel-
    enite, ^{75}Se-selenate, and ^{75}Se-selenomethionine in bovine erythrocytes. Can. J.
    Physiol. Pharmacol. *50,* 927–935.
26. Gasiewicz, T. A., and Smith, J. C.   1978.   The metabolism of selenite by intact rat
    erythrocytes *in vitro.* Chem.-Biol. Interact. *21,* 299–313.

The Spectrum of Selenium and Other Trace Elements Using Computerized Pattern Recognition for Early Detection of Lung Cancer

Xu Hui-Bi
Li De-Hua
Pan Zhong-Ming
Xiang Shou-Xian

Based on epidemiological studies, laboratory experiments and clinical observations, selenium may have anticarcinogenic effects (*1*). The protective effects of selenium against carcinogenesis are counteracted by some other elements. The presence of an antagonistic effect makes the correlation between selenium and cancer difficult, therefore the study of this effect is important both in theory and in practice. In order to consider the total effect of trace elements, the computerized pattern recognition method for trace elements was used (*2*).

The Yunnan tin mine is one of the areas with a high incidence of lung cancer. Because of its environmental conditions, a new method of formulating trace element spectrums by computerized pattern recognition for early detection of lung cancer was designed. The experiments gave promising preliminary results.

RESEARCH METHOD

Sample

Hair was chosen as the sample because it stores a great deal of information on health and the trace elements in hair are relatively stable. Furthermore, hair samples are easy to obtain. In this study, 117 hair samples were taken from 117 workers: 67 healthy miners, 22 early lung cancer patients, and 28 lung cancer patients.

Measurement of Trace Elements

Epidemiological studies in the Yunnan tin mine showed an inverse correlation between the selenium level and cancer mortality (3). It was also shown that one possible cause of the high incidence of lung cancer is the high arsenic level there. Some antagonistic elements of selenium are also considered. Altogether, eight elements were measured (Se, Zn, Cd, Cr, Cu, Pb, As, Sn) by atomic absorption spectrophotometry and polarography.

Extraction of Information

A total of 88 samples were taken stochastically to be nonlinearly mapped (NLM) onto a characteristic plane so that the distribution area of different patients and healthy miners could be determined; the remaining 29 samples were used for predicting early lung cancer. There were three areas which corresponded to healthy, early lung cancer, and lung cancer patients, respectively.

RESULTS AND DISCUSSION

For prediction purposes, 29 samples were taken (17 healthy miners, 7 early lung cancer patients, and 5 lung cancer patients), among which 86% of the early lung cancer patients fell in the early lung cancer area, so the rate of accuracy of diagnosis was 86%.

On the basis of the above-mentioned investigation, it is possible to propose a new method for computerized prediction of otherwise unknown early lung cancer patients.

REFERENCES

1. Passwater, R. A. 1980. Selenium as Food and Medicine , p. 16.
2. Xu, H.-B. 1983. Mol. Sci. *1* (1), 131–133.
3. Zhu, Y.-J. and Yu, S.-Y. 1982. Chin. J. Oncol. *4* (2), 158.

118

Relationship between Selenium in Hair and Cancers of the Digestive Tract, and a New Method for Determining Selenium in Hair

Zhang Fu-Zheng
Ji Lian-Fang
Wu Ting-Guo
Deng Jia-Qi

INTRODUCTION

Selenium is one of the essential trace elements in the human body. Deficiency of Se in the human body can cause diseases. The Se content in hair samples of patients with cancers of the digestive tract and of healthy men has been determined. The difference between the Se content of hair samples of the cancer patients and of healthy men was significant and throws some light on cancer study in clinical medicine.

At present, neutron-activation analysis, atomic fluorescence spectrometry, gas–liquid chromatography, and atomic absorption spectrometry are generally used to determine Se in hair (*1–3*). The instrumentation for these methods is expensive. Fluorimetry is commonly used in China to determine Se. Zhang *et al.* (*4*) have determined Se by using differential anodic stripping voltammetry (DASV) on a gold electrode. The sensitivity reached 0.20 ppb. A new method for determining Se in human hair has been studied by using DASV on a gold electrode. This method offers advantages in terms of sensitivity, simplicity, and rapidity (from the beginning of digestion to the end of determination of one hair sample takes less than 30 min), and its instrumentation is economical [it costs only several thousand Chinese yuan (renminbi)].

EXPERIMENTS AND RESULTS

Relationship between Hair Selenium and Cancers

The Se content in hair samples of 54 male adult patients with cancers of the digestive tract and 50 healthy men has been determined. The differences in the Se content in hair samples from patients and from healthy men have been compared by Student's t test (Table 1).

A New Method for Determining Selenium in Hair

Preparation of Hair Sample.   Soak the hair sample in a solution of 1% (by weight) washing detergent for about 30 min. Keep the temperature of the solution at about 45°C. Wash with tap water. Then wash four times with deionized water. Put the wet hair sample in a beaker in an oven kept at about 70°C for 24 hr. Cut dried hair into small pieces and store in glass bottles.

Digestion of Hair Sample.   Weigh out 0.30 g of the hair sample and place it in a high quartz cell. Add 3.3 ml digestion solution $(H_2SO_4:HClO_4 = 3:4)$ and 0.1 ml 5% $(NH_4)_2MoO_4$ solution. Place the cell on an electroheated plate kept at about 230°C. It is heated until the color of the digestion solution of the hair sample becomes bright yellow green; then continue heating for another 2 min. Remove it and let it cool, then add a little water and heat the solution for a short time. Transfer the solution into a volumetric flask, then dilute to 25 ml with deionized water.

Determination of Selenium.   Pour the above-mentioned solution into a cell. Deposit for a defined time (3–5 min) at −0.20 V, while the solution is stirred by bubbling nitrogen gas. Cease bubbling nitrogen gas for 15 sec and condition the potential of the working electrode at +0.40 V during this period. Scan the potential of the working electrode from +0.40 V to +1.7 V at 100 mV sec^{+1}. Record the Se stripping peak (Fig. 1) by using an X–Y recorder. Condition the potential of the working electrode at +1.7 V for 15 sec. After adding the standard solution,

TABLE 1. Differences in Hair Se Content in Cancer Patients and Healthy Men Compared by Student's *t* Test

Site of cancer	No. of cases	Hair Se of 50 cancer patients (μg g⁻¹)	Hair Se of 50 healthy men (μg g⁻¹)	*t* value	*P*
Esophagus and cardia	16	0.39 ± 0.02	0.74 ± 0.03	5.8 > 2.66	<.01
Stomach	14	0.42 ± 0.01	0.74 ± 0.03	5.0 > 2.66	<.01
Intestine	24	0.43 ± 0.02	0.74 ± 0.03	6.3 > 2.66	<.01
Digestive tract	54	0.41 ± 0.01	0.74 ± 0.03	9.7 > 2.66	<.01

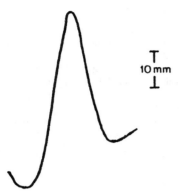

FIG. 1.   Differential anodic stripping curve of Se in human hair. $E_d = -0.20$ V (V·sec·Ag/AgCl); $T_d = 3.0$ min; $V_{scan} = 100$ mV sec^{-1}).

the determination of the Se stripping peak is carried out. The Se content in human hair sample is calculated by means of the standard additions.

Precision and Accuracy

The coefficients of variability and the recovery ratios of the hair samples collected from three men were determined. The results are given in Table 2.

CONCLUSIONS

The difference of the Se content in hair between healthy men and patients with cancers of the digestive tract was highly significant.

An effective method for the analysis of Se content in human hair has been presented. It is suitable to popularize this method for application.

TABLE 2. Coefficients of Variability and Recovery Ratios

Hair sample	Se content in hair sample (μg g^{-1})	Coefficient of variability (%)	Recovery ratio (%)
1	0.64	4.4	98
2	0.50	5.8	98
3	0.54	3.1	96

REFERENCES

1. Morris, J. S., Smith, M. F., Morrow, R. E., Heimann, E. D., Hancock, J. C., and Gall, T. 1982. J. Radioanal. Chem. *69* (1–2), 473–494.
2. Robison, M. F., and Thomson, C. D. 1981. *In* Selenium in Biology and Medicine. J. E. Spallholz, J. L. Martin, and H. E. Ganther (Editors), pp. 283–302. AVI Publishing Co., Westport, CT.
3. Bem, E. M. 1981. Environ. Health Perspect. *37,* 183–200.
4. Zhang, F.-Z., Ji, L.-F., and Deng, J.-Q. 1984. Huaxue Shijie (unpublished research).

119

Observations on Selenium in Human Breast Cancer

Walter M. Lewko
Kenneth P. McConnell

Selenium is an essential dietary trace element. There is good evidence that selenium has anticancer effects (*1,2*). In laboratory animals, dietary selenium inhibits the development of carcinogen- and virus-induced tumors (*3–9*). Other studies suggest that selenium is correlated with a lower incidence of cancer in humans (*1,2,10–17*). The mechanism of selenium's anticancer effect is not clear. Evidence suggests that selenium may be acting systemically to enhance immune response (*18,19*), inhibit the metabolic activation of carcinogens (*20*), or stimulate detoxification and excretion of carcinogens (*21*). Other studies show that selenium influences neoplastic cells directly (*22,23*). Here, we review our findings on serum selenium levels in patients with cancer (*14–17*) and we present results showing direct effects of selenium on the growth of human breast cancer cells in culture (*24*). We also present some preliminary data suggesting that the mechanism of selenium's effect might involve changes in the levels of cellular glutathione (*25*).

MATERIALS AND METHODS

The human breast cancer cell lines used in these studies were obtained from Mason Research Institute, Worchester, Massachusetts through the Breast Cancer Task Force of the National Cancer Institute. The cell lines were maintained at the University of Louisville by the serial passage of initially frozen stocks. The cells were grown in a medium composed of minimum essential medium (MEM, Gibco), 5% bovine calf serum (JR Scientific), and 40 μg/ml gentamycin sulfate (Sigma). Samples of media prepared in this manner contained a basal level of 0.0691 μg Se/ml as determined by the fluorescence method of Hoffman et al. (26) (kindly assayed by Dr. Thomas S. Shearer, University of Oregon). Sodium selenite, selenomethionine, and selenocystine were obtained from Sigma Chemical Company. Trypsinized cells were counted using a hemacytometer.

Protein synthesis was measured in cultured cells by the incorporation of [2-^{3}H]glycine (5 μCi/ml medium, 44 Ci/mmol; Research Products International) into trichloroacetic acid and precipitable protein. The incubations were carried out for 4 hr at 37°C in a 5% CO_2 incubator. Dialyzed calf serum (5%) was added to MEM for the incorporation studies.

Glutathione levels (oxidized and reduced) were measured by the spectrophotometric assay of Tietze (27). Oxothiazolidine carboxylic acid (28) was purchased from Chemical Dynamics Corporation.

Neutron activation analysis for ^{75m}Se (29) was carried out on dialyzed and lyophilized serum samples obtained as previously described (14–17).

Estrogen receptors were measured using a whole-cell ligand exchange assay. Binding studies were carried out on trypsinized cells (75,000) in 1 ml MEM (lacking serum) containing 5 nM [2,4,6,7-^{3}H] estradiol-17β (90.0 Ci/mmol; New England Nuclear). Nonspecific binding was measured in parallel incubations which contained a 200-fold excess of diethylstilbestrol. After incubating for 1 hr at 37°C, the cells were cooled to 4°C, pelleted by centrifugation (800 g, 5 min), and washed twice by resuspension in 2 ml of phosphate-buffered saline (pH = 7.0). Bound radioactivity was extracted into 4 ml of liquid scintillation counting fluor (Budgetsolve, Research Products Internat'l). Samples were counted using a Packard Tri-Carb scintillation counter. Specific binding was calculated as the difference between incubations in the absence and in the presence of excess diethylstilbestrol. Cellular protein was measured in 0.2 M NaOH extracts using the method of Lowry et al. (30) with a bovine serum albumin standard.

RESULTS

Serum selenium levels were measured in dialyzed, lyophilized serum samples by neutron activation analysis. Table 1 shows the serum selenium levels in 18 healthy women compared with 27 women

TABLE 1. Selenium Levels in Serum Samples from Healthy Women and Women with Breast Cancer[a]

Sample (n)	Age mean ± SEM (range)	Se (μg/g dry wt) mean ± SEM (range)
Breast cancer (27)	61.56 ± 1.70 (41–79)	1.287[b] ± 0.046 (0.759–1.734)
Control (18)	62.22 ± 3.02 (43–85)	1.706 ± 0.099 (1.246–4.700)

[a] Adapted from Tables 1 and 2, Ref. 15.
[b] $P = .001$

diagnosed to have breast cancer. The breast cancer patients had a significantly lower concentration of serum selenium. In a separate study (Table 2) which involved 110 patients with carcinoma of many types (Table 3), cancer patients exhibited an average serum selenium level of 1.268 ± 0.034 μg/g dry weight. An average value of 1.481 ± 0.073 μg/g dry weight was observed in the control group of healthy individuals. These studies were extended to analyze the relation between selenium levels and the severity of the disease. The results are presented in Table 4. Patients with low serum selenium levels exhibited higher incidences of metastases, multiple primary tumors, recurrence, and decreased survival time when compared to those patients with higher selenium levels.

In contrast to patients with carcinoma, those afflicted with reticuloendothelial neoplasms had average serum selenium values of 1.76 ± 0.24 μg/g (Table 2). There was a wide range of selenium concentrations in these patients. The mean value was not significantly higher than that of the controls.

In order to determine whether selenium had any direct effects upon tumor cell growth, human breast cancer cells were studied in culture. Figure 1 shows the influence of sodium selenite on the growth of MCF-7 cells. When added to the culture medium, 24 hr after plating, sodium selenite had a biphasic effect on cell growth. Low concentra-

TABLE 2. Selenium Levels in Serum Samples from Healthy Individuals and Patients with Cancer

Sample (n)	Se (μg/g dry wt)
Carcinoma (110)	1.27 ± 0.03[a]
Reticuloendothelial (36)	1.76 ± 0.24
Controls (18)	1.48 ± 0.07

[a] Values represent the mean ± SEM. The difference between carcinoma and controls was significant, $P = .001$.

TABLE 3. Categories of the 110 Carcinomas and the Histologic Types[a]

Pulmonary (37)	
Epidermoid	26
Oat cell	8
Adenocarcinoma	3
Orolaryngeal (24)	
Oral cavity, epidermoid	17
Larynx, epidermoid	7
Gastrointestinal (18)	
Esophagus, epidermoid	4
Pancreas, adenocarcinoma	2
Large intestine, adenocarcinoma	12
Genitourinary (14)	
Kidney, renal cell	3
Ureterovesicle, transitional cell	6
Prostate, adenocarcinoma	4
Testis, seminoma	1
Miscellaneous (17)	
Skin, basal cell	7
Epidermoid	5
Melanoma	1
Breast, adenocarcinoma	1
Thyroid, adenocarcinoma	2
Primary unknown	1

[a] Adapted from Table 1, Ref. 16.

tions, 10^{-8} to 10^{-6} M, stimulated cell growth. Higher concentrations inhibited growth. At 10^{-4} M sodium selenite was cytotoxic and caused cell death. Figure 2 is a photomicrograph showing MCF-7 cells growing in culture. Under the plating conditions used in this particular experiment, the cells were near confluence after 5 days in culture (Fig. 2A). Exposure to 10^{-4} M sodium selenite (Fig. 2B) for 4 days had obvious morphological effects. Cell density was markedly lower. A large percentage of the cells were rounded up, loosely attached to the culture surface or floating in the culture medium. Greater than 90% of

TABLE 4. Serum Selenium Levels and the Biological Behavior of the Tumor[a]

Patient group	Number evaluated	Se range ($\mu g/g$ dry wt)	Metastasis Local	Metastasis Distal	Multiple primaries	Recurrences	Died
I	34	0.426–1.125	21	16	9	7	17
II	31	1.330–1.403	19	14	8	3	7
III	32	1.408–2.432	14	8	1	1	4

[a] Adapted from Tables 2 and 3, Ref. 16. Patients were analyzed in three groups based on serum selenium values at the time of surgery.

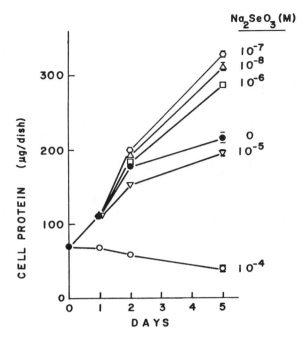

FIG. 1. Influence of sodium selenite on the growth of MCF-7 cells. Twenty-four hours after plating (day 0), sodium selenite was added to the cultures at the indicated concentrations. Protein measurements were carried out on the attached (viable) cells.

the detached cells were dead, as determined by the uptake of Trypan Blue dye. Conversely, more than 90% of the attached cells were still alive.

In the diet and circulating in blood, selenium is generally found as one of several organic derivatives, not as inorganic selenite. We were interested in determining whether organic selenium had any effects on cell growth. Figure 3 shows the influence of selenomethionine and selenocystine. At 5×10^{-4} M, selenomethionine inhibited cell growth while 10^{-3} M was cytotoxic and decreased total cell number. Selenocystine inhibited cell growth and was cytotoxic at 10^{-5} M.

Figure 4 shows the effects of the three forms of selenium on protein synthesis. With increasing concentration, each of the selenium compounds inhibited the incorporation of [³H]glycine into acid-precipitable protein. Selenite was the most potent in terms of effective concentration, followed by selenocystine and selenomethionine, with

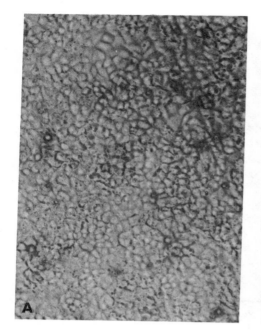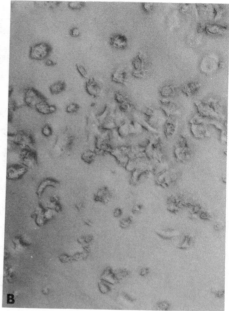

FIG. 2.   Photomicrographs of MCF-7 cells in culture. (A) After 5 days in culture, these cells exhibited near confluent growth. Low concentrations of sodium selenite (less than $10^{-6}\,M$) stimulated growth rate, but did not have any grossly observable effects on cellular morphology. (B) Cells treated for 4 days with $10^{-4}\,M$ sodium selenite were lower in density. Many cells appeared to be detaching from the culture surface or floating in the medium. More than 90% of the detached cells were dead, as determined by the uptake of Trypan Blue dye.

incorporation equaling half control values at concentrations of approximately $10^{-6}\,M$, $10^{-4}\,M$, and $10^{-3}\,M$, respectively.

Table 5 shows the influence of selenium on estrogen binding capacities in ZR75-1 and MCF-7 cells. Both cell types responded similarly to selenite in terms of growth. Low concentrations of selenium stimulated estrogen binding, whereas the higher concentrations which inhibited growth depressed binding.

The effects of selenium on growth and glutathione are shown in Fig. 5. Glutathione levels were elevated in cells treated with the low concentrations of selenium which stimulated growth. At higher concentrations of selenium, cellular glutathione levels were decreased in a dose-dependent manner.

Further evidence for a relationship between selenium and glutathione in cell growth came from experiments with oxothiazolidine

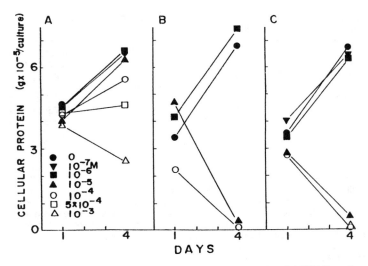

FIG. 3.   Influence of selenium on the growth of MCF-7 cells. Selenium was added to cultures in the form of (A) selenomethionine, (B) selenocystine, or (C) sodium selenite. Cellular protein was measured 4 days later.

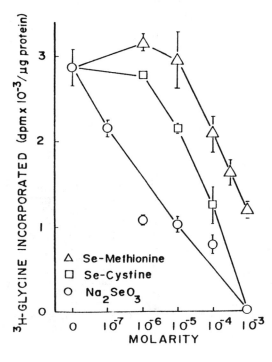

FIG. 4.   Protein synthesis in MCF-7 cells treated with selenium. Selenium was added to the cultures in the form of sodium selenite, selenocystine, or selenomethionine. Twenty-four hours later, protein synthesis was measured by the 4-hr incorporation of [³H]glycine into acid-precipitable protein.

TABLE 5. Influence of Sodium Selenite on Estrogen Binding Capacity in Human Breast Cancer Cells

Cell type	Sodium selenite (M)	Estrogen binding capacity[a] (fmol bd/mg protein)
ZR-75-1	0	98.0 ± 4.0
	5.8×10^{-7}	127.9 ± 11.4[b]
	5.8×10^{-5}	79.5 ± 5.3[c]
MCF-7	0	247
	1×10^{-8}	315
	5×10^{-6}	244
	5×10^{-5}	118

[a] Cultures were exposed to the indicated concentrations of sodium selenite for 24 hr. Viable, attached cells were trypsinized and estrogen receptors were measured by whole cell assay. Values for ZR-75-1 cells represent the mean ± SE for four separate experiments. The values for the MCF-7 cells represent the average of quadruplicate determinations in one experiment.
[b] $P = .046$, relative to control.
[c] $P = .025$, relative to control.

carboxylic acid (OTC). Figure 6 shows that OTC enhances cellular glutathione levels in MCF-7 cells, as has been shown to occur in liver (28). Figure 7 shows the influence of selenium and OTC on MCF-7 cell growth. Increasing concentrations of OTC stimulated growth, with an optimum concentration of 5×10^{-5} M. Sodium selenite, 10^{-4} M, in-

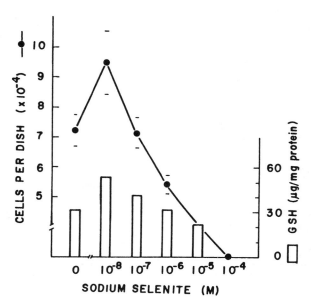

FIG. 5. Influence of selenium on growth and glutathione in MCF-7 cells. Twenty-four hours after plating, sodium selenite was added to the cultures (20,000 cells) at the indicated concentrations. Cell numbers and glutathione levels were determined 4 days later.

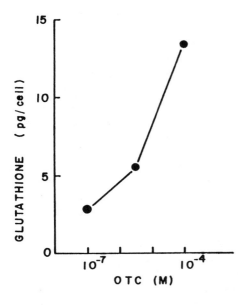

FIG. 6. Influence of oxothiazolidine carboxylic acid on glutathione in MCF-7 cells. Twenty-four hours after plating, oxothiazolidine carboxylic acid was added to cultures at the indicated concentrations. Glutathione levels were measured 24 hr later.

hibited growth. However, increasing concentrations of OTC added to the medium appeared to reverse the inhibitory effects of selenium. This suggested the possibility that selenium might be influencing cell growth by altering glutathione levels.

DISCUSSION

It is an intriguing possibility that low levels of serum selenium measured in cancer patients were responsible for the development and

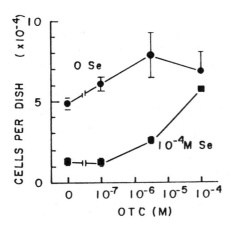

FIG. 7. Influence of sodium selenite and oxothiazolidine carboxylic acid on MCF-7 cell growth. Twenty-four hours after plating, oxothiazolidine carboxylic acid was added to cultures (20,000 cells) alone or in the presence of 10^{-4} M sodium selenite. Cell counts were taken 72 hr later.

severity of the disease. Clearly, animal studies have shown that dietary selenium (albeit abnormally high) has a protective effect in various chemically and virally induced cancers. Our interest in breast cancer prompted us to examine the effects of selenium on the growth of several types of breast cancer cells that grow in culture. We sought to determine whether added selenium might influence tumor development by directly affecting cell growth. We also wanted to use *in vitro* models to determine the molecular mechanism whereby selenium might exert any direct effects.

Three types of human breast cancer cell lines (MCF-7, ZR75-1, MDA-231) were examined. Similar growth effects were realized. Low levels of added sodium selenite, less than 10^{-6} M, tended to stimulate growth, protein synthesis, and DNA synthesis. Higher doses of selenite, greater than 10^{-5} M, tended to inhibit cell growth. Cytotoxicity was observed at concentrations greater than 10^{-4} M.

It is well known that the growth of certain breast cancers is sensitive to the endocrine environment (*31*). Endocrine therapy (e.g., the administration of the antiestrogen tamoxifen) has been used to treat breast cancer. The estrogen receptor is an established marker for endocrine responsiveness (*32*). We sought to determine whether selenium altered estrogen receptors as part of its anticancer effect. Low concentrations of selenite stimulated the binding of estradiol, whereas high concentrations depressed binding. These results suggested that selenium is a potential modifier of response to estrogens. This effect could be of particular importance to the induction of breast cancer, since induction appears to be very sensitive to estrogens (*33*). Nonetheless, altered estrogen receptor levels do not appear to be essential to the selenium response we observed. MDA-231 cells lack estrogen receptors, yet responded similarly with decreased growth when subjected to the high concentrations of sodium selenite.

We measured the influence of selenium on glutathione levels. Low concentrations of selenium which stimulated growth enhanced glutatione, whereas higher doses which inhibited growth depressed glutathione (similar results have been observed in primary cultures of rat mammary tumor cells). The effect of selenium on glutatione could be the result of altered metabolism (changes in the enzymes involved). This remains to be determined. Alternatively, depressed glutathione levels could be the result of the formation of seleno derivatives of glutathione. Ganther (*34*) has studied the synthesis of seleno analogs of glutathione. Vernie *et al.* (*35*) reported that the selenotrisulfide analog of glutathione inhibits protein synthesis. Poirer and Milner (*36*) have proposed that seleno derivatives of glutathione may be responsible for inhibition of tumor cell growth. Glutathione has an important role in cell division and the general metabolic well-being of cells (*37*). Our studies using oxothiazolidine carboxylic acid suggest that glutathione's levels were critical and that the replenishment of glutathione in selenium-treated cells was capable of at least partially reversing selenium's inhibitory effects.

ACKNOWLEDGMENTS

We wish to thank Dr. A. J. Blotcky and W. L. Broghamer for their contributions to these studies and Mr. Danny E. Winn for his excellent technical assistance. This project was funded in part by American Cancer Society Institutional Research grant IN-111 and DHHS NIH grant CA 32240.

REFERENCES

1. Griffin, A. C. 1979. Role of selenium in the chemoprevention of cancer. Adv. Cancer Res. *19*, 419–442.
2. Schrauzer, G. N. 1979. Trace elements in carcinogenesis. Adv. Nutr. Res. *2*, 219–244.
3. Clayton, C. C., and Baumann, C. A. 1949. Diet and azo dye tumors: Effect of diet during a period when the dye is not fed. Cancer Res. *9*, 575–582.
4. Greeder, G. A., and Milner, J. A. 1980. Factors influencing the inhibitory effect of selenium on mice inoculated with Ehrlich ascites tumor cells. Science *309*, 825–827.
5. Milner, J. A., and Hsu, C. Y. 1981. Inhibitory effects of selenium on the growth of L1210 leukemic cells. Cancer Res. *41*, 1652–1656.
6. Ip, C. 1981. Factors influencing the anti-carcinogenic efficacy of selenium on dimethylbenz[*a*]anthracene-induced mammary tumorigenesis in rats. Cancer Res. *41*, 2683–2686.
7. Ip, C., Ip, M. M., and Kim, U. 1981. Dietary selenium intake and growth of the MTW-9B transplantable rat mammary tumor. Cancer Lett. *14*, 101–107.
8. Schrauzer, G. N., White, D. A., and Schneider, C. J. 1976. Inhibition of the genesis of spontaneous mammary tumors in C_3H mice: Effects of selenium and selenium-antagonistic elements and their possible role in human breast cancer. Bioinorg. Chem. *6*, 265–270.
9. Thompson, H. J., and Becci, P. J. 1980. Selenium inhibition of *N*-methyl-*N*-nitrosourea-induced mammary carcinogenesis in the rat. JNCI, J. Natl. Cancer Inst. *65*, 1299–1301.
10. Schrauzer, G. N. 1981. Selenium and cancer: Historical developments. *In* Selenium in Biology and Medicine. J. E. Spallholz, J. L. Martin, and H. E. Ganther (Editors), pp. 98–102. AVI Publishing Co., Westport, CT.
11. Shamberger, R. J., and Frost, D. V. 1969. Possible protective effect of selenium against human cancer. Can. Med. Assoc. J. *100*, 682.
12. Shamberger, R. J., Tytko, S. A., and Willis, C. E. 1976. Antioxidants and cancer. VI. Selenium and age-adjusted human cancer mortality. Arch. Environ. Health *31*, 231–235.

13. Shamberger, R. J., and Willis, C. E. 1971. Selenium distribution and human cancer mortality. CRC Crit. Rev. Clin. Lab. Sci. 2, 211–221.

14. McConnell, K. P., Broghamer, W. L., Blotcky, A. J., and Hart, O. J. 1975. Selenium levels in human blood and tissues and in disease. J. Nutr. 105, 1026–1031.

15. McConnell, K. P., Jager, R. M., Higgins, P. J., and Blotcky, A. J. 1978. Serum selenium levels in patients with and without breast cancer. In Nutrition and Cancer. J. Van Eys, M. S. Seelig, and B. R. Nichols (Editors), pp. 195–197. SP Medical and Scientific Books.

16. Broghamer, W. L., McConnell, K. P., and Blotcky, A. J. 1976. Relationship between serum selenium levels and patients with carcinoma. Cancer 37, 1384–1388.

17. McConnell, K. P., Jager, R. M., Bland, K. I., and Blotcky, A. J. 1980. The relationship of dietary selenium and breast cancer. J. Surg. Oncol. 15, 67–70.

18. Abdullaev, G. B., Gasanou, G. G., Ragimov, R. N., Teplyakova, G. V., Mekhitiev, M. A., and Dzafarov, A. I. 1973. Antineoplastic activity of selenium compounds. Dokl. Akad. Nauk Azerb. SSR 29, 18.

19. Spallholz, J. E. 1981. Selenium: What role in immunity and immune cytotoxicity? In Selenium in Biology and Medicine. J. E. Spallholz, J. L. Martin, and H. E. Ganther (Editors), pp. 103–117. AVI Publishing Co., Westport, CT.

20. Schillaci, M., Martin, S. E., and Milner, J. A. 1982. The effects of dietary selenium on the biotransformation of 7,12-dimethylbenz[a]anthracene. Mutat. Res. 101, 31–37.

21. Jacobs, N. M. 1983. Selenium inhibition of 1,2-dimethylhydrazine-induced colon carcinogenesis. Cancer Res. 43, 1646–1649.

22. Medina, D., and Oborn, C. J. 1981. Differential effects of selenium on the growth of mouse mammary cells in vitro. Cancer Lett. 13, 333–344.

23. Lewko, W. M., and McConnell, K. P. 1982. Biphasic influence of selenium on cell growth and the synthesis of collagen in cultured mammary tumor cells. Fed. Proc., Fed. Am. Soc. Exp. Biol. 41, 623.

24. Lewko, W. M., Winn, D., and McConnell, K. P. 1983. Effect of sodium selenite on growth and protein synthesis in cultured human breast cancer cells. Fed. Proc., Fed. Am. Soc. Exp. Biol. 42, 669.

25. Lewko, W. M., Winn, D. E., and McConnell, K. P. 1984. Influence of selenium on cell growth and glutathione levels in cultured breast cancer cells. Fed. Proc., Fed. Am. Soc. Exp. Biol. 43, 793.

26. Hoffman, I., Westerby, R. J., and Hirdiroglou, M. 1968. Precise fluorometric microdetermination of selenium in agricultural materials. J. Assoc. Off. Anal. Chem. 51, 1039–1042.

27. Tietze, F. 1969. Enzymic method for quantitative determination of nanogram amounts of total and oxidized glutathione. Anal. Biochem. 27, 502–522.

28. Williamson, J. M., Boettcher, B., and Meister, A. 1982. Intracellular cysteine delivery system that protects against toxicity by promoting glutathione synthesis. Proc. Natl. Acad. Sci. U.S.A. 79, 6246–6249.

29. Blotcky, A. J., Arsennault, L. J., and Rack, E. P. 1973. Optimum procedures for the determination of selenium in biological specimens using [77m]Se neutron activation. Anal. Chem. 45, 1050–1060.

30. Lowry, O. H., Roseburgh, N. J., Farr, A. L., and Randall, R. J. 1951. Protein measurement with the Folin phenol reagent. J. Biol. Chem. 193, 205–275.

31. Kennedy, B. J. 1974. Hormonal therapies in breast cancer. Sem. Oncol. 1, 111–130.

32. McGuire, W. L., Carbone, P. O., and Vollmer, E. P. 1975. Estrogen Receptors in Human Breast Cancer. Raven Press, New York.

33. Huggins, C., Grand, L. C., and Brillantes, F. P. 1961. Cancer induced by a single

feeding of polynuclear hydrocarbon and its suppression. Nature (London) *189*, 204–207.

34. Ganther, H. E.   1968.   Selenotrisulfides. Formation by reaction of thiols with selenious acid. Biochemistry 7, 2898–2905.

35. Vernie, L. N., Bant, W. S., and Emmelot, P.   1974.   Inhibition of *in vitro* amino acid incorporation by sodium selenite. Biochemistry *13*, 337–341.

36. Poirier, K. A., and Milner, J. A.   1983.   Factors influencing the antitumorigenic properties of selenium in mice. J. Nutr. *113*, 2147–2154.

37. Kosower, N. S., and Kosower, E. M.   1978.   The glutathione status of cells. Int. Rev. Cytol. *54*, 109–160.

Blood Selenium Level and the Interaction of Copper, Zinc, and Manganese in Stomach Cancer

Kazuo Saito
Takeshi Saito
Toshiyuki Hosokawa
Keizo Ito

Schamberger (1,2) reported the inhibitory effect of Se on carcinogenesis in which Se inhibits a peroxidation that may enhance attachment of the carcinogen to DNA. The significantly lower blood Se level in patients with carcinoma of the colon, liver, pancreas, and stomach, and Hodgkin's disease as compared to normal subjects, which was revealed by Schamberger *et al.* (3) supports his earlier hypothesis. The role of Se for tumorigenicity remains to be clarified; however, it may be very difficult to examine its roles only in epidemiological surveys on cancer patients.

Broghamer *et al.* (4) investigated the histologic type and anatomic sites of the origin of malignant tumors with abnormal serum Se levels. They also evaluated the extent of local and disseminated involvement, the status of hepatic function, the serum protein concentration, the survival time of the patients, the incidence of multiple primary lesions, and the rate of recurrence of the original primary tumor with serum Se levels (5).

Saito *et al.* (6) reported that mean copper (Cu) concentration in blood showed lower levels in the early stages of stomach cancer, but that there were higher Cu levels during the more advanced stages in comparison with normal subjects. Significantly lower levels of zinc (Zn) as well as lower Zn/Cu and Zn/manganese (Mn) ratios were found in the blood of patients with stomach cancer when compared to normal subjects.

The objects of this study were to find out if the blood Se level of patients with stomach cancer varies during the malignant stage, to investigate the interaction of Se to Cu, Zn, and Mn, and to discover what role Se plays in the blood of patients with stomach cancer.

MATERIALS AND METHODS

Thirty-one male stomach cancer patients between the ages of 30 and 80 (mean ± standard deviation, 62.4 ± 11.4 years old) and 14 normal male subjects between the ages of 25 and 48 (mean ± standard deviation, 28.4 ± 6.0 years old) were studied in this investigation. The diagnoses of stomach cancer had been confirmed by histologic examination of a postoperative specimen. The patients were classified into four stages of malignancy, I–IV, according to the rules laid down for surgical and pathologic treatment of stomach cancer by the Studying Committee (6). They consisted of 14 patients at stage I, 4 at stage II, 3 at stage III, and 10 at stage IV. They were also grouped into three "metastatic" categories: nonmetastasis, metastases of the lymph nodes only, and metastases of other organs. The normal male subjects who acted in this investigation as a control group were postgraduate students and staff of the Department of Hygiene and Preventive Medicine, Hokkaido University School of Medicine.

About 7 ml of nonfasting blood samples was drawn by syringe from the vena mediana cubiti of both groups of subjects. Before the operation the stomach cancer patients had not received any special therapy. The samples of whole blood, blood plasma, and erythrocytes were separated from the collected venous blood samples in heparinized tubes and stored at −80°C until use. The Se concentrations in whole blood and in blood plasma were determined essentially by the method of Watkinson (7). Two milliliters of whole blood or blood plasma was dissolved in 10 ml concentrated HNO_3 and 5 ml 70% of $HClO_4$. The mixture was held at 37°C for at least 2 days. A wet digestion with $HClO_4$ and HNO_3 removed the rest of the organic matter. Selenium was then complexed with 2,3-diaminonaphthalene. The complex was extracted and passed into cyclohexane and the fluorescence determined with a spectrofluorophotometer (Shimadzu-RF-500). A standard curve was constructed by measuring the fluorescence of 0, 0.1, 0.2, and 0.4 μg Se. The amount finally measured varied between 0.1 and 0.4 μg, and Se concentrations in the blood were determined by duplicate analyses. When measuring the Cu and Mn of each individual, 0.2 ml of whole blood and blood plasma were diluted 15-fold; for Zn, however, they were diluted 450-fold with 0.1 N HCl. These diluted blood samples were directly analyzed by the standard additions methods, using 10 μl of a sample with a Zeeman-type flameless atomic absorption spectrophotometer (Hitachi Model 170-70), as previously described (8).

RESULTS

Blood Mn, Cu, Zn, and Se Concentrations of Patients with Stomach Cancer

The mean Cu, Zn, Mn, and Se concentrations in 100 ml of whole blood and blood plasma of patients with stomach cancer are presented in Table 1.

The mean whole blood Se concentrations and the standard errors of male stomach cancer patients and normal subjects were 19.3 ± 0.8 and 22.0 ± 0.8 μg/100 ml, respectively. The mean blood plasma Se concentrations and the standard errors of both stomach cancer patients and normal subjects were 11.8 ± 0.5 and 14.1 ± 0.3 μg/100 ml, respectively. The mean whole blood and blood plasma Se levels for the four stages of malignancy and the three groups of metastases are presented in Fig. 1. The Se concentration in the whole blood of the patients showed a significantly lower level than in normal subjects ($P < .01$). The same results were observed at stage I ($P < .05$), at stage II ($P < .05$), and among the nonmetastatic group ($P < .05$). However, no significant difference within the stages of malignancy or between metastatic and nonmetastatic groups was observed. The Se concentration in the blood plasma of the patients showed a significantly lower

TABLE 1. Blood Manganese, Copper, Zinc, and Selenium Concentrations of Male Patients with Stomach Cancer[a]

Element (μg/dl)	Whole blood (31)	Blood plasma (31)
Mn	1.99 ± 0.09	0.72 ± 0.06
Cu	86.2 ± 2.3	90.8 ± 3.2
Zn	669.5 ± 14.3	86.0 ± 3.4
Se	19.3 ± 0.78	11.8 ± 0.46

[a] The results shown are the mean ± SEM. Number of subjects is shown in parentheses.

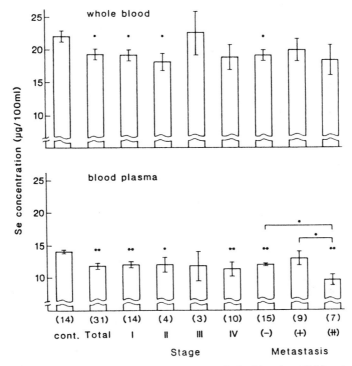

FIG. 1.   Mean Se concentrations in whole blood and blood plasma and standard errors for patients with stomach cancer and normal subjects serving as control. Number of subjects is shown in parentheses. Asterisks above the columns mean a significant difference from the control. Asterisks above the line mean a significant difference between two groups.

level than that of normal subjects ($P < .01$). The same results were observed for stage I ($P < .01$), II ($P < .05$), IV ($P < .01$), the non-metastatic group ($P < .01$), and the group with metastases of other organs ($P < .01$). The group with marked metastases of other organs showed a significantly lower level than those of the nonmetastic group and the group with metastasis of the lymph nodes only ($P < .05$). The correlation of Se concentration present in whole blood with that present in blood plasma of the stomach cancer patients showed a significant tendency at the correlation coefficient of 0.3157 ($P < .10$), which suggests that the Se concentration in the blood plasma reflects the severe state of stomach cancer rather than that of whole blood. In contrast, the correlation of Se concentration in whole blood with that

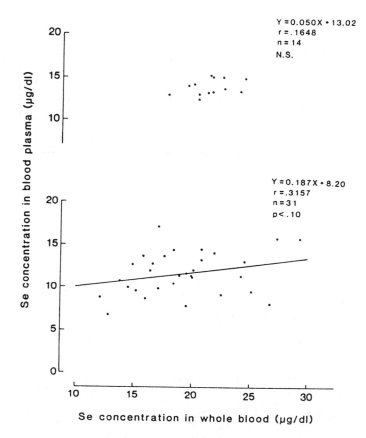

FIG. 2. Relationship of Se concentration in whole blood with that in blood plasma of patients with stomach cancer and of normal subjects.

in blood plasma of normal subjects was not significant at a correlation coefficient of 0.1648 (Fig. 2).

Changes in Blood Mn, Cu, Zn, and Se Concentrations of Patients with Stomach Cancer

Blood Cu, Zn, and Se concentrations decreased during the early stage of malignancy (stage I) in the patients with stomach cancer, but blood Mn concentration increased at the same stage. The blood metal levels during stage II were low, in the order $Zn < Se < Cu < Mn$. The order $Zn < Cu < Mn < Se$ was observed during stage III. These blood metal concentrations during the advanced stage of malignancy (stage IV) showed a decrease in Zn and Se, but an increase in Mn and Cu (Fig. 3).

Relationship of Blood Selenium to Hematocrit and Hemoglobin in Patients with Stomach Cancer

The mean hematocrit and hemoglobin values and their standard errors of the 31 male stomach cancer patients were $40.2 \pm 0.8\%$ and 13.4 ± 0.5 g/dl, respectively. The correlations of Se concentration in both whole blood and blood plasma to the hematocrit and hemoglobin

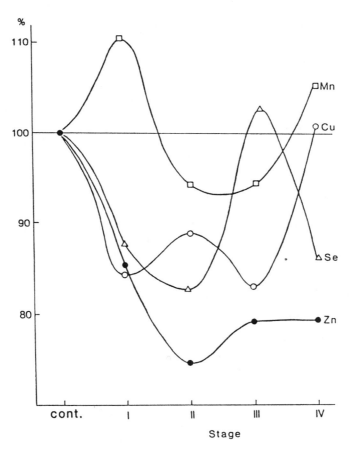

FIG. 3.   Changes in Mn, Cu, Zn, and Se concentration in whole blood of patients with stomach cancer.

of the patients with stomach cancer are shown in Figs. 4 and 5, respectively. A significant correlation between hematocrit and Se concentration in whole blood ($P < .01$) or in blood plasma ($P < .05$) was recognized, but the correlation between hemoglobin and Se concentration in whole blood or in blood plasma was not significant.

Relationship of Blood Se to Blood Mn, Cu, and Zn Concentrations

Correlation coefficients between Se concentration and Mn, Cu, or Zn concentrations in whole blood and in blood plasma are given in Table

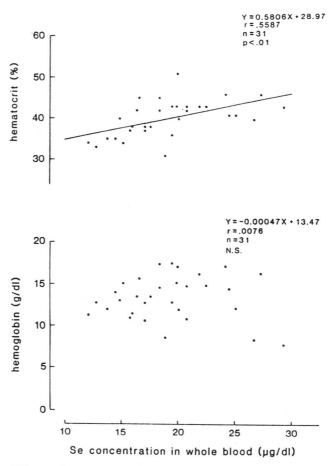

FIG. 4.   Relationship between Se concentration in whole blood and hematocrit or hemoglobin of patients with stomach cancer.

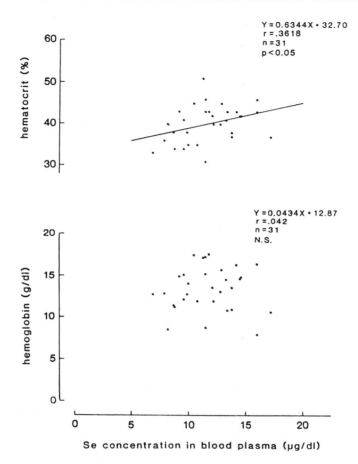

FIG. 5.   Relationship between Se concentration in blood plasma and hematocrit or hemoglobin of patients with stomach cancer.

2. Almost all of the correlation coefficients between blood Se and blood Mn, Cu, or Zn registered negative values, but were not significant except for a significant correlation ($P < .01$) between Se in blood plasma and Mn in whole blood (Fig. 6). We suggest that a low Se concentration in blood plasma with a high Mn concentration in whole blood is one of the characteristics in the patients with stomach cancer.

DISCUSSION

Schamberger *et al.* reported that 12 patients with stomach cancer had a mean whole blood Se level of 15.3 ± 2.10 µg/100 ml and that

TABLE 2. Correlation Coefficients between Blood Selenium Level and Concentrations of Blood Manganese, Copper, or Zinc of Patients with Stomach Cancer[a]

	Whole blood			Blood plasma		
	Mn	Cu	Zn	Mn	Cu	Zn
Se (whole blood)	-0.1794	-0.2754	0.0958	-0.1763	-0.1009	0.2143
Se (blood plasma)	-0.5192*	-0.2285	-0.0085	-0.1204	-0.1700	-0.1072

[a] The values show correlation coefficients. Number of subjects is 31 male stomach cancer patients. Asterisk means a significant correlation between two elements at 1% level.

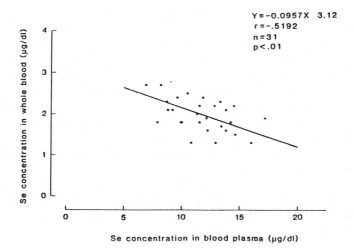

FIG. 6.   Correlation between Se concentration in blood plasma and Mn concentration in patients with stomach cancer.

patients with gastrointestinal cancer or metastases of the gastrointestinal organs had significantly lower blood Se values. They also reported that most patients with carcinoma of the digestive tract (except the rectum) had blood Se values of <18 μg/100 ml, while none of the normals had blood Se values of <18 μg/100 ml. In our results, Se concentrations in whole blood and in blood plasma of the patients with stomach cancer were also significantly lower than those of normal subjects. Of the 14 male normal subjects, 1 showed 17.8 μg/100 ml, and others had blood Se values between 20.0 and 30.8 μg/100 ml. Of 31 male stomach cancer patients, 13 had blood Se values of <18 μg/100 ml, 8 had blood Se values between 18 and 20, and 10 had blood Se values of >20 μg/100 ml. As to Se concentration in blood plasma of the stomach cancer patients, there was a significant difference between the degrees of metastasis. The stomach cancer patients with marked metastases of other organs showed the lowest Se level in blood plasma, and the patients with metastases to the lymph nodes only showed rather high Se values in blood plasma when compared to the groups in the early stage or the advanced severe stage, as Fig. 1 has shown. This characteristic of blood Se is distinct in comparison with Mn, Cu, and Zn, as Fig. 3 has shown. Broghamer et al. (4) pointed out the widely variable serum Se levels in carcinoma. They suggested, first, that low Se levels in patients with carcinoma were more likely to be associated with (1) distant metastasis; (2) multiple primary tumors, which in many instances appeared in different organ systems; (3) multiple recurrences; and (4) a short survival time. Second, they suggested that as selenium levels approach or exceed the mean value for the carcinoma group, then (1) the tumor is more likely to remain confined to the region of its origin; (2) distant metastasis is less likely to occur; and (3) that multiple primary lesions and recurrences will seldom appear. As for the relationship of blood Se levels to blood Mn, Cu, and Zn levels, our study showed that high blood Mn and Cu levels and low blood Zn and Se levels are characteristic of the advanced type of patient with stomach cancer. These blood trace elements at the middle stage of malignancy showed low levels in the stomach cancer patients; i.e., these results suggest that the variability of blood Mn, Cu, Zn, and Se levels is due to the biological behavior of the degrees of malignancy and advance in stomach cancer. Changes in the blood Se and other trace elements shown by the above results may reflect an inherent peculiarity of stomach cancer.

SUMMARY

Blood Se levels in relation to blood Mn, Cu, and Zn levels were measured in male patients with stomach cancer and in normal male subjects. Mean Se concentrations and the standard errors were 19.3 ± 0.8 µg/100 ml for whole blood and 11.8 ± 0.5 µg/100 ml for blood plasma in the patients with stomach cancer. Those of normal subjects were 22.0 ± 0.8 µg/100 ml for whole blood and 14.1 ± 0.3 µg/100 ml for blood plasma. The blood Se levels of the stomach cancer patients were significantly lower than those of normal subjects and these levels could be observed in groups classified according to malignancy and metastasis. Significantly positive correlations between blood Se concentrations and hematocrit, no correlation between blood Se concentration and hemoglobin, and a significantly negative correlation between plasma Se level and Mn concentration in whole blood were noted in the patients with stomach cancer. As to the relationship of blood Se to blood Mn, Cu, and Zn, variation in the inherent pattern of the blood Mn, Cu, Zn, and Se during the advance of stomach cancer was observed, and the changes in these trace elements in blood may possibly reflect the peculiarity of stomach cancer.

ACKNOWLEDGMENTS

The authors are grateful to Dr. Danjo of the Kaisei Hospital for supplying the blood of patients with stomach cancer and to Dr. Yamamoto of the Hokkaido Institute of Public Health for his technical assistance with blood Se measurement.

REFERENCES

1. Shamberger, R. J.   1970.   Relationship of selenium to cancer. I. Inhibitory effect of selenium on carcinogenesis. J. Natl. Cancer Inst. (U.S.) *44*, 931–936.
2. Shamberger, R. J.   1972.   Increase of peroxidation in carcinogenesis. J. Natl. Cancer Inst. (U.S.) *48*, 1491–1497.
3. Shamberger, R. J., Rukovena, E., Longfield, A. K., Tytko, S. A., Deodhar, S., and Willis, C. E.   1973.   Antioxidants and cancer. I. Selenium in the blood of normals and cancer patients. J. Natl. Cancer Inst. (U.S.) *50*, 863–870.
4. Broghamer, W. L., McConnell, K. P., and Blotcky, A. L.   1976.   Relationship between serum selenium levels and patients with carcinoma. Cancer (Philadelphia) *37*, 1384–1388.
5. Shamberger, R. J., Tytko, S. A., and Willis, C. E.   1976.   Antioxidants and cancer. VI. Selenium and age-adjusted human cancer mortality. Arch. Environ. Health *31*, 231–235.
6. Saito, K., Saito, T., Hosokawa, T., Fujimoto, S., and Sasaki, T.   1984.   Changes of blood copper, zinc, and manganese in stomach cancer. Trace Elem. Med. *1*, 24–28.
7. Watkinson, J. H.   1966.   Fluorometric determination of selenium in biological material with 2,3-diaminonaphthalene. Anal. Chem. *38*, 92–97.
8. Saito, K., Sasaki, T., Sato, Y., and Yasuda, H.   1979.   Distribution of trace metals in snowfall and human blood in the northern region of Japan. *In* Trace Substances in Environmental Health—XIII. D. D. Hemphill (Editor), pp. 68–86. Univ. of Missouri Press, Columbia.

Selenium in the Blood of Japanese and American Women with and without Breast Cancer and Fibrocystic Disease

Gerhard N. Schrauzer   Klaus Kuehn
Tammy Schrauzer   Hiroshi Yamamoto
Sherri Mead   Eiji Araki

Selenium suggested itself as a possible breast cancer-protecting micronutrient after its efficacy in preventing virally induced mammary tumorigenesis was demonstrated in animal experiments, and it was shown that the mortalities from female breast cancer in the United States population are lower in regions of the United States naturally high in selenium (Schrauzer and Ishmael 1974). Subsequently, statistically significant inverse associations were reported between the calculated dietary selenium intakes or blood selenium levels and the age-corrected breast cancer mortalities in different countries (Schrauzer and White 1978). To further corroborate this evidence, we decided to compare the selenium concentrations in the blood of Japanese and American women with and without breast cancer. Japanese breast cancer patients represent an interesting group in view of the relatively low incidence of breast cancer among Japanese women, which is commonly attributed to their low dietary fat intakes. However, the higher dietary selenium intakes in the general Japanese population may also be important within this context.

Blood selenium concentrations from patients with newly diagnosed breast cancer and with recurrent disease were analyzed separately.

During the study, blood samples from patients with benign fibrocystic disease became available and were also analyzed.

MATERIALS AND METHODS

Samples of whole blood were obtained from patients with newly diagnosed breast cancer, recurrent breast cancer, and benign fibrocystic disease, and from healthy women at the National Cancer Center, Tokyo and the University of California Cancer Center, San Diego, stabilized with EDTA, and stored in vacuum tubes. For selenium determination, 1-ml samples of homogenized blood were wet-ashed with $HNO_3/HClO_4$. Selenium was determined by the fluorimetric method with 2,3-diaminonaphthalene (Olson 1969), as modified by Chan (1976).

RESULTS

Blood selenium concentrations of healthy Japanese women aged 21–72 years old were significantly higher than those of women with newly diagnosed and recurrent breast cancer (ages 26–77 years) and also higher than the blood selenium concentrations of healthy American women from San Diego, California in similar age groups. However, the blood selenium levels of Japanese women with breast cancer were similar to those of our San Diego breast cancer cases and controls. Blood selenium concentrations of the cases with benign fibrocystic disease were between those of women with and without breast cancer, but the differences were statistically insignificant (Table 1). There was no consistent dependence of blood selenium levels on age in any of the groups.

DISCUSSION

Blood Selenium Concentrations of Healthy Subjects

The observed selenium concentrations in whole blood of the healthy Japanese women of our study are similar to the 0.27 µg/ml previously reported for healthy Japanese adults (Schrauzer et al. 1977). They are somewhat higher than the mean of 0.223 µg/ml reported by other authors (Kurahashi et al. 1980), but these workers wet-ashed their

TABLE 1. Blood Selenium Concentrations of Japanese and American Women with and without Breast Cancer and Fibrocystic Disease

Sample	Japanese			American		
	n	Mean (µg/ml ± SD)	Range	n	Mean (µg/ml ± SD)	Range
Controls	25	0.285	0.230–0.332	14	0.183	0.134–0.211
Fibrocystic disease	10	0.200[a] ± 0.046	0.089–0.233	8	0.142[d] ± 0.010	0.078–0.168
Breast cancer, new	79	0.195[b] ± 0.057	0.081–0.334	11	0.164[d] ± 0.039	0.086–0.238
Breast cancer, recurrent	14	0.188[c] ± 0.061	0.122–0.320	10	0.164[d] ± 0.039	0.078–0.168

[a] Mean different from control group with $P = .05$, Student's t test, one tail.
[b] Mean different from control group with $P = .001$.
[c] Mean different from control group with $P < .01$.
[d] Mean not significantly different from control group.

samples only with HNO_3 rather than with $HNO_3/HClO_4$, resulting in lower than the actual selenium values in biological samples.

In healthy subjects, blood selenium concentrations reflect dietary selenium intakes. The blood selenium levels of our American controls, all from San Diego, are similar to those previously reported for California residents (Schrauzer and White 1978). From the observed linear relationship between the blood selenium concentrations and the selenium intakes (Schrauzer and White 1978),

$$[Se]_{intake}(\mu g/day) = 1104 \times [Se]_{blood}(\mu g/ml) - 55$$

the daily selenium intakes of the Japanese women was calculated to 255 $\mu g/day$ (range 126–300 $\mu g/day$), in reasonable agreement with the 287 $\mu g/day$ estimated from food consumption data (Schrauzer et al. 1977). For the American controls of our study, selenium intakes of 144 $\mu g/day$ (range 100–177 $\mu g/day$) are calculated, in close agreement with the 130 $\mu g/day$ (range 92–160 $\mu g/day$) previously determined for 10 California residents (Schrauzer and White 1978).

Blood Selenium Concentrations in Breast Cancer and Fibrocystic Disease

Blood selenium concentrations of the Japanese breast cancer patients are lower than those of healthy Japanese women, and the difference is statistically significant (see Table 1). In newly diagnosed breast cancer patients, blood selenium concentrations may be expected to reflect dietary intakes more likely than effects of disease. Such patients are as a rule not cachectic, and selenium sequestration by the tumor is unlikely to deplete selenium body stores significantly, as the primary lesions are usually small. No significant differences between the blood selenium levels exist between patients in early stages of the disease and those with recurrent malignant breast disease, both for the Japanese and American patients. This may be due to the fact that our samples were generally obtained from nonterminal patients. For terminal patients, low blood selenium levels are to be expected in view of generally deteriorating health status. In one patient with recurrent breast cancer, the blood Se level of 0.236 $\mu g/ml$ dropped to 0.133 $\mu g/ml$ when she became terminal.

Cystic proliferative changes of the breast, referred to as fibrocystic or chronic cystic disease, are generally regarded as premalignant conditions, leading to about a 3-fold increase of the breast cancer risk (Spratt and Donegan 1968). There is at present no published evidence on a possible role of selenium in the etiology of this disease. Although our data base is small, the results suggest that low dietary selenium intakes may increase the risk of fibrocystic disease development.

DIET AND CANCER RISK

The traditional Japanese diet provides approximately twice the amount of selenium than that of the typical American diet; major dietary sources of selenium are rice and seafoods. During the past 30 years, however, the proportions of meat, eggs, milk, fats, and sugar have increased significantly (Insull *et al.* 1968). This westernization of the Japanese diet may lead to a diminution of the selenium intakes and could influence the cancer risk in such a manner that cancers occurring with relatively high incidence in the United States becomes more common in Japan. Such effects of dietary change were previously observed only among Japanese immigrants to the United States. In Japanese women living in California and Hawaii, for example, a rapid increase of the incidence was observed and has since also become noticeable in the Japanese home population itself (Hirayama *et al.* 1980). A detailed dietary analysis of Japanese breast cancer patients would be of interest to assess the influence of dietary changes involving selenium, selenium antagonists, dietary fat, and other macro- and micronutrients on breast cancer risk. As may be estimated from a previously published empirical relationship (Schrauzer, 1976), a lowering of the per capita dietary selenium intake from 250 to 125 µg/day could cause a 4- to 5-fold increase of the breast cancer risk. Accordingly, relatively minor dietary modifications of the traditional Japanese diet should have noticeable effects on the incidence of cancer of the breast in the Japanese population and thus explain current morbidity and mortality trends. Similar conclusions may be reached for cancers of other organs, notably the intestine, rectum, lung, prostate, ovary, skin, pancreas, bladder, and urinary organs, and for leukemia.

ACKNOWLEDGMENTS

This work was supported in part by grant CA 23100 of the National Cancer Institute and the National Fisheries Institute.

REFERENCES

Chan, C. Y. 1976. Improvement in the fluorimetric determination of selenium in plant material with 2,3-diaminonaphthalene. Anal. Chim. Acta *82*, 213–215.

Hirayama, T., Waterhouse, J. A. H., and Fraumeni, J. F., Jr.   1980.   Cancer Risks by Site, pp. 128–129. UICC (International Union Against Cancer), Geneva.

Insull, W., Jr., Oiso, T., and Tsuchiya, K.   1968.   Diet and nutritional status of Japanese. Am. J. Clin. Nutr. *21,* 753–777.

Kurahashi, K., Inoue, S., Yonekura, S., Shimoishi, Y., and Toei, K.   1980.   Determination of selenium in human blood by gas chromatography with electron-capture detection. Analyst *105,* 690–695.

Olson, O. E.   1969.   Fluorimetric analysis of selenium in plants. J. Assoc. Off. Anal. Chem. *52* (3), 627–634.

Schrauzer, G. N., and Ishmael, D.   1974.   Effects of selenium and of arsenic on the genesis of spontaneous mammary tumors in inbred female C_3H mice. Ann. Clin. Lab. Sci *4,* 441–444.

Schrauzer, G. N., and White, D. A.   1978.   Selenium in human nutrition: Dietary intakes and effects of supplementation. Bioinorg. Chem. *8,* 303–318.

Schrauzer, G. N., White, D. A., and Schneider, C. J.   1977.   Cancer mortality correlation studies. III. Statistical associations with dietary selenium intakes. Bioinorg. Chem. *7,* 23–34.

Spratt, J. S., Jr., and Donegan, W. L.   1978.   Cancer of the Breast, pp. 28–30. W. B. Saunders Co., Philadelphia, PA.

Nonmelanoma Skin Cancer and Plasma Selenium: A Prospective Cohort Study

Larry C. Clark   *Bruce W. Turnbull*
Gloria F. Graham   *Barbara S. Hulka*
John Bray   *Carl M. Shy*

In order to test the hyothesis that patients with low plasma selenium levels are at an increased risk of developing new skin cancer, we conducted a prospective study of patients at high risk of developing new nonmelanoma skin cancer (NMSC). All patients in the cohort had a previous diagnosis of NMSC, either a basal cell epithelioma (BCE) or a squamous cell carcinoma (SCC). These patients were originally examined in our 1980 case control study of the selenium and cancer hypothesis (1).

We conducted our case control study of plasma selenium and skin neoplasms (1) because previous studies of the hypothesis that selenium status affects the risk of cancer were difficult to interpret (2–10). In particular, the sampling procedures for the selection of the cases and the controls in the earlier studies were poorly defined, and the poorer health status of the cancer patients may have influenced their selenium status. The patients in our initial study were essentially healthy individuals whose plasma selenium status was unlikely to be affected by the previous diagnosis of an NMSC, because NMSCs are usually small localized tumors that have an initial cure rate of over 95%.

The results of our initial case control study indicated that patients in the lowest decile of plasma selenium levels were 4.4 times more likely to have been diagnosed with either a BCE or an SCC than patients in the highest decile of plasma selenium. In a nested case control study which used prediagnostic plasma samples from the Hypertension Detection and Follow-up Program (HDFP), patients in the lowest quintile of plasma selenium were two times more likely to develop cancer than patients in the highest quintile (11). Remarkably, both the HDFP case control study and our skin cancer case control study estimated very similar logistic regression coefficients, indicating a similar strength for the association between plasma selenium and cancer in both studies. Subsequent case control studies of the selenium and cancer hypothesis have been recently reviewed (12).

Numerous animal experiments have used skin cancer as a model of carcinogenesis; only one experiment, however, has investigated the protective effect of selenium on skin carcinogenesis in mice (13). In a recent experiment, the inflammatory response from ultraviolet (UV) radiation in hairless mice was reduced by selenium supplementation (14).

METHODS

The 240 patients who comprise this cohort of patients at high risk of skin cancer were originally examined in 1980 as part of a case control study of nonmelanoma skin cancer and plasma selenium. The patients all had a history of one or more BCE or SCC prior to the initial study clinic visit in 1980. At the initial study visit, informed consent was obtained from each patient prior to conducting the personal interview, clinical examination, and obtaining samples of hair and plasma. The clinical examination included the diagnosis of new skin cancers and the assessment of the degree of sun damage of the forearm and malar surfaces. The personal interview collected information on potential skin cancer risk factors.

All of the eligible patients were invited to revisit the clinic during 1983 for an annual follow-up examination. New NMSCs, which were diagnosed 3 or more months after the initial examination, were considered incident tumors, while NMSCs diagnosed within 3 months of the initial examination were considered prevalent tumors. All clinically diagnosed tumors were biopsied and examined histologically to confirm the diagnosis.

The analysis of the plasma retinol and total carotenoids in 1980 was done using a standard trifluoroacetic acid method (15). The plasma selenium samples were analyzed with neutron activation analysis at the Burlington Laboratory of North Carolina State University. Each sample was irradiated for 24 hr with 1.5×10^{13} neutrons/cm^2/sec. A further description of this study has been published previously (1).

The incidence of NMSC presented here is the annual cumulative incidence calculated from incidence density using person years of observation as the denominator. The stratified analysis of the risk of NMSC for plasma selenium by level of sun damage and total carotenoids used the Mantel Haenszel test for person time denominators available for the programmable calculator (16). The 90% test-based confidence intervals are presented for the estimates of the relative risk. The high and low groups for variables are dichotomized at the median.

Separate Cox Proportional Hazard models were developed for patients with total plasma carotenoid levels above and below the median level of 120 μg/dl in order to investigate the association between plasma selenium status and the time to first new skin cancer (3 or more months after the case control examination), after adjusting for potential confounding variables. The parameter estimates for the Cox Proportional Hazard model were estimated using program P2L in BMDP on a VAX 11/750 microcomputer.

RESULTS

A total of 223 of the 240 persons in the cohort were reexamined 3 or more months after the initial examination, 92.3% of the original cohort. Of the 17 patients not reexamined, 4 died in 1980, 2 refused to participate, and 11 patients were lost to follow-up. In addition, 5 patients were excluded from the analysis because of incomplete biochemical data. The cohort represents a total of 6784 person months of observation, during which 75 patients had at least one new BCE or SCC, for a cumulative incidence rate of 12.4/100 per year.

A number of potential risk factors for skin cancer identified in the initial case control study are important univariate predictors of future skin cancer incidence. These variables include plasma selenium, retinol and total carotenoids, a clinical index of sun damage, age, sex, and the number of large and small arsenical keratoses. The incidence for each of these dichotomized risk factors and the relative risk for the low vs high comparison are presented in Table 1 for the 177 patients who were not prescribed vitamin A supplements. (During the follow-up period 41 patients were prescribed vitamin A supplements by the clinic's dermatologist and were excluded from further analysis of this cohort. These excluded patients differed from other cohort patients because of higher levels of sun damage, plasma selenium, plasma total

TABLE 1. Incidence and Relative Risk of Nonmelanoma Skin Cancer for Potential Skin Cancer Risk Factors Dicotomized at Their Median Value

Exposure	Incidence[a]		Relative risk	Mean	Units
	Low	High			
Selenium	12.2	9.3	0.77	0.137	µg/g
Sun damage	7.7	17.2	2.2	0.37	% high
Sex (F, M)	7.2	14.3	2.0	0.51	% male
Age	8.2	13.6	1.7	58.3	Years
Total carotenoids	9.3	12.2	1.3	126.0	µg/dl
Retinol	8.7	13.0	1.5	42.7	µg/dl
Small arsenical keratoses[b]	8.6	13.2	1.5	8.8	Number
Prescribed vitamin A supplement	10.7	20.8	1.9	0.19	(n = 218)

a Cumulative incidence/100 for the 177 patients who did not use prescribed supplements of vitamin A. The risk factors are all dicotomized at their median value.
b Small arsenical keratoses less than 3 mm in diameter.

carotenoids, and a higher incidence of new skin cancers.) Statistical significance testing is not presented because the risk factors are not independent of one another.

The correlation matrix of the risk factors in Table 2 indicates a significant correlation between several of these important skin cancer risk factors. For example, sun damage is worse in older patients, in patients with lower levels of total carotenoids, in males, and in patients with multiple arsenical keratoses.

Table 3 presents a stratified analysis of plasma selenium levels controlling for the effect of sun damage and total plasma carotenoids. The overall relative risk adjusted for total plasma carotenoids and sun damage is 0.74, with 90% confidence limits of 0.43 and 1.26. However, an examination of the cell-specific relative risk suggests that the relative risk for selenium could be dependent on the level of total carotenoids. A similar effect was observed in our earlier case control study, where a protective effect for selenium was evident if either total carotenoid or retinol levels were low, but not if both were high. For patients with plasma levels of carotenoids lower than 120 µg/dl, the relative risk for high vs low levels of selenium is 0.53, with 90% confidence intervals of (1.03, 0.26), suggesting a possible protective effect for a 0.057-µg/g difference in plasma selenium levels. When plasma levels of total carotenoids are above 120 µg/dl, the relative risk for selenium is essentially unity, indicating that the incidence of skin cancer was not affected by plasma selenium status when carotenoid levels were high.

The clinical assessment of sun damage is an important univariate predictor of future skin cancer incidence, with a crude relative risk of 2.2 for high vs low levels of sun damage. Examining the relative risk for plasma selenium levels separately for high and low levels of sun damage indicates that the relative risks are similar, 1.53 for patients with higher levels of sun damage, and 1.26 for patients with lower levels of sun damage. This suggests a lack of interaction between the level of sun damage and plasma selenium level for the risk of NMSC.

The Cox Proportional Hazard model can increase the statistical power for this study by treating selenium as a continuous variable rather than a dichotomized one and by using the time to tumor diagnosis as the dependent variable.

Two separate Cox Proportional Hazard models were developed, one for low plasma levels of total carotenoids (120 µg/dl or below), shown in Table 4, and one for high levels of total carotenoids (above 120 µg/dl), shown in Table 5. In the low carotenoid model, the negative β coefficient for selenium indicates a protective effect of plasma selenium levels. The relative risk for the 75th percentile vs the 25th percentile is 0.46

TABLE 2. Correlation Coefficients for Potential Skin Cancer Risk Factors[a]

Factors	Selenium	Carotenoids	Retinol	Sun	Age	Sex
Selenium	1.000	NS[b]	NS	0.113	NS	NS
	0.000	NS	NS	0.134	NS	NS
Carotenoids	NS	1.000	0.218	-0.328	NS	-0.320
	NS	0.000	0.003	0.001	NS	0.001
Retinol	NS	0.218	1.000	NS	0.098	NS
	NS	0.003	0.000	NS	0.192	NS
Sun damage	0.113	-0.328	NS	1.000	0.220	0.215
	0.134	0.001	NS	0.000	0.003	0.004
Age	NS	NS	0.098	0.2198	1.000	NS
	NS	NS	0.192	0.003	0.000	NS
Sex	NS	-0.320	NS	0.215	NS	1.000
	NS	0.001	NS	0.004	NS	0.000

[a] n = 177 patients.
[b] Not significant.

TABLE 3. Relative Risk of Skin Cancer for Plasma Selenium Level Stratified by Sun Damage and Total Carotenoid Levels

	Low carotenoids			High carotenoids		
	Cases	PRYR[a]	Incidence[a]	Cases	PRYR	Incidence
Low sun						
Se +	3	84.8	3.5	7	79.3	8.5
Se −	6	67.1	8.6	10	95.5	9.9
Relative risk			0.42			0.83
High sun						
Se +	6	48.1	11.7	8	32.7	21.7
Se −	9	44.5	18.3	5	23.1	19.5
Relative risk			0.63			1.11
Adjusted relative risk						
Marginal[b]	RR = 0.53 (1.03, 0.26)[c]			RR = 0.95 (1.96, 0.45)		
Overall	RR = 0.74 (1.25, 0.43)					

[a] Cumulative incidence/100 calculated using person years (PRYR) of observation.
[b] Relative risk adjusted for level of sun damage.
[c] Relative risk and 90% confidence intervals.

1128

TABLE 4. Cox Proportional Hazard Model for Patients with Total Plasma Carotenoids 120 µg/dl or Below Regression Coefficients, Estimated Risk Ratio, and Confidence Interval (CI) for High vs Low Values of Each Potential Risk Factor[a]

Factor	β	β/std	Mean	High	Low	RR	(90% CI)
Sun	0.268	1.97	5.1	6.5	3.7	2.12	(1.13, 4.76)
Sex	−0.629	0.23	0.27	Male	Female	0.28	(0.12, 0.96)
Age	0.026	1.31	57.8	66.7	50.1	1.54	(0.89, 2.39)
Se	−13.733	2.05	0.140	.166	.110	0.46	(0.25, 0.96)
Retinol	−0.006	0.28	40.8	52.0	33.5	0.89	(0.49, 1.27)
Carotenoids	0.003	0.32	88.6	103.8	73.3	1.10	(0.67, 1.48)
Previous tumor	1.051	2.18	0.22	1.0	0.0	2.86	(1.30, 10.8)

[a] Cases = 24, censored = 67 (73.6%); global chi square = 18.70, P = .0092.

TABLE 5. Cox Proportional Hazard Model for Patients with Total Plasma Carotenoids Above 120 μg/dl Regression Coefficients, Estimated Risk Ratio, and Confidence Interval (CI) for High vs Low Values of Each Potential Risk Factor[a]

Factor	β	β/std	Mean	High	Low	RR	(90% CI)
Sun	0.278	0.14	4.3	6.5	3.7	2.18	(1.13, 5.07)
Sex	0.581	2.72	−0.26	Male	Female	3.20	(1.58, 19.16)
Age	0.072	2.47	58.7	66.7	50.1	3.29	(1.49, 19.21)
Se	6.982	1.37	0.135	0.166	0.110	1.48	(0.92, 2.22)
Retinol	0.029	1.50	44.6	52.0	33.5	1.71	(0.95, 2.91)
Carotenoids	0.007	1.08	165.6	193.5	138	1.45	(0.82, 2.16)
Previous tumor	0.351	0.56	0.05	Yes	No	1.42	(0.51, 2.02)

[a] Cases = 30, censored = 56 (65.2%); global chi square = 22.9, P = .0017.

(1/2.17), with 90% confidence limits of 0.25 and 0.95, and 95% confidence limits of 0.22 and 0.99. This relative risk represents a relative risk adjusted for the other important factors for skin cancer. The two most important predictors of the risk of new skin cancers are the diagnosis of a new skin cancer at the entry examination (RR = 2.86) and the average degree of sun damage on the malar surfaces and forearms (RR = 2.12).

A 16-year difference in the age of a patient predicts an RR of 1.54; however, this RR is not statistically significant, since its confidence interval includes 1.0. None of the other estimated regression coefficients approaches statistical significance.

The Cox Proportional Hazard model for the high carotenoid group has a global chi square of 22.9, which is significant at $P = .0017$. The signs of the regression coefficients for sex, selenium, and retinol change from negative in the low carotenoid model to positive in the high carotenoid model. Sex is a significant predictor of risk, with a predicted risk ratio of 3.29 (1.49, 19.21) for males compared to females. Both retinol and selenium are of marginal statistical significance, with lower 90% confidence bounds that include 1. Age and sun damage both have significant relative risks of 3.20 (1.58, 19.16) and 2.18 (1.13, 5.07). The incidence rate for new skin cancer is 1.3 times higher for the patients in the high carotenoid model than in the low carotenoid model, which may explain some of the differences in the parameter estimates in the two models.

DISCUSSION

The results of this study are intriguing because they show that a number of potential risk factors can predict the incidence of new skin cancers and that micronutrient status is an important determinant of risk.

The follow-up for this cohort of individuals at high risk of developing new skin cancer was excellent, with over 90% of the patients being reexamined 3 or more months after the entry examination. However, the interpretation of the results from this cohort study is constrained by the small sample size available for analysis and by the exclusion of 41 patients who were prescribed vitamin supplements during the follow-up period.

The use of 5-fluorouracil (5-FU), topical retinoic acid, and sun screens may also reduce the incidence of new skin cancer, but their effect has not been analyzed for the current data, since the analysis would need to be of particular exposed epithelial surfaces rather than of individual patients. The omission of these factors may increase the misclassification of the predicted risk of new skin cancers, resulting in a biased estimate of the relative risk for the exposure; however, this usually makes it more difficult to observe a statistically significant effect when the misclassification is nondifferential. There is no reason to suspect that the use of any of these agents is more likely to occur in a particular selenium status group, since plasma selenium levels are essentially independent of the other risk factors. However, the use of 5-FU is correlated with both the degree of sun damage and age and may affect the interpretation of carotenoid status.

Both our case control and prospective cohort study defined selenium status by measuring plasma selenium concentration at one point in time. The validity of this approach is limited by the relative stability of plasma selenium levels and selenium status with time. If plasma selenium levels fluctuate widely on a weekly or daily basis, then the resulting misclassification of exposure should make it extremely difficult to observe a strong association. The results of our two studies would suggest that in this population a single plasma selenium level is predictive of the risk of developing NMSC. However, neither of these studies suggests the length of the latency period for a decrease in the risk of cancer after an improvement in selenium status. This question is best addressed in a double-blind randomized clinical trial.

The results of the Cox Proportional Hazard models and the stratified analysis suggest that the observed plasma retinol and carotenoid status do not have a strong protective effect for NMSC. A detailed analysis of these associations will be published in the near future.

CONCLUSION

This is the first prospective cohort study to investigate whether plasma selenium status affects the risk of cancer. The results of this study, although limited by a small sample size, suggest that the protective effect of selenium status for NMSC is strongest for patients with low total plasma carotenoid levels, which is consistent with the results of our previous case control study of NMSC. The HDFP nested case control study suggests that vitamin E status, an antioxidant, may also affect the magnitude of the observable protective effect of selenium status.

Despite the lack of statistical power compared to our previous study, our cohort study has the major methodological advantage of being a prospective study, with the selenium status of each patient being determined prior to the start of patient follow-up. We believe that the results of this study are consistent with a causal interpretation of the hypothesis that elevated selenium status is associated with a decreased risk of cancer and provides the justification for a conclusive test of the hypothesis, which is a double-blind clinical trial of nutritional levels of selenium supplementation for the prevention of NMSC.

REFERENCES

1. Clark, L. C., Graham, G. F., Crounse, R. G., Hulka, B., and Shy, C. M. 1984. Plasma selenium and skin cancer: A case control study. Nutr. Cancer 6, 13–21.
2. Shamberger, R. J., Rukovena, E., Longfield, A. K., Tytko, S. A., Deodhar, S., and Willis, C. E. 1973. Antioxidants and cancer. I. Selenium in the blood of normals and cancer patients. J. Natl. Cancer Inst. (U.S.) 50, 867–870.
3. McConnell, K. P., Broghamer, W. L., and Blotcky, A. J. 1975. Selenium levels in human blood and tissue in health and disease. J. Nutr. 105, 1026–1031.
4. Broghamer, W. L., McConnell, K. P., and Blotcky, J. L. 1976. Relationship between serum selenium levels and patients with carcinoma. Cancer (Philadelphia) 37, 1384–1388.
5. Broghamer, W. L., Jr., McConnell, K. P., Grimaldi, M., and Blotcky, A. J. 1978. Serum selenium and reticuloendothelial tumors. Cancer (Philadelphia) 41, 1462–1466.
6. Calautti, P., Moschini, G., Steivano, B. M., Tomio, L., Calzavara, E., and Perona, G. 1980. Serum selenium levels in malignant lymphoproliferative diseases. Scand. J. Haematol. 23, 63–66.
7. McConnell, K. P., Jayer, R. M., Bland, K. I., and Blotcky, A. J. 1980. The relationship of dietary selenium and breast cancer. J. Surg. Oncol. 15, 67–70.
8. Vernie, L. N., De Vries, M., Benckhuijsen, C., De Goeij, J. J. M., and Zegers, C. 1983. Selenium levels in blood and plasma, and glutathione peroxidase activity in blood of breast cancer patients during adjuvant treatment with cyclophosphamide, methotrexate, and 5-fluorouracil. Cancer Lett. 18, 283–289.
9. Goodwin, W. J., Lane, H. W., Bradford, K., Marshall, M. V., Griffin, A. C., Geopfert, H., and Jesse, R. H. 1983. Selenium and glutathione peroxidase levels in patients with epidermoid carcinoma of the oral cavity and oropharynx. Cancer (Philadelphia) 51, 110–115.
10. Robinson, M. F., Godfrey, P. J., Thomson, D. D., Rea, H. M., and van Rij, A. M. 1979. Blood selenium and glutathione peroxidase activity in normal subjects and in surgical patients with and without cancer in New Zealand. Am. J. Clin. Nutr. 32, 1477–1485.
11. Willett, W., Polk, B. F., Morris, S., Stampfer, M. J., Pressel, S., Rosner, B., Taylor, J. O., Schneider, K., and Hames C. G. 1983. Prediagnostic serum selenium and risk of cancer. Lancet 8342, 130–134.
12. Clark, L. C. 1985. The epidemiology of selenium and cancer. Fed. Proc., Fed. Am. Soc. Exp. Biol. (in press).
13. Shamberger, R. J., and Rudolf, G. 1966. Protection against cocarcinogenesis by antioxidants. Experientia 22, 116–118.
14. Thorling, E. B., Overvad, K., and Bjerring, P. 1983. Oral selenium inhibits skin

reactions to UV light in hairless mice. Acta Pathol. Microbiol. Immunol. Scand., Sec A 91A, 81–83.

15. Bauer, J. D., Ackermann, P. G., and Toro, G.   1974.   Clinical Laboratory Methods, 8th Edition. C. V. Mosby Co., St. Louis, MO.

16. Rothman, K. J., and Boice, J. D.   1979.   Epidemiologic Analysis with a Programmable Calculator, USDHEW NIH-NO 79-1644. USDHEW, Washington, DC.

Index